Boston Studies in the Philosophy and History of Science

Volume 334

The series *Boston Studies in the Philosophy and History of Science* was conceived in the broadest framework of interdisciplinary and international concerns. Natural scientists, mathematicians, social scientists and philosophers have contributed to the series, as have historians and sociologists of science, linguists, psychologists, physicians, and literary critics.

The series has been able to include works by authors from many other countries around the world.

The editors believe that the history and philosophy of science should itself be scientific, self-consciously critical, humane as well as rational, sceptical and undogmatic while also receptive to discussion of first principles. One of the aims of Boston Studies, therefore, is to develop collaboration among scientists, historians and philosophers.

Boston Studies in the Philosophy and History of Science looks into and reflects on interactions between epistemological and historical dimensions in an effort to understand the scientific enterprise from every viewpoint.

More information about this series at http://www.springer.com/series/5710

Mario Piazza • Gabriele Pulcini
Editors

Truth, Existence and Explanation

FilMat 2016 Studies in the Philosophy
of Mathematics

 Springer

Editors
Mario Piazza
Classe di Lettere e Filosofia
Scuola Normale Superiore
Pisa, Italy

Gabriele Pulcini
Department of Mathematics
Universidade Nova de Lisboa
Caparica, Portugal

ISSN 0068-0346   ISSN 2214-7942   (electronic)
Boston Studies in the Philosophy and History of Science
ISBN 978-3-030-06643-7   ISBN 978-3-319-93342-9   (eBook)
https://doi.org/10.1007/978-3-319-93342-9

Preface

The philosophy of mathematics aims at understanding the wonderful phenomenon of mathematics, and its immense resonance, from a philosophical point of view. This multifaceted intellectual enterprise is large enough to require a wide vista of present-day mathematics and its practice and to be constrained by matters of epistemology, metaphysics, logic, history of mathematics, and philosophy of language, mind, and science. However, it is also precisely because of its distinctive capacity of crossing disciplinary boundaries that the philosophy of mathematics has reached a sophisticated level of autonomy under the mantle of philosophy, with its own goals and standards of success. Nevertheless, autonomy does not mean insularity. Indeed, the outcome of the application of formal methods and techniques to a repertoire of vexed ontological and epistemological questions about mathematics encroaches on philosophy *sans phrase* as well as it deserves mathematical audience. This is what the present volume essentially aspires to.

The chapters here collected, in particular, zoom in on assorted themes concerning the triumvirate *truth*, *existence*, and *explanation* in mathematics, trying to move forward current debates in the literature along different axes of point of view. Needless to say, the intimate relation between truth, existence, and explanation propels another batch of interrelated issues for the philosopher: truth serves explanatory purposes, and successful mathematical explanations of the physical reality have been traditionally considered as a path toward arguing for the existence of mathematical objects.

Most of these contributions originate from the Second FilMat Conference that took place at the University of Chieti-Pescara from May 26 to 28, 2016. This conference hosted 19 talks and received 36 submissions from 32 international universities and research institutions from Austria, Belgium, Brazil, Canada, France, Germany, Hungary, India, Israel, Italy, Mexico, Poland, USA, and UK. FilMat conferences are organized under the aegis of the Italian Network for the Philosophy of Mathematics, FilMat Network (www.filmatnetwork.com). The network now counts almost 70 Italian scholars worldwide in the philosophy of mathematics and closely related disciplines. The present book follows the volume *Objectivity, Realism, and Proof* edited by F. Boccuni and A. Sereni in 2016 in the same Springer series, which selected contributions from the First FilMat Conference held at San Raffaele University, Milan, in May 29–31, 2014.

There is no doubt that the "space" of a collective volume is continuous, not discrete. So, in subdividing this kind of book into distinct parts there always remains a residue of arbitrariness on the editor's part. The chapters of this volume are self-contained, for the most part. The chapters in Part I deal with a group of issues concerned with mathematical truth, touching on problems such as the question of how to conceptualize the truth value of undecidable sentences, the intricacies of the interaction between absolute provability and truth theories, the motivation for a deflationary account according to which there is no substantial property of truth, and the important but elusive role of intensionality in mathematics.

This is a synopsis of the individual chapters.

Enrico Moriconi's contribution "Some Remarks on *True* Undecidable Sentences" considers the classical problem of assigning a truth value to Gödel's independent sentences. In particular, Moriconi's proposal is to examine the trajectory of this question in relation to three different levels: ordinary mathematics, informal theories, and formal theories such as Peano Arithmetic. The author argues that these three levels should not be treated separately inasmuch as it is only when one considers their constant interplay that a proper solution to the problem of the truth value of Gödel's sentences emerges.

Johannes Stern, "Penrose's New Argument and Paradox," examines the contemporary assessments of Penrose's New Argument for the claim that the human mind cannot be mechanized. In particular, Stern shows that Penrose's argument cannot be formalized within a theory of truth and absolute provability, in the sense that there is no consistent theory that allows for a formalization of the argument in a straightforward way. In a second step, the author sets up a reasonable theory of truth and absolute provability in which Penrose's argumentative strategy leads to a sound argument but, as a more general conclusion, he shows that the argument relies on a pathological feature of the theory.

Carlo Nicolai in "On Expressive Power Over Arithmetic" examines the relationship between a syntactical base theory and extensions by modal predicates and/or truth predicates. For the syntactical base theory, the author takes into account sequential theories which fall into two different kinds, either inductive or finitely axiomatizable. As for the relationship between the modal theory and its base theory, Nicolai considers three notions that have been suggested in the literature: conservativeness, relative interpretability, and speed up. He shows that many of the results known are highly dependent on the choice of the base theory. The author therefore questions some of the philosophical conclusions that have been too hastily drawn arguing that a more fine-grained analysis is needed.

Jaroslav Peregrin, in his chapter "Intensionality in Mathematics," addresses a key question in the philosophy of mathematics: Can we make room for *intensions* in mathematics? That question is specifically treated in the light of the fact that only *extensions* seem to be meaningfully associated with mathematical expressions, according to the standard possible-worlds based account of intensions. However, the author claims that we cannot give up intensionality in the metamathematical talk and so the possible-worlds account needs to be suitably modified. Peregrin's final proposal is that intensions should be taken as corresponding to rules regulating how (much) extensions are allowed to change without infringing the "boundaries of identity."

Denis Bonnay and Henri Galinon in "Deflationary Truth Is a Logical Notion" examine the thesis according to which "truth" should be taken as a *logical* notion. Such a thesis was inaugurated by Quine and now endorsed by some contemporary deflationists such as Horwich and Field. In their paper, the authors supply new arguments in favor of the logical nature of the notion of "truth" by focusing on Tarski's invariance approach to logicality.

Andrea Strollo's chapter "Making Sense of Deflationism from a Formal Perspective: Conservativity and Relative Interpretability" purports to import and interpret technical results in the study of formal theories of truth into the classical philosophical debate in which the notion of "truth" is taken metaphysically. In particular, the author discusses the role of the formal notions of *conservativity* and *relative interpretability* in the evaluation of the deflationary nature of "truth". The final result of Strollo's approach is a new taxonomy of axiomatic theories of truth.

The chapters in Part II are more heterogeneous. They explore a variety of issues concerning the ubiquitous notion of structure in mathematics, the nature of mathematical understanding, the explanatory role of completeness theorem and inductive proofs, the interplay between the informal side of computability and the formal one, and the effectiveness and applicability of mathematics. The last chapter of the book is meant to be a measure of the sophistication of ancient topics in the philosophy of mathematics, such as Aristotelian view of continuity.

Reinhard Kahle, in "Structure and Structures," takes the view that mathematical truth makes sense only insofar as it refers to structures. However, the author observes that it is far from being obvious how these structures, together with their internal structure, are given. By taking a different point of view with respect to the standard debate on mathematical structuralism, he addresses the question of how the informal talk about "structures" in ordinary mathematics is included in the notion of "structure" as it is technically defined in mathematical logic. To this purpose, the author distinguishes between *first-order*, *primitive*, and *abstract* structures.

Janet Folina in "Towards a Better Understanding of Mathematical Understanding" examines the general notion of *understanding* in mathematics. This chapter incorporates two main thesis. According to the first, mathematical understanding should be considered as a "family of resemblance" type of concept, and this explains why it is in general very hard to frame it into a well-defined theory. Secondly, the notion of "mathematical structure" can help in filling the conceptual gap between the different epistemological meanings associated with "understanding."

In the chapter "The Explanatory Power of a New Proof: Henkin's Completeness Proof," **John Baldwin** embarks on a detailed comparison of the Henkin and Gödel proofs of the completeness theorem. The author's main purpose here is to underline the explanatory power of Henkin's argument as both a proof of the completeness of first-order logic and, more in generality, show how this argument allowed for the entire recasting of model-theory as an autonomous and flourishing mathematical discipline.

In her chapter "Can Proofs by Mathematical Induction Be Explanatory?," **Josephine Salverda** argues against the received view according to which proofs by induction usually lack explanatory value. In particular, Marc Lange recently maintained that inductive proofs can never be explanatory. Against this latter strong claim, Salverda's chapter aims to provide examples of explanatory inductive proof, together with a positive suggestion for making sense of these examples on Lange's account.

In the chapter "Ontological Commitment and the Import of Mathematics," **Daniele Molinini** critically focuses on a recent article by Alan Baker in which the author argues that there is a class of explanations in science, namely optimization explanations, for which the use of more general mathematical resources leads to a reduction of concrete commitments and to a boost in the explanatory power. Molinini challenges this thesis in the context of Baker's preferred example involving periodical cicadas, thus raising doubts against the alleged import of Baker's considerations in the Platonism vs. Nominalism debate.

In his chapter "Applicability Problems Generalized," **Michele Ginammi** aims to lay the foundation for a general philosophical account of *applicability*. To this end, the author provides interesting case studies concerning all the possible relevant applicability configurations involving mathematics and physics. Finally, Ginammi tries to "extract" a general pattern common to each one of the math-to-physics, math-to-math, and physics-to-math applications. The kind of generality he finally achieves allows for a better understanding of the complex interaction between physics and mathematics.

Luca San Mauro's chapter "Church-Turing Thesis, in Practice" provides the reader with a philosophical account of the notion of "proof by Church's Thesis"; this is the notion which allows mathematicians to resort to informal methods when working in computability theory. Firstly, the author offers a detailed reconstruction of the historical development of this notion, from Post to Rogers. Then, in contrast with the received view on the topic, San Mauro shows how the *informal* side of computability is not completely reducible to its *formal* counterpart.

The last chapter in this volume is devoted to Aristotle's conception of the relationship between existence and conceivability. In "Existence vs. Conceivability in Aristotle: Are Straight Lines Infinitely Extendible?," **Monica Ugaglia** argues that Aristotle's philosophy is consistent with the mathematics of its time. In particular, the chapter shows how Aristotle, resorting to the infinite procedure of division of the continuum, accounted for the infinity of numbers and asymptotic properties, while at the same time maintaining that mathematical entities are immanent in physical ones and that the physical world is finite.

Pisa, Italy   Mario Piazza
Caparica, Portugal   Gabriele Pulcini
September 2018

Acknowledgments

We would like to thank Pierluigi Graziani who co-organized with us the FilMat Conference in Chieti, as well as the Department of Philosophical, Pedagogical and Economic-Quantitative Sciences of the University of Chieti-Pescara.

We are also grateful to the referees for the help in ensuring the quality of the papers.

Gabriele Pulcini thankfully acknowledges the support from the Portuguese Science Foundation, FCT, through the project "Hilbert's 24th Problem" PTDC/MHCFIL/2583/2014.

Contents

Part I
Truth and Expressiveness

Chapter 1
Some Remarks on *True* Undecidable Sentences

Enrico Moriconi

Abstract In this paper I try to discuss the question of the truth-value of Gödel-type undecidable sentences in a framework which keeps into due account the idea that mathematical inquiry develops in a three-level framework: informal (or pre-formal) mathematics, (informal) theories, and formal theories. Moreover, it is to be stressed that no phase deletes the other ones; all of them, so to speak, live together.

Keywords Truth · Proof · Gödel's incompleteness theorems · Undecidable sentences

1.1 Introduction

At the very root of the question I wish to discuss there is the opposition between the notion of truth and that of provability. This has been lucidly focused by W. Tait in (1986) by stressing that

> An arithmetical proposition A, for example, is about a certain structure, the system of natural numbers. It refers to numbers and relations among them. If it is true, it is so in virtue of a certain fact about this structure. And this fact may obtain even if we do not or (e.g., because of its relative complexity) cannot know that it does (p. 61).

On the other hand, he adds that

> we learn mathematics by learning how to do things – for example, to count, compute, solve equations, and more generally, to prove. Moreover, we learn that the ultimate warrant for a mathematical proposition is a proof of it. In other words, we are justified in asserting A – and therefore, in any ordinary sense, the truth of A – precisely when we have a proof of it (ivi).

I wish to thank my colleague Luca Bellotti for many helpful discussions on these topics.

E. Moriconi (✉)
Dipartimento di Civiltà e Forme del Sapere, Pisa, Italia
e-mail: enrico.moriconi@unipi.it

© Springer International Publishing AG, part of Springer Nature 2018　　　　　　　3
M. Piazza, G. Pulcini (eds.), *Truth, Existence and Explanation*,
Boston Studies in the Philosophy and History of Science 334,
https://doi.org/10.1007/978-3-319-93342-9_1

The former quotation represents of course a simple and rough expression of the so-called *Platonist* (or *realist*) point of view towards mathematics. Against this perspective, largely widespread among mathematicians, various perplexities have been raised, hinging notably on the fact that we lack any direct apprehension of the *abstract* structure of natural numbers. A structure with which, moreover, we cannot have any interaction of causal type.

Thus, for many people it has been natural to wonder about what could link the practice of proving, what we *learned* to do, with what *holds* in the structure of the integers. And one could ask whether it is available any evidence that proof procedures accord with what holds in the structure (i.e., that they are, so to speak, *sound*). Bringing to the fore the practice of proving, many people have hence tried to provide a meaning for mathematical propositions in a way which, hopefully, can do without (what would this way result to be) the illusory reference to mathematical objects and structures. This is the general framework which is shared by various (foundational) proposals: constructivism, operationalism, formalism, ...

1.2   A "Different" Perspective

Previously mentioned themes are no doubt fundamental and have been accordingly strongly and widely debated.[1] Nonetheless, it could be interesting to frame matters adopting a different point of view. That is, stressing the *fact* that a very large part of mathematical activity develops at an *informal* or *pre-formal* level. Which means that there is a "common language" of mathematical practice where conceptual constructions are intertwined with proof procedures, and both are open to stimuli possibly coming from other sciences or from outer reference. At this level researchers believe that there are specific and largely agreed criteria about how and which propositions have to be held true.

As the italian mathematician Ennio De Giorgi said in a lecture of 1981

> All mathematicians share the firm belief that mathematical sentences and mathematical proofs have and exhibit specific qualities. It is absolutely beyond any doubt for the mathematicians the meaning of assertions like "a theorem is meaningful" (that is, that it abides by the internal rules of mathematics), or "a proof is correct" or "a proof is mistaken". This is a question which all mathematicians agree on (Bassani et al. 2001, p. 157).[2]

---

[1]One can find the various options adequately developed in, for instance, Benacerraf and Putnam (1983). Historically, very important texts are Hempel (1945) and Benacerraf (1973).

[2]We report here the italian text:

> Ciò che viene accettato concordemente da tutti i matematici è il carattere specifico di tutti gli enunciati matematici, delle dimostrazioni matematiche. È assolutamente fuori discussione fra i matematici che cosa voglia dire che "un teorema è sensato", ossia scritto in modo coerente alle regole interne della matematica e che "una dimostrazione è giusta" o "è sbagliata". Su questo esiste una totale unanimità fra i matematici.

A wonderful exploration of how mathematics evolves at this level was provided by the now classic *Proofs and Refutations* by I. Lakatos (see Lakatos 1976). There we find an incisive exposition of how regimenting proofs in order to clarify their assumptions and the nature of the accomplished steps – the process which formalization was going to idealize – is just one phase in the complex process that leads to the growth of mathematical knowledge. Within this "common language" certain research areas are circumscribed with a variable level of precision, sometimes gathering their characteristics in a theory. The birth of a theory, as Hilbert wrote in 1918, in *Axiomatisches Denken*, presupposes that a certain set of proving procedures and conceptual constructions in a given field of research had reached a point of sufficient ripeness.

> When we assemble the facts of a definite, more-or-less comprehensive field of knowledge, we soon notice that these facts are capable of being ordered. This ordering always comes about with the help of a certain *framework of concepts* in the following way: a concept of this framework corresponds to each individual object of the field of knowledge, and a logical relation between concepts corresponds to every fact within the field of knowledge. The framework of concepts is nothing other than the *theory* of the field of knowledge (Hilbert 1918, p. 1107–8.).

We have to take in due account that producing a *theory* contains a precise link to a concrete historical situation of the considered field of research, and that it is an operation which does not close that field off from scientific development. Otherwise, we risk to misunderstand the very nature of relationship between the notions of truth and proof. I think that two other quotations from the same Hilbert's text are appropriate:

> If we consider a particular theory more closely, we always see that a few distinguished propositions of the field of knowledge underlie the construction of the framework of concepts, and these propositions then suffice by themselves for the construction, in accordance with logical principles, of the entire framework (p. 1108).

> [The fundamental propositions] regarded from an initial standpoint as the *axioms of the individual fields of knowledge* [provide a solution to] the problem of grounding the individual field of knowledge; but this solution was only temporary. In fact, in the individual fields of knowledge the need arose to ground the fundamental axiomatic propositions themselves. [So one realized that the "proofs" so acquired] basically only make it possible to trace things back to certain deeper propositions, which in turn are now to be regarded as new axioms instead of the propositions to be proved. [...] The procedure of the axiomatic method, as it is expressed here, amounts to a *deepening of the foundations* of the individual field of knowledge (p. 1108–9).

1.2.1   Formal Theories

The previous theoretical framework underwent a remarkable change between the end of nineteenth century and the Thirties of the twentieth century, when formalization came to the fore. Frege's and other people's discovery of the possibility to formalize mathematical knowledge produced a new theoretical subject: *formal theories*. A couple of clarifications concerning the use of the adjective "formal" are in order:

- On the one hand, "formal" means that a given part of mathematical knowledge undergoes a process of rebuilding which aims at making its logical structure explicit, so that the justification procedure of informal mathematical proofs can be fully understood.
- On the other hand, "formal" means that we give up using natural languages and adopt conventional artificial ones, without any particular meaning assigned to (strings of) the signs of the language.

It would be a sheer misunderstanding to consider the construction of a formal theory as a simple work of *decoration*. Indeed, it produces new questions, new knowledge, and new standpoints.

For instance, one could wonder whether when a *formal* proof of a mathematical sentence is available, we know something more than knowing that that sentence is provable (in a pre-formal theory, for instance).

Moreover, it becomes possible to appreciate the difference between speaking, both at the first level of pre-formal mathematical enquiry, and at the second level when (informal) theories are produced, of the *truth of a mathematical sentence*, and speaking, at the third level, of the *truth in a model of a sentence belonging to a formal mathematical theory*. In the latter case we mean a *mathematical* definition of the notion of satisfiability of a formula in a model, that is, in a set-theoretically defined structure.

Notably, only with respect to the formal language of a formal theory it has been possible to develop a formal theory of the way in which *truth* can be attributed to a sentence of a formal mathematical theory. *Formal Semantics* built by Tarski in the Thirties of the last century was a(n implicit) definition given *in* mathematics of the notion of *truth-in-a-model* with respect to a given *formal* language. And it is to be stressed again that both the notion of formal language and that of the structures which can be its models are mathematical structures.

1.2.2   A Three-Level Framework

This way, we get a three-level framework:

1. informal, or pre-formal, mathematics,
2. (informal) theories, and
3. formal theories.

It is to be stressed that no phase deletes the previous ones; all of them, so to speak, live together. Formal theories cannot be studied separately from all the non-formal, or pre-formal, background behind them. In respect of this, I think appropriate to point out that in *Der Wahrheitsbegriff in den formalisierten Sprachen* Tarski emphasized that he dealt with formal languages insofar as they were *formalizations of informal languages*. That is to say: not just pure and simple, abstract, formal languages, waiting for a successive attribution of meaning, but, in a sense, always already interpreted languages.

> It remains perhaps to add that we are not interested here in *formal* languages and sciences in one special sense of the word *formal*, namely sciences to which no meaning is attached. For such sciences the problem here discussed has no relevance, it is not even meaningful. We shall always ascribe quite concrete and, for us, intelligible meanings to the signs which occur in the languages we shall consider. [...] The sentences which are distinguished as axioms seem to us materially true, and in choosing rules of inference we are always guided by the principle that when such rules are applied to true sentences the sentences obtained by their use should also be true. [my emphasis] (Tarski 1983, p. 166–7).

The (set-theoretical) structures we build to provide a meaning to our formal languages are objects we build "in mathematics"; that is, in the pre-formal mathematics which is the setting where the theories, both pre-formal and formal, come into the world and grow. And the fact that we are able to work with these structures presupposes, first, that we understand the mathematical propositions which deal with those structures, and, second, that we are capable to assess, at least for some of them, that they hold in those structures.

In other words, the presupposition is that it is within the compass of our capacities to acknowledge the truth of the propositions whose truth is involved in the very construction of the structure. This kind of truth, which could be called plain *semantic truth*, is what we hinted at when we spoke about the "truth of a mathematical proposition". It is also the kind of truth we assume one is already able to master when (s)he defines the notion of "truth in a model with respect to the sentences of a formal mathematical theory". Obviously, in fact, mastering the latter notion presupposes a preliminary understanding of what it means to say that a certain proposition is true, without any further qualification.

1.3   *Marginalia* on Gödel's Incompleteness Results

It would be useless to try to retrace here all of the threads of Gödel's argument, claiming also to say something new (and correct, of course); so, we shall simply list some *marginal* items which seem more relevant for us.[3]

Gödel's theory involved a three-layered framework:

* the formal language of **PA**, say $\mathcal{L}(\mathbf{PA})$,
* its semi-formal metatheory (faithful, in this respect, to the spirit of Hilbert's *inhaltliche Metamathematik*), and
* the structure $\mathcal{N} =< \omega, =^{\mathcal{N}}, +^{\mathcal{N}}, \times^{\mathcal{N}}, s^{\mathcal{N}}, 0^{\mathcal{N}} >$ constituting the so-called *standard model* of the natural numbers.

---

[3] Very interesting and up-to-date expositions of the relevant matters are available in Beklemishev (2010) and Buldt (2014). Both texts offer plenty of further references.

The formal theory **PA** can refer to elements of ω basically in two ways: by means of *pronouns* or by means of proper names.

- In the former case we have the individual variables of $\mathcal{L}(\mathbf{PA})$: $x_1, x_2, \ldots$; or metatheorical expressions like $\sigma_{x_k}^n$, where x_k is an individual variable, $n \in \omega$, and σ is an assignment.
- In the latter case, beside 0, $\mathcal{L}(\mathbf{PA})$ can produce a representation for each $n \in \omega$ by applying n times the function s to 0: $s(\ldots s(0) \ldots)$. We'll denote the last figure, called the *numeral* of n, by **n**.

In the metatheory it is possible to define a function, say g, which assigns a number, or *code*, to any term t, formula A or derivation $\mathcal{D}$ of **PA**. We'll write, respectively, $g(t)$, $g(A)$, and $g(\mathcal{D})$. This is the so-called "arithmetization of metamathematics", or "gödelization".

1.3.1 Representation

As a consequence of gödelization, any metatheorical assertion becomes the statement of a numerical relation. And Gödel was able to show that in many cases such *metatheorical* assertions become relations *definable* in $\mathcal{N}$ which can be expressed or *represented* in (**PA**) (so going this time in opposite direction to what was Hilbert's intention). To say that a numerical relation, say $R(n, m)$, is *representable* in **PA** means that there exists a formula of $\mathcal{L}(\mathbf{PA})$, say $\underline{R}(x_1, x_2)$, such that if n is R to m, that is, if $\mathcal{N} \models R(n, m)$, then $\mathbf{PA} \vdash \underline{R}(x_1/\mathbf{n}, x_2/\mathbf{m})$, and if n is not R to m, that is, if $\mathcal{N} \not\models R(n, m)$, then $\mathbf{PA} \vdash \neg\underline{R}(x_1/\mathbf{n}, x_2/\mathbf{m})$.

This representation consisted of two steps.

1. First, Gödel showed that, with the notable exception of the notion of theorem, the numerical sets, operations and relations associated, thanks to the arithmetization, to metatheorical properties, operations and relations, turn out to be all *effective*, and more precisely all **primitive recursive**.
2. Second, he proved that even though in **PA** only the symbols $0, +, \times, s$ are available, all the primitive recursive operations and relations are representable in **PA**.

As it is obvious, but anyway worth to be stressed, this notion of "representability" depends on both the richness of $\mathcal{L}(\mathbf{PA})$, and on the power of the proof system of **PA**.

1.3.2 Formalization of Metatheory

The notion of representability is at the core of Gödel's procedure, which proceeded by comparing and contrasting the notions of informal metatheory – notably the proof relation – and the properties formally expressible, thanks to the arithmetization, within the formal system.

The formalization of metatheory is a typical question of adequacy: one has to make explicit the conditions which the formal theory has to satisfy in order to be able to *adequately* express the informal metatheory. Such conditions are three-fold:

1. logical conditions;
2. encoding conditions;
3. derivability conditions.

In the first class, I include also the *consistency* and *ω-consistency* of the formal theory.

1.3.3 *ω-Consistency*

Although belonging to the class of logical conditions, *ω-consistency* is also relevant for encoding conditions.

In fact, if $Prov_{\mathbf{PA}}(\mathbf{n}, \mathbf{m})$ represents in **PA** the encoding of the (metatheorical) assertion that n is the code of a proof of the formula encoded by m, then we can define the set of natural numbers which code a provable proposition, say $Th_{\mathbf{PA}}$, as the set $\{n \mid \exists m \, Prov_{\mathbf{PA}}(m, n)\}$.

This guarantees the recursive enumerability of $Th_{\mathbf{PA}}$, but we need *ω-consistency* to prove that the formula $\exists x_n \, Prov_{\mathbf{PA}}(x_n, x_m/\mathbf{m})$ weakly, or semi-, represents $Th_{\mathbf{PA}}$ in **PA**.

- Let's assume that $Th_{\mathbf{PA}}(n)$. Then for a given m we have $Prov_{\mathbf{PA}}(m, n)$. Since $Prov_{\mathbf{PA}}(m, n)$ is a primitive recursive relation, it is representable in **PA** by a formula $Prov_{\mathbf{PA}}(x_n, x_m)$, which in turn means that

$$\mathbf{PA} \vdash Prov_{\mathbf{PA}}(x_n/\mathbf{n}, x_m/\mathbf{m}).$$

  From this fact, by logic, it follows $\mathbf{PA} \vdash \exists x_n \, Prov_{\mathbf{PA}}(x_n, x_m/\mathbf{m})$. Note that, besides the consistency of **PA**, we exploited only the assumption that for any $n \in \omega$ there exists a correspondent numeral.

- Let's now assume that $\mathbf{PA} \vdash \exists x_n \, Prov_{\mathbf{PA}}(x_n, x_m/\mathbf{m})$. From the *ω-consistency* of **PA** it follows that there exists at least an n such that

$$\mathbf{PA} \not\vdash \neg Prov_{\mathbf{PA}}(x_n/\mathbf{n}, x_m/\mathbf{m}).$$

From this fact, being $Prov_{\mathbf{PA}}$ a Δ_0-formula, it follows that

$$\mathbf{PA} \vdash Prov_{\mathbf{PA}}(x_n/\mathbf{n}, x_m/\mathbf{m}).$$

Because $Prov_{\mathbf{PA}}(x_n, x_m)$ represents $Prov_{\mathbf{PA}}$ in **PA**, from the last fact it follows that $Prov_{\mathbf{PA}}(m, n)$ holds, which in turn means that $Th_{\mathbf{PA}}(n)$.

The property of ω-consistency, of course, expresses a sort of numerical soundness, and is in close relationship with the property of *categoricity*. In fact, ω-consistency requires that the denotations of *numerals*, which are closed terms, *proper names* of integers, offer no significant difference with respect to provability, if compared with the denotations of individual variables which, working as *pronouns*, have as the range of variation all of the individuals of the domain.

At odds, in a sense, with what we have just now remarked, and calling G Gödel's undecidable sentence, it could be interesting to note that the unprovability of $\neg G$, which is obtained by essentially exploiting the ω-consistency of the theory, entails immediately that **PA** is *not* categorical. If **PA** is ω-consistent, in fact, G is neither provable nor refutable in **PA**. Of course, our assumption entails that **PA** is consistent, and such are also $\textbf{PA} \cup \{G\}$ and $\textbf{PA} \cup \{\neg G\}$. By semantic completeness of first-order theories, both theories will get a model, say $\mathcal{M}$ and $\mathcal{N}$. Moreover, $\mathcal{M}$ and $\mathcal{N}$ will of course be models also of **PA**. Since they can not be isomorphic to each other, we get immediately the non-categoricity of **PA**.

1.4   The Truth of Undecidable Sentences

Just to see what we are dwelling upon, let's go through the construction steps of the undecidable sentence G.

1. Let $G(x)$ the formula $\neg \exists y \, Prov_{\textbf{PA}}(y, \underline{st}(x, x))$.
2. Remark: this formula contains only the free variable x; and $\underline{st}$ represents in **PA** the so-called *substitution function*, that is the primitive recursive binary function which, applied to a pair of natural numbers, say (n, p), outputs the code of the formula obtained by substituting in the formula coded by n any occurrences of its free variable by $\textbf{p}$, if n codes a formula with just one free variable; otherwise, st outputs n.
3. Let $g(\neg \exists y \, Prov_{\textbf{PA}}(y, \underline{st}(x, x))) = k$.
4. Gödel's undecidable sentence G will be then $\neg \exists y \, Prov_{\textbf{PA}}(y, \underline{st}(\textbf{k}, \textbf{k}))$.
5. Being in $\mathcal{L}(\textbf{PA})$, G is a number-theoretic sentence. However, it has been built within the framework of gödelization, and by exploiting the key of the encoding algorithm we can reach the *meta*theoretical level and in this way we can realize that G speaks about the (*object language*)-sentence obtained from the sentence whose code is k by substituting its only free variable x with $\textbf{k}$, and asserts that the latter sentence is unprovable in **PA**.
6. First, one can note that G is exactly the result of this substitution, which means that it holds: $\textbf{PA} \vdash g(\textbf{G}) = \underline{st}(\textbf{k}, \textbf{k})$.
7. From this fact, one then usually moves to emphasize that it holds: $\textbf{PA} \vdash G \leftrightarrow \neg \exists y \, Prov_{\textbf{PA}}(y, \textbf{g}(\textbf{G}))$, or also $\textbf{PA} \vdash G \leftrightarrow \neg Th_{\textbf{PA}}(\textbf{g}(\textbf{G}))$.
8. Before starting to draw consequences from the equivalence of previous item 7, from which we know that the two sides of equivalence are mutually interchangeable with regard to provability within **PA**, it could be useful to spell

it out carefully:

$$\mathbf{PA} \vdash \neg \exists y \underline{Prov}_{\mathbf{PA}}(y, \underline{st}(\mathbf{k}, \mathbf{k})) \leftrightarrow \neg \exists y \underline{Prov}_{\mathbf{PA}}(y, \mathbf{g(G)}).$$

9. Now, by applying again the decoding algorithm to this equivalence, and so considering it again a *meta*-sentence, we can realize that the two sides of the equivalence attribute the same property (that is, the unprovability in **PA**), to two sentences which have got, respectively, the codes $\underline{st}(\mathbf{k}, \mathbf{k})$, and $\mathbf{g(G)}$.

10. The previous remark 9, of course, just points to the role played at the metatheorical level by the equation, provable inside **PA**, of item 6. Watching again from the point of view of the metalevel, in fact, that equation makes us know that the two sides of the equivalence speak of the same syntactic object, and being equivalent, they are true in exactly the same situations. Not, however, that any of them speaks of itself.

11. The undeniable flavour of self-reference which surrounds G comes from its construction. However,

    - it's one thing to say that G asserts of itself to be unprovable within **PA**,
    - and another to say that G asserts the unprovability in **PA** of the sentence of $\mathcal{L}(\mathbf{PA})$ whose Gödel number is $st(k, k)$.

12. To speak of *self-reference* one has to go one step further in the metalevel, and to check that $g(G)$ is actually the very result of calculating $st(k, k)$.[4]

Referring to G, and foreseeing in a sense the subsequent debate, Gödel himself stressed in the footnote 15 of his 1931 paper that

> Contrary to appearances, such a proposition involves no faulty circularity, for initially it [only] asserts that a certain well-defined formula (namely, the one obtained from the qth formula in the lexicographic order by a certain substitution) is unprovable. Only subsequently (and so to speak *by chance*) does it turn out that this formula is precisely the one by which the proposition itself was expressed [my emphasis].

It is maybe opportune to remark that G, *in itself*, is a huge authentic number theoretic sentence, too long to be extensively written down, and probably of scarce interest from the point of view of number theory. On the other hand, *for us*, G comes into the world within the arithmetized metatheory of **PA**, as the fixed point of the (arithmetized) predicate $\neg(Th_{\mathbf{PA}}(x))$. We didn't obtain G as the arithmetization of a given number theoretic sentence, as could be with, e.g., $1 + 1 = 2$, that is $s(0) + s(0) = s(s(0))$, which gödelization transforms into something like:

$$2^{\mathbf{g(s)}} \times 3^{\mathbf{g((}} \times 5^{\mathbf{g(0)}} \times 7^{\mathbf{g))}} \times 11^{\mathbf{g(+)}} \times 13^{\mathbf{g(s)}} \times 17^{\mathbf{g((}} \times 19^{\mathbf{g(0)}} \times 23^{\mathbf{g))}} \times$$

$$29^{\mathbf{g(=)}} \times 31^{\mathbf{g(s)}} \times 37^{\mathbf{g((}} \times 41^{\mathbf{g(s)}} \times 43^{\mathbf{g((}} \times 47^{\mathbf{g(0)}} \times 53^{\mathbf{g))}} \times 59^{\mathbf{g))}}$$

which, in its turn, is another number-theoretic sentence.

---

[4]For a thorough treatment of the question of self-reference one can see Halbach and Visser (2014).

We did obtain G by proceeding the other way round, that is after the gödelization was done, and the huge number theoretic formula expressing the provability relation could be considered as *given*, in a highly idealistic sense of "given", of course. At the metalevel, we can afterwards baptize that formula $Prov_{PA}(x, y)$, and on this basis proceed to the construction of the sentence we baptized G. Of course, G is also the name of an arithmetical sentence, even though writing it down is an improbable task. It makes no sense to speak of G *before* the arithmetization, even though it is a truly arithmetical sentence. A sentence, however, which we can only imagine – and not write down – as the result of the application of the decoding function to G. Roughly speaking, as the result of calculating $g^{-1}(G)$.

1.4.1 Truth of G

After *self-reference*, the question which soon came to the fore was the question of the truth-value of G. The subject is of course remarkable since if G could be meaningfully acknowledged as true, then we would be given an explicit separation between what is provable and what is true, without hope to reduce the latter notion to the former one.

Also in this case, I think, it is worth recalling Gödel's remark in the 1931 paper (with notation adapted to the present context), since there he hints at the direction to follow:

> From the remark that G says about itself that it is not provable, it follows at once that G is true, for G *is* indeed unprovable (being undecidable). Thus, the proposition that is undecidable *in the system* **PA** still was decided by metamathematical considerations (Gödel 1986, p. 151).

The previous quotation contains the only reference to the truth of G made by Gödel in the 1931 essay; there he just added that

> The precise analysis of this curious situation leads to surprising results concerning consistency proofs for formal systems (ib.).

Here, of course, he hints at the possibility to internalize in **PA** the argument which, in the first half of the first incompleteness theorem, from the *assumption* of the consistency led to the unprovability of G, this way getting:

$$\vdash_{PA} \neg Th_{PA}(g(0 = 1)) \rightarrow \neg Th_{PA}(g(G))$$

or

$$\vdash_{PA} Cons_{PA} \rightarrow G \text{ or, equivalently, } Cons_{PA} \vdash_{PA} G$$

which is his second incompleteness result.

Much later, in a letter to H. Wang of 19/XII/1967, Gödel addressed again the question, but with a different and much stronger disposition:

[I]t should be noted that the heuristic principle of my construction of undecidable number theoretical propositions in the formal systems of mathematics is the highly transfinite concept of "objective mathematical truth", as opposed to that of "demonstrability", with which it was generally confused before my own and Tarski's work. Again the use of this transfinite concept eventually leads to finitarily provable results, e.g., the general theorems about the existence of undecidable propositions in consistent formal systems (Gödel 2003, p. 398).

1.4.2 Classical Truth

Reasoning according to the classical perspective, the notion of "truth" is ruled by a bunch of principles, including identity, non-contradiction, and another principle which I think better to split in the following way:

1. the *principle of bivalence*, to which also intuitionists agree: two truth-values are available; that is, no sentence *may* be neither true nor false;
2. the *principle of the excluded middle*, which is the distinctive feature of the classical point of view: any sentence *must* get at least and at most one of the two truth-values.

Henceforth, it follows the paradoxical nature of the Liar sentences; but what about G?

1.4.3 Truth of G: How and Where

Before trying to answer the last question, it is better to frame how and where this question can be answered.

- On the one hand, being an instance of the principle of the excluded middle, we get

$$\vdash_{PA} (G \lor \neg G)$$

thus, reasoning on the basis of classical logic, $(G \lor \neg G)$ must be true.
- However, and reasoning on the basis of the same assumptions, because both are unprovable, nothing can be stated about the truth value of both G and $\neg G$.

Things are quite different if we consider instead the assertion

"G is true or G is false"

that is,

"G is true or $\neg G$ is true"

which, being an instance of the principle of bivalence, is of course true, still reasoning on the basis of classical logic. The difference consists in the fact that, for any $A \in \mathcal{L}(\mathbf{PA})$, the assertion "$A$ is true" $\notin \mathcal{L}(\mathbf{PA})$. Hence, the latter assertion cannot be assessed within $\mathbf{PA}$.[5]

The object of our reflection is no more "the formal sentence G", but, so to speak, its "semantic content".

After "how", let's now consider "where" the truth of G could be stated, and I would like to stress that in this case the argument cannot be developed within a formal theory, where by *truth* we mean

truth in a structure

with respect to a sentence of a formal language. Here the focus is on the notion of

truth of a mathematical sentence

and so we are arguing within the pre-formal domain which we acknowledged as the necessary background of the building of informal and formal theories.

From both Gödel's Incompleteness Theorems we know that the truth of G cannot be the outcome of a proof *in* $\mathbf{PA}$: to this aim, in fact, it would be necessary *a proof of the unprovable*.

On the other hand, Tarski taught us that no predicate belonging to $\mathcal{L}(\mathbf{PA})$ can represent *in* $\mathbf{PA}$ the truth predicate relative to $\mathcal{L}(\mathbf{PA})$.

Thus, in order to *express* the very sentence "G is true" we have to move from $\mathbf{PA}$ to, say, $\mathbf{PA}^{\mathrm{TTT}}$, which is the theory we get when to $\mathbf{PA}$ we adjoin Tarskian Theory of Truth relative to $\mathbf{PA}$. $\mathbf{PA}^{\mathrm{TTT}}$ is obtained from $\mathbf{PA}$ by enriching the language with the new truth predicate $Tr_{\mathbf{PA}}(x)$, and by strengthening the deductive apparatus by means of the Tarski-biconditionals for managing the new predicate. Therefore, within $\mathbf{PA}^{\mathrm{TTT}}$ it is possible to express the Tarski-biconditional relative to G.

1.4.4   Truth of G: Closing the Circle

Note that Tarskian Theory of Truth *must* be formulated in a metatheory which, in order not to generate an infinite hierarchy, is to be treated in a semi-formal way. The right hand side of the relevant biconditional, that G *cannot be proved in* $\mathbf{PA}$, does not belong to $\mathcal{L}(\mathbf{PA})$, but to $\mathcal{L}(\mathbf{PA}^{\mathrm{TTT}})$. Better and, in a sense, closing the circle of our argument, this means that when we assert "G is true" we utter a sentence of our pre-formal mathematical language.

---

[5] What I mean is the impossibility to have in $\mathbf{PA}$ the uniform assertion of its soundness. It is well known, on the other hand, that it is possible to build *partial* truth definitions which, given, say, an arithmetical sentence $A \in \Sigma_n$, allow to formalize in $\mathbf{PA}$ the obvious intuitive implication from provability to truth with respect to sentences of *that* complexity.

In other words: I mean that in order to be able to acknowledge that the formalized (in PA^{TTT}) notion of truth (with respect to $\mathcal{L}(PA)$) is genuine, real, truth, it is necessary to reason adopting the standpoint of the pre-formal level of inquiry.[6]

References

Bassani, F., A. Marino, and C. Sbordone, eds. 2001. *E. De Giorgi. Anche la scienza ha bisogno di sognare*. Pisa: Pisa University Press.
Beklemishev, L.D. 2010. Gödel incompleteness theorems and the limits of their applicability. I. *Russian Mathematical Surveys* 65: 857–899.
Benacerraf, P. 1973. Mathematical truth. *The Journal of Philosophy* 70: 661–679.
Benacerraf, P., and H. Putnam, eds. 1983. *Philosophy of mathematics. Selected readings*. Cambridge: Cambridge University Press.
Buldt, B. 2014. The scope of Gödel's first incompleteness theorem. *Logica Universalis* 8: 499–552.
Ewald, William, ed. 1996. *From Kant to Hilbert. A source book in the foundations of mathematics*, vol. II. Oxford: Clarendon Press.
Gödel, K. 1986. *Collected works. Volume I (Publications 1929–1936)*, ed. S. Feferman, et al. New York/Oxford: Oxford University Press/Clarendon Press.
Gödel, K. 2003. *Collected works IV–V: Correspondence*, ed. S. Feferman, et al. New York: Oxford University Press.
Halbach, V., and A. Visser. 2014. Self-reference in arithmetic I and II. *The Review of Symbolic Logic* 7(4): 671–712.
Hempel, C.C. 1945. On the nature of mathematical truth. *American Mathematical Monthly* 52: 543. Reprinted in Feigl, H., and W. Sellars, *Readings in philosophical analysis*, New York, 1949.
Hilbert, D. 1918. Axiomatisches denken. *Mathematische Annalen* 78: 405–415. Reprinted in Hilbert, 1935. Eng. trans. in Ewald, 1996.
Lakatos, I. 1976. *Proofs and refutations*. Cambridge: Cambridge University Press. The articles were originally published in the *British Journal for the Philosophy of Science*, 1963–1964.
Pantsar, M. 2009. *Truth, proof and Gödelian arguments: A defence of Tarskian truth in mathematics*. Philosophical studies from the University of Helsinki, vol. 23. Department of Philosophy, University of Helsinki.
Piazza, M., and G. Pulcini. 2013. Strange case of Dr. soundness and Mr. consistency. In *Logica yearbook*, 161–172. College Publications.
Piazza, M., and G. Pulcini. 2015. A deflationary account of the truth of the Gödel sentence *G*. In *From logic to practice. Italian studies in the philosophy of mathematics*, ed. G. Lolli, M. Panza, and G. Venturi, 71–90. Heidelberg: Springer.
Piazza, M., and G. Pulcini. 2016. What's so special about the Gödel sentence *G*? In *Objectivity, realism, and proof. FilMat studies in the philosophy of mathematics*. Boston studies in the philosophy and history of science, vol. 318, ed. F.B.A. Sereni, 245–263. Cham: Springer.
Tait, W. 1986. Truth and proof: The platonism of mathematics. *Synthese* 69: 341–370.
Tarski, A. 1983. *Logic, semantics, metamathematics. Papers from 1923 to 1938*, 2nd revised ed. Indianapolis: Hackett.

---

[6]The question of (the sense in which one can speak about) the truth of *G* has been vastly debated: we limit to mention the very recent (Piazza and Pulcini 2013, 2015, 2016); each of them contains many pertinent references. For an interesting survey of the entire debate concerning the assessment of the (possible) truth-value of *G*, one can profitably see also (Pantsar 2009).

Chapter 2
Penrose's *New Argument* and Paradox

Johannes Stern

Abstract In this paper we take a closer look at Penrose's *New Argument* for the claim that the human mind cannot be mechanized and investigate whether the argument can be formalized in a sound and coherent way using a theory of truth and absolute provability. Our findings are negative; we can show that there will be no consistent theory that allows for a formalization of Penrose's argument in a straightforward way. In a second step we consider Penrose's overall strategy for arguing for his view and provide a reasonable theory of truth and absolute provability in which this strategy leads to a sound argument for the claim that the human mind cannot be mechanized. However, we argue that the argument is intuitively implausible since it relies on a pathological feature of the proposed theory.

2.1 Introduction

Gödel's Incompleteness Theorems are beyond doubt amongst the greatest and most interesting results in mathematical logic of the twentieth century and have attracted interest far beyond the frontiers of logic and mathematics. Gödel's theorems even found application outside of logic and mathematics, and, in fact, several non-mathematical propositions were claimed to be consequences of the incompleteness theorems. One particularly striking such proposition was the claim that *the human mind cannot be mechanized*, which was defended by several authors, most prominently Lucas (1961) and Penrose (1989, 1994, 1996). As a matter of fact the origins of this debate are with Gödel himself. Gödel in his celebrated Gibb's lecture, entitled *Some basic theorems on the foundations of mathematics and their implications* (Gödel 1995), explored some philosophical consequences

J. Stern (✉)
Department of Philosophy, University of Bristol, Bristol, UK
e-mail: johannes.stern@bristol.ac.uk

© Springer International Publishing AG, part of Springer Nature 2018   17
M. Piazza, G. Pulcini (eds.), *Truth, Existence and Explanation*,
Boston Studies in the Philosophy and History of Science 334,
https://doi.org/10.1007/978-3-319-93342-9_2

of his incompleteness theorems. Famously, his reflections led him to a disjunctive conclusion, which is known as Gödel's Disjunction:

> Either mathematics is incompletable in this sense, that its evident axioms can never be comprised in a finite rule, that is to say, the human mind (even within the realm of pure mathematics) infinitely surpasses the powers of any finite machine, or else there exist absolutely unsolvable diophantine problems (...). (Gödel 1995, p. 310)

According to the first disjunct the set of mathematical theorems that the (idealized) human mind can produce, that is prove, cannot be recursively axiomatized, indeed, the set is not recursively enumerable and therefore cannot be the output of a Turing machine or any other effective algorithm. The second disjunct asserts that there are sentences of the mathematical language such that neither the sentence nor its negation are among the mathematical theorems the human mind can produce. If this is paired with a classical, that is, bivalent view according to which each sentence is either true or false, then there are sentences that are absolutely unprovable.[1] The notion of absolute provability we have just employed is meant to stand for some intuitive notion of proof that is supposed to capture the process in which the human mind produces mathematical theorems.

In our terminology Gödel's Disjunction (henceforth GD) thus asserts that *either the set of absolutely provable sentences cannot be recursively enumerated or there exists a true mathematical sentence which is not absolutely provable*. Indeed, putting aside worries concerning the obscure notion of absolute provability for the moment, GD follows from the incompleteness theorems if it is understood in this way. However, Gödel made it clear, that he did not think that either one of the disjuncts of the disjunction could be established solely by appeal to the incompleteness theorems. In contrast, Lucas, Penrose and other authors thought differently and argued that the incompleteness theorems imply the first disjunct of GD. In other words, Lucas and Penrose thought that they could refute mechanism, that is, the idea that the outputs of the human mind can be produced by an effective algorithm.[2] There are basically two arguments for the first disjunct of GD. The first argument was popularized by Lucas (1961), while the second argument, Penrose's so-called *New Argument*, was introduced by Penrose (1994, 1996).

Lucas argued that while any recursively enumerable system F can never prove its Gödel sentence, that is, a sentence saying of itself that it is not provable in F, the human mind can always perceive it to be true: the Gödel sentence for F is absolutely provable but not provable in F. According to Lucas this shows that the human mind exceeds the notion of proof of every recursively axiomatizable mathematical theory.

---

[1]In fact, assuming bivalence is not necessary to "derive" Gödel's Disjunction. It suffices to argue that there are certain sentences that are either true or false but which are absolutely unprovable. Gödel (1995) argues convincingly that this situation also arises if non-classical mathematics is assumed.

[2]Mechanism can also be understood as the stronger thesis that the human mind functions like some particular algorithm, rather than that the human mind and some algorithm have the same outputs. We confine ourselves to the latter understanding and refer the reader to the Lindström (2001), Shapiro (2003, 2016) and Koellner (2016) for a discussion of this issue.

However, to draw this conclusion Lucas needs the additional assumption that the human mind can perceive that F is consistent, i.e., that it is absolutely provable that F is consistent. Otherwise, Lucas would have argued only for the conditional statement asserting that the Gödel sentence of F is true, if F is consistent. But this latter claim is provable in F itself and therefore not sufficient for establishing the first disjunct of GD. In general, it seems to be an implausible assumption that the— even idealized—human mind can perceive the consistency (and inconsistency) of every recursively axiomatizable theory.[3] Moreover, as Koellner (2016) points out, this assumption is also problematic for more principled reasons because it implies that we are in possession of a Π^0_1-oracle, which is tantamount to asserting that the outputs of a human mind cannot be produced by a finite machine—but this is precisely what Lucas' argument was meant to show. All of this suggests that Lucas' argument does not make a convincing case for the first disjunct of GD.

In this paper we shall chiefly be concerned with Penrose's so-called *New Argument*. As we shall see Penrose added a further twist to the Lucas-style arguments for the first disjunct of GD by introducing some basic assumptions concerning the notion of absolute provability. This brings the obscure notion of absolute provability back into the limelight and one might reasonably expect Penrose to provide some clarification. After all without further explanation this notion remains entirely mysterious and opaque, and it is impossible to evaluate claims that are made on behalf of this notion. Moreover, by introducing specific assumptions on behalf of absolute provability he also provides his opponent with an opportunity for resisting such an argument for the first disjunct of GD. While it would be an unreasonable move to deny Gödel's incompleteness theorems, as they are part of core mathematics, it is perfectly acceptable to resist some particular assumption concerning absolute provability[4]: there is no generally agreed theory of absolute provability Penrose could rely on. However, in this paper we will not discuss these issues any further and refer the reader interested in a more thorough discussion of the notion of absolute provability to the work by Myhill (1960) and Leitgeb (2009).[5,6] Rather following the lead of Koellner (2016), we will pursue a charitable approach throughout this paper and grant Penrose the assumptions he makes concerning absolute provability. We shall query whether under Penrose's assumptions his *New Argument* is a coherent argument for the first disjunct of GD.

---

[3] See also Benacerraf (1967) or Lindström (2001) for remarks along these lines.

[4] For remarks along these lines see, e.g., Shapiro (2003).

[5] Absolute provability and GD have also been studied in so-called *epistemic arithmetic*. See, for example, Shapiro (1985), Reinhardt (1986), and Carlson (2000).

[6] Ultimately, we are very skeptical whether anything interesting and coherent can be said on behalf of absolute provability but as indicated in the main text we bracket this worry for the purpose of this paper.

2.2 Penrose's New Argument

Penrose (1996) summarizes his *New Argument* in the following way, where F is some recursively axiomatizable system

> Though I don't know that I am necessarily F, I conclude that if I were, then system F would have to be sound, more to the point, F' would have to be sound, where F' is F supplemented by the further assertion "I am F". I perceive that it follows from the assumption that I am F that the Gödel statement $G(F')$ would have to be true and, furthermore, that it would not be a consequence of F'. But I have just perceived that if "if I happened to be F, then $G(F')$ would have to be true", and perceptions of this nature is precisely what F is supposed to achieve. Since I am F, I deduce that I cannot be F after all. (Penrose 1996)

Let us try to reconstruct the argument in an explicit way.[7] This will also help us to assess which assumptions concerning absolute provability are required for carrying out the argument. To start the reconstruction we need to introduce some additional terminology and to say a little bit more about what these *systems* are that Penrose alludes to in the argument. From now on we take a system to be a recursively enumerable set that is closed under modus ponens and extends Peano Arithmetic (PA).[8] We assume a system to be closed under modus ponens because modus ponens is the standard rule of proof in logic. This assumption thus guarantees that a system qua set of sentences does not deviate from the set of theorems that the system can prove.[9,10]

To reconstruct Penrose's argument we also need a sentential predicate K that expresses absolute provability. The intended interpretation of this predicate is the set of mathematical theorems the human mind can produce. Finally we need to clarify what it means for a system F to be sound and how this is expressed. A system F is sound iff all its theorems are true. If we avail ourselves to a metalinguistic perspective then the soundness of the system F can be expressed by the schemata

$$F\ulcorner\phi\urcorner \to \phi,$$

which needs to holds for all sentences of the language.[11] Throughout the paper $\ulcorner\phi\urcorner$ is the numeral of the Gödel number of ϕ and thus acts as name of ϕ. Notice

---

[7]See Shapiro (2003) for a very similar reconstruction of Penrose's argument.

[8]Both assumption are plausible and frequently assumed in the literature. See, e.g., Koellner (2016) for discussion. A weaker arithmetical theory would do as well as long as it is sufficient for proving Gödel's incompleteness theorems.

[9]Everything we say at this point would go through even if we do not assume a system to be closed under modus ponens. In this case we would need to distinguish throughout the argument between the system and what the system proves.

[10]Of course, there are logics in which modus ponens is not a sound rule of proof. But in such systems there will be alternative rules of proof and a similar problem will arise. Moreover, we would be surprised, if Penrose were to argue against modus ponens.

[11]Such a metalinguistic perspective is not unproblematic since it does not seem to be available to the anti-mechanist for he would need to step outside the system F, i.e. their "mind", to assume it. At this point we ignore this problem but it is yet another reason why formulating Penrose's *New Argument* in a rigorous way seems to require a truth predicate in the language (see below).

that by Gödel's second incompleteness theorem no recursively axiomatizable theory can prove its own soundness.

We now carry out the reconstruction in a step-by-step manner:

1. if I were [the system F], then system F would have to be sound.[12]

$$\forall x(Fx \leftrightarrow Kx) \to (F\ulcorner\phi\urcorner \to \phi) \qquad \text{for all } \phi$$

2. F' would have to be sound, where F' is F supplemented by the further assertion "I am F".

$$\forall x(Fx \leftrightarrow Kx) \to (F'\ulcorner\phi\urcorner \to \phi) \qquad \text{for all } \phi$$

3. I perceive that it follows from the assumption that I am F that the Gödel statement $G(F')$ would have to be true

$$\forall x(Fx \leftrightarrow Kx) \to G(F')$$

4. furthermore, that it [$G(F')$] would not be a consequence of F'

(i) $\qquad\qquad \forall x(Fx \leftrightarrow Kx) \to \neg F'\ulcorner G(F')\urcorner$

(ii) $\qquad\qquad \forall x(Fx \leftrightarrow Kx) \to \neg F\ulcorner\forall x(Fx \leftrightarrow Kx) \to G(F')\urcorner$

5. I have just perceived that if "if I happened to be F, then $G(F')$ would have to be true"

$$K\ulcorner\forall x(Fx \leftrightarrow Kx) \to G(F')\urcorner$$

6. Since I am F, I deduce that I cannot be F after all

(i) $\qquad\qquad \forall x(Fx \leftrightarrow Kx) \to F\ulcorner\forall x(Fx \leftrightarrow Kx) \to G(F')\urcorner$

(ii) $\qquad\qquad \forall x(Fx \leftrightarrow Kx) \to \bot.$

To get the argument off the ground we need to assume that notion of absolute provability is a sound notion: every sentence that is absolutely provable, i.e. every mathematical theorem the human mind produces, is true. In other words we need to assume for all sentences ϕ of the language:

(T) $\qquad\qquad\qquad\qquad\qquad\qquad K\ulcorner\phi\urcorner \to \phi.$

If (T) is assumed, Line 1 is a valid assumption and we may proceed from there. Line 2 then follows by weakening (and modus ponens). Notice that by the deduction theorem we have the following equivalence

---

[12]Throughout the paper we take $\forall x\phi$ to be short for $\forall x(\mathsf{Sent}(x) \to \phi(x))$ unless we explicitly mention the restriction of the quantifier. Sent is a predicate representing the set of sentences of the language.

(†) $F'(\ulcorner \phi \urcorner) \leftrightarrow F(\ulcorner \forall x(Fx \leftrightarrow Kx) \rightarrow \phi \urcorner),$

which will be important later in the argument.

We obtain Line 3 by instantiating the schematic Line 2 to the Gödel sentence of F'. By the very construction of the Gödel sentence Line 3 is equivalent to Line 4(i) and by (†) the latter is equivalent to 4(ii).[13] Line 5 requires a further assumption on behalf of absolute provability: if the human mind has produced a theorem, i.e., a sentence has been shown to be absolutely provable, then this fact itself is absolutely provable. This is nothing but the rule of necessitation for absolute provability:

(Nec) $\dfrac{\phi}{K \ulcorner \phi \urcorner}.$

Given the rule of necessitation we obtain Line 5 from Line 3. By classical logic Line 5 implies Line 6(i) but Line 6(i) together with Line 4(ii) yields Line 6(ii), which establishes the first disjunct of GD, namely, that the mathematical theorems the human mind can produce do not coincide with the theorems of any recursively axiomatizable system. Notice that in order to draw this conclusion it was important that no special assumption concerning F was made, since we need to use the rule of universal generalization in the metalanguage.[14] Thus assuming the principle (T) and the rule (Nec) each step of Penrose's *New Argument* is sound and consequently one might think that Penrose has succeeded in providing a coherent argument for the first disjunct of GD. Unfortunately though, the assumptions concerning absolute provability, that is the principle (T) and the rule (Nec), are jointly inconsistent as Myhill (1960) and Montague (1963) have taught us.[15]

Theorem 1 (Myhill/Montague) *Let Σ be a theory extending Robinson Arithmetic such that for all sentences ϕ of the language:*

(i) $\Sigma \vdash K \ulcorner \phi \urcorner \rightarrow \phi$
(ii) $\Sigma \vdash \phi \Rightarrow \Sigma \vdash K \ulcorner \phi \urcorner$

for some primitive or complex predicate K. Then Σ is inconsistent.

This inconsistency result points to a further complication in providing an argument for the first disjunct of GD: it is not sufficient to suggest plausible principles for absolute provability but one also has to guarantee that these principles are jointly consistent. Moreover, the result by Myhill and Montague is only one of many inconsistency results and it is not an easy task to provide a satisfactory account

---

[13]The Gödel sentence $G(F')$ is a sentence for which, using the diagonal lemma, we can prove $\neg F'\ulcorner G(F')\urcorner \leftrightarrow G(F')$.

[14]By the argument we obtain $\neg \forall x(Fx \leftrightarrow Kx)$ and by universal generalization in the metalanguage $\forall F \neg \forall x(Fx \leftrightarrow Kx)$. This yields $\neg \exists F \forall x(Fx \leftrightarrow Kx)$: there exists no recursively axiomatizable system F that coincides with the set of absolutely provable sentences.

[15]The fact that the principles Penrose appeals to in formulating his *New Argument* are jointly inconsistent was already pointed out by Chalmers (1995) and Shapiro (2003), amongst others.

or theory of absolute provability.[16] For this reason the aforementioned inconsistency results are also known as intensional paradoxes.

Indeed, according to Wang (1996), Gödel thought that once we were in possession of a satisfactory solution to the intensional paradoxes, we could successfully establish the first disjunct, that is, we could show that the human mind surpasses any finite machine:

> If one could clear up the intensional paradoxes somehow, one would get a clear proof that mind is not [a] machine. (Wang 1996, p. 187)

Even though it is perhaps safe to say that now, almost half a century later, we still lack an entirely satisfactory solution to the intensional paradoxes, it is interesting to apply the existing proposals to Penrose's *New Argument* and check whether we thereby obtain a coherent argument in favor of the first disjunct of GD. If this were possible, this would be a partial vindication of Penrose but also of Gödel's conviction that a satisfactory solution to the intensional paradoxes would lead to an argument for the first disjunct of GD.

2.3   Truth and Intensional Paradox

The intensional paradoxes, which affect notions such as proof, knowledge, or belief, have not received quite as much attention as the semantic paradoxes such as the Liar paradox. But at least from a technical point of view the intensional and the semantic paradoxes are closely related. In both cases paradox arises due to the application of these notions to themselves. As a consequence, in both cases we have roughly the same options how paradox can be avoided. The first option is to restrict the characteristic schemes of these notions to a salient set of sentences that in some way or another singles out the "good" instances from the "bad" ones in such a way that paradox can no longer arise. For example, in the case of absolute provability this would amount to restricting the scheme (T) and the rule (Nec) to specific sentences so that Montague's theorem no longer applies. The second, alternative option would be to adopt weaker principles or schemata for these self-applicable notions; principles that even if applied in full generality are jointly consistent.

This latter option of adopting weaker schemata does not seem to be a promising option if one is interested in resurrecting Penrose's *New Argument*. As we have seen, we need (T) and (Nec) in order to carry out Penrose's reasoning and, at least as long as no further resources are added, weaker principles will simply not do. So we are left with the former option of restricting the scope of the schemata. However, as far as we can see there is no obvious restriction that would enable us to vindicate Penrose's *New Argument*. The most immediate one that comes to mind is to restrict the schemata to arithmetical sentences. But this restriction

---

[16]See Egré (2005) or Stern (2016, Chap. 3) for an overview of the various inconsistency results.

would make the transition from Line 1 to Line 2 and from Line 4 to Line 5 of the argument invalid.[17] A further and maybe more plausible strategy might be to divide the sentences of the language into the paradoxical and the non-paradoxical ones. Unfortunately, this distinction has proven rather elusive in the past and difficult to pin down. Moreover, Penrose needs the classification, or at least a principled sub-classification thereof, to be decidable for otherwise the reasoning leading to the first disjunct could not be carried out in a recursively axiomatizable system.[18] This would undermine the reductio-strategy of his argument, which is based on the assumption that the human mind is recursively axiomatizable. As a consequences the chances of finding an acceptable restriction become even smaller since the more promising classifications proposed in the literature are not decidable, indeed they are of much greater complexity. Now, even if, against all odds, we manage to restrict the schemata in a recursive way that divides the sentences of the language into the "good", non-paradoxical and the "bad", paradoxical ones, we still need to make sure that the sentence (Nec) is applied to in Line 5 is such a "good" case. Unfortunately, the work by Koellner (2016) shows that this will not generally be the case.

These remarks suggest that the usual options for dealing with the paradoxes of self-applicable notions may not yield a formal framework in which the first disjunct of GD can be proved along the lines of Penrose's *New Argument*. Fortunately, there is a further option that arises for certain intensional and semantic notions. The idea is that at the root of all the paradoxes there is only one paradoxical notion, namely the notion of truth. This idea would tie the intensional paradoxes to the semantic, i.e. truth-theoretic, paradoxes in such a way that the paradoxicality of the intensional notion, e.g. absolute provability, depends solely on the paradoxicality of the notion of truth. The central idea of the strategy is to formulate the principles of absolute provability using the truth predicate. For example, the principle (T) we have appealed to in reconstructing Penrose's reasoning would then be formulated as

$$(\mathsf{T_K}) \qquad\qquad\qquad \forall x(Kx \to Tx).$$

If these revised principles of absolute provability are combined with a consistent theory of truth, we obtain a consistent theory of truth and absolute provability. Such a strategy has been recently developed by Stern (2014a,b, 2016) and, independently, Koellner (2016). Moreover, Koellner's work is directly motivated by Gödel's Disjunction and the evaluation of Penrose's *New Argument*. To implement this strategy we need to introduce a truth predicate to the framework that is allowed to interact with the absolute provability predicate. But Penrose's *New Argument* makes

---

[17]Introducing a hierarchy of typed absolute provability predicates would not be of any help here. Rather at each level we would face the problem anew and could never draw the desired conclusion. See Shapiro (2003) for remarks along these lines.

[18]The restriction has to be decidable only if we assume the schemata to be characteristic of the notion of absolute provability, that is, if they are assumed to be the axioms of the recursively axiomatizable system. Otherwise, the restriction could be recursively enumerable.

explicit use of the notion of truth at several places[19] and thus an evaluation of his argument in a framework where we have both, an absolute provability predicate *and* a truth predicate, seems highly desirable and independently motivated. In the remainder of this paper we shall therefore adopt this framework and investigate whether in such a framework we can provide a coherent argument for the first disjunct of GD.

2.3.1 *Penrose's* New Argument *Reconsidered*

Koellner (2016) extracts three principle of truth and absolute provability he takes to be crucial for reconstructing Penrose's argument in a language with a truth and an absolute provability predicate. These principles are the rule (Nec) and the principles

$$(T_K) \qquad\qquad \forall x(K(x) \to Tx)$$

$$(T\text{-In}) \qquad\qquad \phi \to T\ulcorner\phi\urcorner$$

Using these principles Penrose's argument can be formalized, roughly following Koellner (2016), as follows:

1. $\forall x(Fx \leftrightarrow Kx) \to \forall x(Fx \to Tx)$ by (T_K)
2. $\forall x(Fx \leftrightarrow Kx) \to \forall x(F'x \to Tx)$ by $(T\text{-In})$
3. $\forall x(Fx \leftrightarrow Kx) \to G(F')$ by 2 and definition of $G(F')$
4. $\forall x(Fx \leftrightarrow Kx) \to \neg F'\ulcorner G(F')\urcorner$ by definition of $G(F')$
5. $K\ulcorner\forall x(Fx \leftrightarrow Kx) \to G(F')\urcorner$ 4, by (Nec)
6. $\forall x(Fx \leftrightarrow Kx) \to \neg F\ulcorner\forall x(Fx \leftrightarrow Kx) \to G(F')\urcorner$ 4, by definition of F'.
7. $\forall x(Fx \leftrightarrow Kx) \to F\ulcorner\forall x(Fx \leftrightarrow Kx) \to G(F')\urcorner$ 5
8. $\forall x(Fx \leftrightarrow Kx) \to \bot$ 6, 7.

Shortly, we will take a closer look at this reconstruction and investigate whether the different steps of the argument are valid, given the principles of truth and absolute provability Koellner puts forth. But before we do so, it is worth pointing out that even in the presence of a truth predicate we need a self-applicable absolute provability predicate to carry out the reasoning. For example, in Line 5 the absolute provability predicate is applied to a sentence containing the predicate itself. What about the truth predicate; need it be self-applicable? On the face of it at no point in the argument is the truth predicate applied to a sentence in which it explicitly occurs and, consequently, one might think that a typed truth predicate, that is, a truth predicate that can only be applied to sentences in which the truth predicate does not occur, will be sufficient for the argument. Eventually we will have a look at the possibility of carrying out Penrose's *New Argument* within a typed theory

---

[19]See Line 1, 2, 3 and 5.

of truth but ultimately opting for a typed truth predicate is a cheat. After all in the present set up the (idealized) human mind is meant to reflect about its proofs and capacities. For example, the human mind is supposed to perceive that it is sound, that is, the principle (T_K) should be absolutely provable.[20] But then it follows from (T_K) itself that we may apply the truth predicate to sentences in which it occurs.[21] Therefore, it is essentially right when Koellner (2016) concludes that *"any formal system in which the above argument can be implemented will be one involving a type-free theory of K and a type-free theory of T"*. This, of course, leads to the question whether there are such theories of truth and absolute provability in which Penrose's *New Argument* or, more generally, arguments for the first disjunct of GD, can be carried out. Koellner (2016) seems to be skeptical in this respect but detects a general problem for establishing such a negative conclusion:[22]

> It would be ideal if we could quantify over "all possible type-free theories of truth" and show that no such theory yields a system that provided a convincing argument for the first disjunct. But given the open-endedness of the notion of a possible "type-free theory" it is hard to see how to do this. (Koellner 2016, p. 166)

As a consequence Koellner focuses on one sample theory of truth and absolute provability he deems to be particularly attractive and shows that in this theory no argument for the first disjunct can be provided. We shall take a somewhat different approach to Koellner and propose to split up Koellner's initial question into two distinct questions: (1) Can Penrose's *New Argument* be carried out in some consistent theory of truth and absolute provability? (2) Are there reasonable theories of truth and absolute provability that yield an argument for the first disjunct of GD? The first question focuses entirely on whether one can coherently argue for the first disjunct of GD following the outlines of Penrose's *New Argument*. The second question, however, asks for a coherent argument for the first disjunct of GD tout court. There is no restriction on the structure of the argument and, in particular, we don't need to respect the outlines of Penrose's *New Argument*. While it seems difficult to provide a conclusive answer for the second question, which is the question Koellner addresses, such a conclusive answer might be possible for the first question.

Indeed we think that for the first question a rather strong conclusion is possible provided the reconstruction of Penrose's *New Argument* we employ is accepted: there seems to be no coherent theory of truth and absolute provability in which the *New Argument* can be carried out. In contrast, for the second question, we show

---

[20]Penrose (1996) explicitly agrees with this claim.

[21]It is possible to block this conclusion by restricting the principle (T_K) to sentences of the language without the truth predicate. However, this would not really affect the argument: after all we would have sentences in which the truth predicate is applied to sentences which the knowledge predicate occurs. The knowledge predicate, in turn, may be applied to sentences in which the truth predicate occurs. So the truth predicate is applied to sentences in which it implicitly occurs. In order to maintain a coherent picture it should be possible to apply the truth predicate to sentences containing itself.

[22]See Koellner (2016, p. 184/185) for a clear expression of his skepticism.

that, under certain assumptions, there are reasonable theories of truth and absolute provability in which a sound argument for the first disjunct can be given. However, as we shall see, even though the argument is sound it is not very convincing for it exploits a pathological feature of the theories of truth and absolute provability we shall be considering. Since we are dealing with paradoxes, all solutions to these paradoxes, i.e. theories of truth and absolute provability, will have certain pathological features. We think that if an argument for the first disjunct relies on the pathological part rather than the intuitively motivated part of the theory then it will fail to be convincing. After all it will lack any intuitive support and will remain a technical peculiarity. Ultimately, we are skeptical as to whether a *convincing* argument for the first disjunct can be given in a reasonable theory of truth and absolute provability.

We now return to the first question. To this end we need to take a closer look at the reconstruction of Penrose's argument Koellner proposes. Koellner (2016) does not investigate the argument in detail for he can show that in the theory of truth DT (Feferman 2008) he is considering the first disjunct of GD is independent: it can be neither proved nor refuted. However, if all steps of the argument were sound given the assumptions Koellner puts forth, then there would be consistent theories of truth and absolute provability in which the *New Argument* could be carried out.[23] However, we already find ourselves in trouble when trying to reconstruct the inference from Line 1 to Line 2. To see this, it is worth recalling that F is closed under modus ponens.

If we add to the system F the sentence $\forall x(Fx \leftrightarrow Kx)$ we obtain a new recursively enumerable set. Let us call this set Σ. Σ may not be a system in our sense since it may not be closed under modus ponens. It is by closing Σ under modus ponens that we obtain the system F'. Maybe surprisingly, it is precisely this closure under *modus ponens* that creates a problem in the step from Line 1 to Line 2. Even though we know by Line 1 and $(T\text{-In})$ that all members of Σ are true, we do not know without further assumption that all the members F' are true because we do not know whether the truth predicate is closed under *modus ponens*. For all we know there could be sentences ϕ and ψ such that $T\ulcorner\phi\urcorner$, $T\ulcorner\phi \rightarrow \psi\urcorner$ but $\neg T\ulcorner\psi\urcorner$. To guarantee that the truth predicate will be closed under *modus ponens* we need to add this requirement as a further assumption:

$(T\text{-Imp})$ $\qquad \forall x, y(\mathsf{Sent}(x \rightarrow y) \rightarrow (T(x \rightarrow y) \rightarrow (Tx \rightarrow Ty)))$.[24]

If $(T\text{-Imp})$ is assumed, then the inference from Line 1 to Line 2 follows from $(T\text{-In})$, $(T\text{-Imp})$ and an induction on the length of a proof in F'.

---

[23] There is an interpretation of the resulting theory in the theory KF together with the completeness axiom. The interpretation would translate the absolute provability predicate as the truth predicate but hold the remaining vocabulary fixed. See Halbach (2011) for more on the truth theory KF and the completeness axiom.

[24] Notice that it is of no help to replace F' by Σ, which is not closed under modus ponens, throughout the argument. To carry out the argument we need to show that whatever can be *proved*

There is a further small lacuna in the proposed reconstruction of the argument, namely, in order to derive Line 3 we need to assume that no false arithmetical sentence is true (in the object-linguistic sense).[25] There are several ways this can be achieved, but one straightforward and rather plausible way is to require the truth predicate to be an adequate truth predicate for the arithmetical language. That is, the Tarski biconditionals should hold for sentences of the arithmetical language: for arithmetical sentences ϕ

(TB) $T\ulcorner\phi\urcorner \leftrightarrow \phi.$

Assuming (TB) we can derive Line 3 as illustrated below where $G_{F'}$ stands for the Gödel sentence of F', i.e., a sentence such that

$$\neg F'(\ulcorner G(F')\urcorner) \leftrightarrow G(F')$$

can be proved via the diagonal lemma. We start our reasoning from Line 2.

2. $\forall x(Fx \leftrightarrow Kx) \rightarrow \forall x(F'x \rightarrow Tx)$

   a. $\forall x(Fx \leftrightarrow Kx) \rightarrow (F'\ulcorner G(F')\urcorner) \rightarrow T\ulcorner G(F')\urcorner$ by Line 2.

   b. $\forall x(Fx \leftrightarrow Kx) \rightarrow (F'\ulcorner G(F')\urcorner \rightarrow G(F'))$ by Line 2(a)., (TB)

3. $\forall x(Fx \leftrightarrow Kx) \rightarrow G(F')$ by Line 2(b) and the definition of $G_{F'}$.

This closes the final gap in the reconstruction of Penrose's *New Argument*. If (T_K), $(T\text{-In})$, $(T\text{-Imp})$, (TB) and (Nec) are assumed then every step of the *New Argument* is sound. But, unfortunately, we find ourselves essentially in the same situation as in the reconstruction of the argument without the truth predicate: the assumptions needed for carrying out the argument are jointly inconsistent.

Theorem 2 (Folklore) *Let Σ be a theory extending Robinson arithmetic. Then*

$$(T\text{-In}), (T\text{-Imp}), (TB) \vdash_\Sigma \bot.[26]$$

The fact that the principles of truth required for carrying out Penrose's reasoning are jointly inconsistent suggests a deep flaw with his reasoning: Penrose seems to rely on a *naive* notion of truth. But this is something we cannot do because *naive* conceptions of truth, or absolute provability for that matter, lead to paradox. This lesson has been taught to us by Tarski and Gödel a long time ago. Admittedly, at this point such a strong conclusion seems a bit premature for two reasons. First, so far we have not proved that the first disjunct of GD cannot be derived using the

---

from Σ is true. To show this we need to assume the truth predicate to be closed under modus ponens.

[25] Actually, we only need to assume $\neg T\ulcorner 0 = 1\urcorner$. However, this would not change the general situation. In particular Theorem 2 would still hold if (TB) were replaced by $\neg T\ulcorner 0 = 1\urcorner$.

[26] See Friedman and Sheard (1987) for a proof of Theorem 2.

principles Koellner puts forth, that is, we have not shown that the first disjunct of GD is independent of these principles. However, as we show in an appendix to this paper, even if we assume the truth predicate to be arithmetically sound, the first disjunct does not follow from these principles.

Now, the second reason why, at this point, one should be careful with drawing too strong conclusions from the inconsistency result is that, as we noted earlier, from a technical point of view it might be possible to carry out Penrose's *New Argument* using typed principles of truth, that is, principles of truth in which the truth predicate is not applicable to sentences in which it occurs. Earlier in this paper we dismissed such a typing restriction for philosophical and systematic reasons but if we could carry out Penrose's *New Argument* using typed principles that are jointly consistent this would show that Penrose's reasoning does not necessarily appeal to a *naive* (and inconsistent) notion of truth.

2.3.2 Penrose's **New Argument** *and Simplistic Typing*

Looking back at Koellner's reconstruction of the *New Argument* and our slight amendment thereof it seems that at least the principles $(T\text{-In})$, $(\mathsf{T_K})$ and (Nec) can be restricted to sentences in which the truth predicate does not occur and Penrose's reasoning would still to go through. Notice that the typing restriction we have in mind is rather crude and simplistic because it blocks the use of these principles with respect to sentences in which the truth predicate occurs while it allows for the application of these principles to sentences in which the truth predicate is mentioned. From this point of view the typing restriction we propose is unprincipled and ad hoc but, as we have pointed out, at this point we are only interested whether the restriction turns the *New Argument* into a valid argument.

To this end we let $\mathscr{L}_{\mathcal{H}}$ be the language without the truth predicate, $\mathsf{Sent}_{\mathscr{L}_{\mathcal{H}}}$ a predicate the set of sentences of the language and replace the principles $(T\text{-In})$, $(\mathsf{T_K})$ and (Nec) by the following variants:

$(\mathsf{T_K^R})$ $\qquad\qquad\qquad \forall x(\mathsf{Sent}_{\mathscr{L}_{\mathcal{H}}}(x) \rightarrow (Kx \rightarrow Tx))$,

$(T\text{-In}_K)$ $\qquad\qquad\qquad \phi \rightarrow T\ulcorner\phi\urcorner$ $\qquad\qquad\qquad\qquad\qquad\qquad \phi \in \mathscr{L}_{\mathcal{H}}$,

(Nec_K) $\qquad\qquad\qquad \dfrac{\phi}{K\ulcorner\phi\urcorner}$ $\qquad\qquad\qquad\qquad\qquad\qquad\quad \phi \in \mathscr{L}_{\mathcal{H}}$.

Unfortunately, the simple minded typing restriction of the principles at play won't be sufficient to block a variant of Theorem 2. In other words the principles required for deriving the first disjunct are still jointly inconsistent.

Theorem 3 *Let Σ be a theory extending Robinson arithmetic. Then*

$$(\mathsf{T_K^R}), (T\text{-In}_K), (T\text{-Imp}), (\mathsf{Nec}), \mathsf{TB} \vdash_\Sigma \bot.^{27}$$

Proof Let δ be a sentence of $\mathscr{L}_{\mathscr{K}}$ such that

$(\ddagger)$ $\Sigma \vdash \delta \leftrightarrow \neg K \ulcorner \delta \urcorner.$

We reason as follows:

1. $K \ulcorner \delta \urcorner \to T \ulcorner \delta \urcorner$ (T_K)
2. $K \ulcorner \delta \urcorner \to T \ulcorner \neg \delta \urcorner$ $(\ddagger), (T\text{-In}_K)$
3. $K \ulcorner \delta \urcorner \to T \ulcorner \neg \delta \wedge \delta \urcorner$ $1, 2, (T\text{-In}_K), (T\text{-Imp})$
4. $K \ulcorner \delta \urcorner \to \bot$ $3, (T\text{-In}_K), (T\text{-Imp}), \mathsf{TB}$
5. $\neg K \ulcorner \delta \urcorner$ 4
6. δ $5, (\ddagger)$
7. $K \ulcorner \delta \urcorner$ $6, (\mathsf{Nec}_K)$
8. $\bot$ $5, 7$
                                                                                                                    □

Examining the reconstruction of Penrose's *New Argument* there does not seem to be an alternative possible restriction of the principles at play that would make the principles jointly consistent while facilitating Penrose's reasoning. Theorem 3 should therefore put an end to all attempts of vindicating the *New Argument* along the lines of our proposed reconstruction. Of course, Penrose could try a similar strategy to the one we discussed at the beginning of Sect. 3. That is, he could argue that there is some restriction that blocks the use of, say, (Nec) in the "bad" cases like in Line 7 in the proof of Theorem 3 but allows the use in the "good" cases like Line 5 in Koellner's reconstruction of the *New Argument*. But all the critical remarks we made a the beginning of Sect. 3 will carry over to the present case. In conclusion it thus seems fair to say that Penrose's *New Argument* is simply incoherent. The problem is that its reasoning makes essential use of a *naive* notion of truth (or absolute provability), which in a context where a certain amount of self-applicability is required irrevocably leads to paradox.

2.4 Global Reflection and the First Disjunct

Under the proposed formalization, Penrose's *New Argument* cannot be turned into a coherent argument but, of course, there might be alternative arguments for the first disjunct of GD. This leads to the second question we outlined in Sect. 3, i.e., the

---

[27]In order to carry out Penrose's reasoning we cannot restrict $(T\text{-Imp})$ to sentences of $\mathscr{L}_{\mathscr{K}}$. However, in the formulation of Theorem 3 we could use the restricted version of the principle as its proof should make clear.

question of whether there are reasonable theories of truth and absolute provability that yield an argument for the first disjunct of GD. As Koellner (2016) points out, given the plentitude of theories of truth and absolute provability it seems difficult to deal with all possible such theories. In contrast to Koellner, however, we do not focus on one particular theory of truth.[28] Instead we investigate the prospects of providing a successful argument for the first disjunct using Penrose's reductio strategy. To this end, we shall now make stronger assumptions on what a formal system F is, that is, from now on we will take systems F to be *classical* recursively axiomatizable theories. Under these presuppositions the assumption that "I am F" can be reformulated as $\forall x(\mathsf{Pr}_F(x) \leftrightarrow Kx)$, where Pr_F is a natural provability predicate of the theory F. This amounts to a strengthening of the Mechanistic thesis as we now only consider Turing machines whose outputs are closed under classical logic. In other words we now focus on arguments for the claim that *the human mind cannot be mechanized if classical logic is assumed* rather than arguments for the claim that *the human mind cannot be mechanized* simpliciter.

Throughout we shall assume the principle (T_K) and look for consistent theories of truth and absolute provability in which the assumption that "I am F" leads to a contradiction. Under this assumption (T_K) implies the so-called Global Reflection principle for F:

$$(\mathsf{GRef}_F) \qquad\qquad \forall x(\mathsf{Pr}_F(x) \rightarrow Tx).$$

Of course, no recursively axiomatizable theory F that is only remotely plausible can prove the Global Reflection principle for itself because by Gödel's second incompleteness theorem we have

$$(*) \qquad\qquad F \vdash \forall x(\mathsf{Pr}_F(x) \rightarrow Tx) \quad\Rightarrow\quad F \vdash \perp.^{29}$$

But then, since the assumption that "I am F" together with (T_K) implies the Global Reflection Principle, which in turn implies a contradiction it might seem that we have already provided a reductio ad absurdum of the assumption and thereby established the first disjunct of GD.

This would be a rather premature conclusion in at least two ways: First, while we can prove that if F proves the Global Reflection Principle, then F is inconsistent we cannot prove that F proves that the Global Reflection Principle implies a contradiction, i.e., F does not necessarily prove

$$(**) \qquad\qquad \forall x(\mathsf{Pr}_F(x) \rightarrow Tx) \rightarrow \perp.$$

---

[28] Koellner (2016) constructs the theory DTK, which extends Feferman's theory DT (Feferman 2008) and shows that in this theory the first disjunct of GD is independent: it can be neither proved nor refuted.

[29] We take it that any remotely plausible theory prevents the truth predicate from being entirely trivial. In other words it should rule out the truth of false arithmetical sentences, i.e., $\neg T\ulcorner 0 = 1\urcorner$ should be a theorem of F.

But a successful argument for the first disjunct via the reductio strategy requires a proof of (∗∗). The weaker claim (∗) is not sufficient for deriving the first disjunct of GD.

Second, and more importantly, in conducting the reductio argument we may not assume that we actually reason in F for this would not only presuppose that "I am F" but also that *I know which system F I am*. This point goes back to Benacerraf (1967) and was further discussed by, e.g., Reinhardt (1986). Reinhardt showed, at least under a particular formalization, that while it is inconsistent to assume that "I am F" and that "I know which system F I am", it is consistent, to assume that "I am F", while *not knowing which particular system F I am*.[30] In more technical terms this means even though the idealized human mind might coincide with a particular system F, it may not recognize the provability predicate Pr_F as "its" provability predicate, that is the provability predicate of F. As a consequence, in the reductio argument we cannot assume the Löb derivability conditions and, in particular, we may not infer $Pr_F(\ulcorner\phi\urcorner)$ whenever ϕ has been derived.[31] This observation is crucial since otherwise Penrose's *New Argument* would have a coherent reconstruction. By using Löb's derivability conditions we could dispense of the rule (Nec) in the reconstruction of the argument in Sect. 2.2 and thereby restrain ourselves to a consistent set of principles of absolute provability. The argument would then run as follows (we start with Line 3):

3. $\forall x(Fx \leftrightarrow Kx) \rightarrow G(F')$
4. $\forall x(Fx \leftrightarrow Kx) \rightarrow \neg F\ulcorner\forall x(Fx \leftrightarrow Kx) \rightarrow G(F')\urcorner$
5. $F\ulcorner\forall x(Fx \leftrightarrow Kx) \rightarrow G(F')\urcorner$ 3, by Löb's derivability conditions
6. $\forall x(Fx \leftrightarrow Kx) \rightarrow \bot$.

Bearing the previous remarks in mind this argument is clearly not a convincing argument for the first disjunct. It fails to establish that there is no recursively axiomatizable theory whose theorems coincide with the absolutely provable sentences but shows that there is some particular system F that falls short of absolute provability: the system F in which line 3 can be proven and the human mind, that is F, knows that this is so. In other words, the argument establishes that it is impossible that *I am F and I know that I am F* but fails to establish the first disjunct of GD. Recapitulating our discussion, the question of whether we can successfully argue for the first disjunct of Gödel's Disjunction, or rather our modified version thereof, using the reductio strategy depends on whether, given a reasonable theory of truth and absolute provability, the Global Reflection Principle

---

[30]The discussion took place in the framework of epistemic arithmetic (Reinhardt 1986; Shapiro 1985) where paradoxical sentences can not be formed due to syntactic restrictions of the language. In Reinhardt (1985b) showed that "I am F", suitably formalized, was consistent *contra* Lucas and Penrose. Reinhardt (1985a) establishes the inconsistency of *I am F and I know that I am F*. In a paper that somewhat completed this line of research Carlson (2000) showed that the so-called *strong mechanistic thesis*, that is, the proposition that *I am F and I know that I am some recursively axiomatizable system* was consistent. See, e.g., Koellner (2016) for discussion.

[31]See Shapiro (2003) for similar remarks.

leads to a contradiction without introducing any particular assumption about which recursively axiomatizable theory we are working in. The theory and particular its proofs are completely opaque to the (idealized) human mind: it does not *know* its axioms or rules of proof. This severely restricts the force of the Global Reflection Principle: even if, assuming that "I am F", I were to derive (GRef_F), I would not realize that I have established my own soundness and as a consequence I might be unable to derive my inconsistency. Indeed, the fact that I do not realize that I have established my own soundness may save me from becoming inconsistent. As it were, from my perspective, (GRef_F) could reflect about *any* recursively axiomatizable system—I just don't know which one. Given the dialectical situation it seems difficult to derive a contradiction from the Global Reflection Principle because what it tells us under these circumstances is that whatever is provable in every recursively axiomatizable theory (perhaps extending some basic arithmetical theory) is true. At least prima facie this seems to be a rather unproblematic and uncontroversial claim for it only means that logical truths should in fact be true (in the object-linguistic sense). It thus may come as a surprise that even under these weak assumptions there exist reasonable theories of truth and absolute provability in which a Penrose-style reductio argument for the first disjunct can be carried out.

2.4.1   A New Argument Against Mechanism[32]

The theory **KFC**, i.e. *Consistent Kripke-Feferman*, is a compositional theory of truth in the sense that it commutes with all logical connectives with exception of negation, for which we may eliminate double negation inside the scope of the truth predicate.[33] Moreover, **KFC** as opposed to **KF** simpliciter asserts the consistency of the truth predicate, which in **KF** is equivalent to the principle

$(T\text{-Out})$ $$T\ulcorner\phi\urcorner \to \phi.$$

By results in Stern (2014b) we know that there are plenty of reasonable consistent theories of truth and absolute provability that extend **KFC** and prove $(\mathsf{T_K})$. The argument below shows that the proponent of such **KFC**-style theories of truth and absolute provability is committed to accepting the first disjunct of GD, that is, they have to deny mechanism provided they accept all the background assumptions. Let λ be the standard Liar sentence, that is, a sentence such that the theory F under consideration proves:

(L) $$\neg T\ulcorner\lambda\urcorner \leftrightarrow \lambda.$$

---

[32]For an alternative presentation and a very different philosophical interpretation of the argument see Stern (2018).

[33]See Halbach (2011) for more details about **KFC**.

1. $F \vdash \forall x(K(x) \leftrightarrow \mathrm{Pr}_F(x)) \to \forall y(\mathrm{Pr}_\sigma(y) \to T(y))$ $\mathsf{T_K}$
2. $F \vdash \forall x(K(x) \leftrightarrow \mathrm{Pr}_F(x)) \to (\mathrm{Pr}_F(\ulcorner \lambda \vee \neg \lambda \urcorner) \to T(\ulcorner \lambda \vee \neg \lambda \urcorner))$ 1
3. $F \vdash \mathrm{Pr}_\varnothing(\ulcorner \lambda \vee \neg \lambda \urcorner)$
4. $F \vdash \forall x(K(x) \leftrightarrow \mathrm{Pr}_F(x)) \to T(\ulcorner \lambda \vee \neg \lambda \urcorner)$ 2, 3
5. $F \vdash \forall x(K(x) \leftrightarrow \mathrm{Pr}_F(x)) \to T(\ulcorner \lambda \urcorner) \vee T(\ulcorner \neg \lambda \urcorner)$ 4, KF
6. $F \vdash \forall x(K(x) \leftrightarrow \mathrm{Pr}_F(x)) \to \neg \lambda$ 5, (L), (T-Out)
7. $F \vdash \forall x(K(x) \leftrightarrow \mathrm{Pr}_F(x)) \to T\ulcorner \lambda \urcorner$ 6, (L)
8. $F \vdash \forall x(K(x) \leftrightarrow \mathrm{Pr}_F(x)) \to \lambda$ 7, (T-Out)
9. $F \vdash \neg \forall x(K(x) \leftrightarrow \mathrm{Pr}_F(x))$ 6, 8

The crucial step in the argument is Line 3, which says that $\lambda \vee \neg \lambda$ is provable without premises, that is, in logic alone. This implies that for all classically recursively axiomatizable theories F, $\mathrm{Pr}_F(\ulcorner \lambda \vee \neg \lambda \urcorner)$. The argument therefore establishes the first disjunct of GD since no special assumption concerning F was made. It is noteworthy that the argument, even though it is a reductio argument, is very different to Penrose's *New Argument*. It exploits the fact that in KFC-style theories of truth and absolute provability the so-called internal logic, that is the logic within the scope of the truth predicate, and the external logic diverge. The external logic is just classical logic while the internal logic of KFC-style theories is strong Kleene logic. As a consequence classical tautologies are not generally true in the object-linguistic sense and cannot be for sake of consistency. However, we have assumed classical logic throughout and, in particular, our formalization of "I am F", i.e. $F \vdash \forall x(K(x) \leftrightarrow \mathrm{Pr}_F(x))$, reflects this fact since the axioms of classical logic are built into the standard provability predicate. This implies that we are only considering theories formulated in classical logic and thus no matter which particular system F we consider, F will prove the classical tautologies. As a consequence ($\mathsf{T_K}$) has the effect of adding the classical tautologies to the internal logic of the theory. But, as our argument shows, this leads to a contradiction in KFC-style theories of truth and absolute provability and we may conclude that for no recursively axiomatizable system F, it can be established that "I am F".

Is this new argument for the first disjunct of GD a good argument? That depends on whether one takes KFC-style theories to be suitable theories for discussing Gödel's Disjunction and related issues, and, moreover, whether one takes the argument to be intuitively plausible. We take it that the answer to both questions is negative. First, KFC-style theories violate one of the fundamental assumptions underlying the debate surrounding GD, namely, that provability in recursively axiomatizable theories of mathematics implies truth. But in KFC-style theories there exist sentences ϕ, e.g. $\lambda \vee \neg \lambda$, such that $\mathrm{Pr}_F(\ulcorner \phi \urcorner)$ and $\neg T\ulcorner \phi \urcorner$. In other words these theories are provably unsound according to their own standards. Indeed, they are provably unsound with respect to the standard of *every* recursively axiomatizable theory. Second, it is precisely this feature that our argument for the first disjunct of GD relies on because (GRef_F) asserts that provability implies truth which it cannot in KFC-style theories. The argument thus exploits a pathological feature of the theory and this undermines the credibility of the argument. In order to be convincing there should be a plausible and intuitive rationale to the argument but there is none and this is at least partly due to the fact the argument relies on the pathological aspects of the theory. In light of these remarks even the convinced anti-mechanist, that is the proponent of the first disjunct of GD, should grant that KFC-style theories of truth and absolute provability might just not be a solution to the intensional paradoxes that allows us to establish the first disjunct of GD. However, the problem goes beyond KFC-style theories of truth and absolute provability because the only possibility to derive a contradiction on the basis of (GRef_F) is precisely if such a clash between the external and the internal logic of the theory arises: in such a situation (GRef_F) forces the internal logic to coincide with the external logic, which may provoke a contradiction. If we are right, then our skeptical remarks concerning our argument extend to any argument based on the reductio strategy, at least if classical logic is assumed.

2.5 Conclusion

The present investigation was devoted to arguments for the first disjunct of GD. We put all conceptual worries aside and assumed that there was an intelligible notion of absolute provability where a sentence was supposed to be absolutely provable iff it was the output of, i.e. proved by, an idealized human mind. But even pursuing this charitable course we were unable to provide a valid version of Penrose's *New Argument* and we do not see how the problems with the argument can be fixed. In our view the argument is deeply flawed because it assumes *naive* conceptions of self-applicable notions like truth and absolute provability. But Tarski and Gödel told us that these naive notions lead to paradox: ultimately, this is why Penrose's *New Argument* cannot be fixed.

We then looked at Penrose's reductio strategy and, using this strategy, provided an argument for the first disjunct that was at least technically sound. However, we argued that the argument exploited a pathological feature of the underlying theory which undermined the plausibility of the argument and also caused the argument to lack an intuitive rationale. Moreover, the shortcomings of our argument are most likely to affect all successful arguments for the first disjunct based on the reductio strategy as we presented it in Sect. 2.4.

Of course, none of our findings show that there cannot be an argument for the first disjunct based on the incompleteness theorems and a theory of truth and absolute provability. But combining our findings with the observations of Benacerraf (1967), Reinhardt (1986), Lindström (2001, 2006), Shapiro (2003), Koellner (2016) and others we slowly seem to run out of options how such a plausible and successful argument could look like.[34] It might be time for the anti-mechanist to come up with

---

[34]So far the discussion has been entirely focused on classical logic. Since a very common reaction to the paradoxes is to give up classical logic it is a rather immediate thought to explore Penrose's

an entirely different type of argument for the first disjunct of Gödel's Disjunction, a type of argument that does not rely on Gödel's incompleteness theorems.

Acknowledgements This work was supported by the European Commission through a Marie Sklodowska Curie Individual Fellowship (TREPISTEME, Grant No. 703529). I wish to thank Catrin Campbell-Moore, Martin Fischer, Leon Horsten, Peter Koellner, Carlo Nicolai, and an anonymous referee for helpful comments on the content of this paper. Earlier versions of the paper were presented at the *FSB Seminar in Bristol*, the *Fourth New College Logic Meeting*, the University of Malaga and the *Third Leuven-Bristol Workshop*. I thank the audiences of these talks for their feedback.

Appendix

In this appendix we show that the first disjunct of GD cannot be proved on the basis of the principles (T_K), $(T\text{-In})$, (TB), and (Nec).[35] We introduce some new terminology to state the result in a precise way. We write W_e to denote the output of the Turing machine with index e. Using this terminology the first disjunct of GD can be formulated as

$$(*) \qquad \neg \exists e \forall x (W_e(x) \leftrightarrow Kx).$$

We will show that $(*)$ is not a consequence of the principles (T_K), $(T\text{-In})$, (TB), and (Nec).

Theorem 4 *Let* PATK *be the extension of* PA *in the language containing a truth and an absolute provability predicate* $(\mathscr{L}_{PATK})$. *Then*

$$(T_K),\ (T\text{-In}),\ (TB),\ (Nec) \nvdash_{PATK} \neg \exists e \forall x (W_e(x) \leftrightarrow Kx).$$

Proof The proof of Theorem 4 is an compactness argument: if we can prove $(*)$ in PATK on the basis of (T_K), $(T\text{-In})$, (TB), and (Nec), then there must be a finite proof of $(*)$ in PATK and, in particular, a proof with only finitely many applications of the rule (Nec). The argument we give shows that there cannot be such a proof of finite length for $(*)$. To this end we define a family of theories by recursion:

$$\Sigma_0 := Cn[PATK + (T_K) + (T\text{-In}) + (TB)]$$
$$\Sigma_{n+1} := Cn[\Sigma_n + \{K\ulcorner\phi\urcorner : \Sigma_n \vdash \phi\}]$$

New Argument or, more generally, arguments for the first disjunct of GD assuming some non-classical logic.

[35] A similar, almost trivial argument shows that the first disjunct of GD can also not be obtained using the principles (T_K), $(T\text{-In})$, $(T\text{-Imp})$, and (Nec). A proof is left to the reader.

where Cn denotes the operation that closes these sets of sentences under logical consequence. Σ_n allows for proofs with n-many applications of the rule (Nec). Notice also that by construction $\Sigma_{n+1} \subseteq \Sigma_{n+1}$. By our compactness argument it follows that if

$$(T_K), (T\text{-In}), (TB), (Nec) \vdash_{PATK} \neg \exists e \forall x (W_e(x) \leftrightarrow Kx),$$

then there must be a proof of $(*)$ in some Σ_n for $n \in \omega$. We now construct a suitable model M_n for each Σ_n with $n \in \omega$ such that

$$M_n \models \exists e \forall x (W_e(x) \leftrightarrow Kx).$$

Let $M_0 = (\mathbb{N}, \|T\|_0, \|K\|_0)$ with

$$\|T\|_0 := \{\phi : (\mathbb{N} \models \phi \, \& \, \phi \in \mathscr{L}_{PA}) \text{or} (\phi \in \mathscr{L}_{PATK} \, \& \, \phi \notin \mathscr{L}_{PA})\}$$

$$\|K\|_0 := \{\phi : PATK \vdash \phi\}.$$

Clearly, $M_0 \models \Sigma_0$ and $M_0 \models \forall x (Pr_{PATK}(x) \leftrightarrow Kx)$. But the theorems of PATK are recursively enumerable and we thus know that $M_0 \models \exists e (W_e(x) \leftrightarrow Kx)$. Now, for M_n we define

$$\|T\|_{n+1} := \|T\|_n$$

$$\|K\|_{n=1} := \{\phi : \Sigma_n \vdash \phi\}.$$

It trivially follows that $M_{n+1} \models \Sigma_{n+1}$ and $M_{n+1} \models \forall x (Pr_{\Sigma_n}(x) \leftrightarrow Kx)$, which implies $M_n \models \exists e (W_e(x) \leftrightarrow Kx)$. This concludes the proof of Theorem 4. There cannot be a proof establishing the first disjunct of GD on the basis of the principles $(T_K), (T\text{-In}), (TB),$ and (Nec).

References

Benacerraf, P. 1967. God, the Devil, and Gödel. *The Monist* 51(1): 9–32.

Carlson, T.J. 2000. Knowledge, machines, and the consistency of Reinhard's strong mechanistic thesis. *Annals of Pure and Applied Logic* 105: 51–82.

Chalmers, D.J. 1995. Mind, machines, and mathematics. A review of shadows of the mind by Roger Penrose. *Psyche* 2(9): 11–20.

Egré, P. 2005. The knower paradox in the light of provability interpretations of modal logic. *Journal of Logic, Language and Information* 14: 13–48.

Feferman, S. 2008. Axioms for determinateness and truth. *Review of Symbolic Logic* 1: 204–217.

Friedman, H., and M. Sheard. 1987. An axiomatic approach to self-referential truth. *Annals of Pure and Applied Logic* 33: 1–21.

Gödel, K. 1995. Some basic theorems on the foundations of mathematics and their implications. In *Kurt Gödel: Collected works*, vol. 3, ed. S. Feferman, J.W. Dawson Jr, W. Goldfarb, C. Parsons, and R.M. Solovay, 304–323. Oxford: Oxford University Press. Manuscript written in 1951.

Halbach, V. 2011. *Axiomatic theories of truth*. Cambridge: Cambridge University Press.

Koellner, P. 2016. Gödel's disjunction. In *Gödel's disjunction. The scope and limits of mathematical knowledge*, ed. L. Horsten, and P. Welch, 148–188. Oxford: Oxford University Press.

Leitgeb, H. 2009. On formal and informal provability. In *New waves in philosophy of mathematics*, ed. O. Linnebo, and O. Bueno, 263–299. Basingstoke/New York: Palgrave Macmillan.

Lindström, P. 2001. Penrose's new argument. *Journal of Philosophical Logic* 30: 241–250.

Lindström, P. 2006. Remarks on Penrose's "New Argument". *Journal of Philosophical Logic* 35: 231–235.

Lucas, J.R. 1961. Minds, machines, and gödel. *Philosophy* 36: 112–127.

Montague, R. 1963. Syntactical treatments of modality, with corollaries on reflexion principles and finite axiomatizability. *Acta Philosophica Fennica* 16: 153–167.

Myhill, J. 1960. Some remarks on the notion of proof. *The Journal of Philosophy* 57: 461–471.

Penrose, R. 1989. *The emperor's new mind: Concerning computers, minds, and the laws of physics*. Oxford: Oxford University Press.

Penrose, R. 1994. *Shadows of the mind: In search for the missing science of consciousness*. Oxford: Oxford University Press.

Penrose, R. 1996. Beyond the doubting of a shadow. *Psyche* 2(23): 89–129.

Reinhardt, W.N. 1985a. Absolute versions of incompleteness theorems. *Nous* 19: 317–346.

Reinhardt, W.N. 1985b. The consistency of a variant of church's thesis with an axiomatic theory of an epistemic notation. In: *Proceedings of the Fifth Latin American Symposium on Mathematical Logic held in Bogota*, Revista Columbiana de Matematicas, ed. X. Caicedo, 177–200.

Reinhardt, W.N. 1986. Epistemic theories and the interpretation of gödel's incompleteness theorems. *The Journal of Philosophical Logic* 15(4): 427–474.

Shapiro, S. 1985. *Intensional mathematics*. Amsterdam: North-Holland.

Shapiro, S. 2003. Mechanism, truth, and Penrose's new argument. *Journal of Philosophical Logic* 32: 19–42.

Shapiro, S. 2016. Idealization, mechanism, and knowability. In *Gödel's disjunction. The scope and limits of mathematical knowledge*, ed. L. Horsten, and P. Welch, 189–207. Oxford: Oxford University Press.

Stern, J. 2014a. Modality and axiomatic theories of truth I: Friedman-Sheard. *The Review of Symbolic Logic* 7(2): 273–298.

Stern, J. 2014b. Modality and axiomatic theories of truth II: Kripke-Feferman. *The Review of Symbolic Logic* 7(2): 299–318.

Stern, J. 2016. *Toward predicate approaches to modality, trends in logic*, vol. 44. Cham: Springer.

Stern, Johannes. 2018. Proving that the mind is not a machine? *Thought. A Journal of Philosophy* 7(2): 81–90.

Wang, H. 1996. *A logical journey: From Gödel to philosophy*. Cambridge: MIT Press.

Chapter 3
On Expressive Power Over Arithmetic

Carlo Nicolai

Abstract The paper is concerned with the fine boundary between expressive power and reducibility of semantic and intensional notions in the context of arithmetical theories. I will consider three notions of reduction of a theory characterizing a semantic or a modal notion to the underlying arithmetical base theory – relative interpretability, speed up, conservativeness – and highlight a series of cases where moving between equally satisfactory base theories and keeping the semantic or modal principles fixed yields incompatible results. I then consider the impact of the non-uniform behaviour of these reducibility relations on the philosophical significance we usually attribute to them.

3.1 Introduction

The claim that a reasonable arithmetical theory, in its twofold role of a mathematical and syntactic theory, cannot express satisfactory notions of truth, necessity, knowledge, belongs to the basic toolkit of the logically informed philosopher. Such a claim is certainly true under a specific reading of *expressing* in terms of explicit definitions: due to the well-known paradoxes of Tarski and Montague, there is no arithmetical (syntactic) formula satisfying the characterizing principles of these (and closely related) modal notions.

These limitative phenomena impose that adequate theories of such notions result in *extensions* of the arithmetical base theory. If much is known about truth-theoretic extensions of a reasonable base theory,[1] much less investigated are extensions of a

---

[1] The reader may consult Halbach (2014) for a comprehensive overview. A proviso: first, the theories in Halbach (2014) are all constructed over Peano Arithmetic PA and it is not always immediate to extend the results there to other base theories.

C. Nicolai (✉)
King's College London, Strand, London, UK

© Springer International Publishing AG, part of Springer Nature 2018
M. Piazza, G. Pulcini (eds.), *Truth, Existence and Explanation*,
Boston Studies in the Philosophy and History of Science 334,
https://doi.org/10.1007/978-3-319-93342-9_3

reasonable syntactic base with further intensional notions.[2] An initial study in this direction is the monograph Stern (2016), where it is considered the interaction of truth and predicates for necessity and possibility.

In this paper we are concerned with the fine boundary between expressive power and reducibility of intensional notions to the arithmetical or syntactic base theory. Over the years several authors have imposed adequacy conditions for the reducibility of intensional predicates to the base theory. For a recent example Fischer and Horsten (2015) consider a theory of truth to be adequate if it is semantically conservative over the base theory – i.e. any model of the base theory can be expanded to a model of the theory of truth – and at the same time expressively irreducible to it in the sense of being not relatively interpretable in it and having additional emerging expressive features such as non-elementary speed-up over it.[3]

In the following sections, we will show that there are non-uniformity phenomena concerning the combinations of these notions of reduction, i.e. conservativeness, relative interpretability, non-elementary speed-up, that cast serious doubts on the very possibility of analyzing intensional notions in terms of their reducibility to the base theory. In a nutshell, we will mainly highlight the fact that by moving from a reasonable base theory to another one may unintentionally go from a reducible to an irreducible intensional notion. If the adequacy of the truth or modal theory is evaluated in terms of its reducibility, as Horsten and Fischer propose, the variation of seemingly innocuous assumptions on the bearers of semantic and modal ascriptions may result in an incoherent body of information. This may seriously compromise any attempt of extracting philosophical information from modal and semantic theories. In the concluding section, after presenting the situation more clearly, we will suggest a possible way out of this riddle.

Note to the reader. To highlight the main conceptual points and enhance readability, I have tried to reduce the formal details to the bare bones. I have nonetheless kept the main points of the proofs of the results I refer to by describing them in the body of the text. The reader interested in full proofs may refer to the bibliography accompanying the informal arguments. Also, the reader may find certain conceptual moves in Sects. 3.2, 3.3, and 3.4 rather hasty. This is because I have decided to condense technicalities and the necessary preliminaries for a proper discussion in these sections and then reconstruct the main line of reasoning, with more care for the philosophical points, in the final section.

---

[2]Note on terminology: In what follows, we employ the adjective 'intensional' and 'modal' in an inclusive sense, familiar to medieval authors such as Ockham, which comprehend also truth and propositional attitudes.

[3]Further references abound: we go from the well-known conservativeness argument discussed in Shapiro (1998), Ketland (1999), Field (1999) to more recent discussions of deflationism Horsten (2012) and Nicolai (2015).

3.2 Bearers of Modal Ascriptions and the Three Case Studies

In what follows, we will consider base theories B that are sufficiently strong to develop a nice theory of the bearers of truth and modal ascriptions. Such theories are generally required to articulate the properties of expression types – that will be in what follows the objects to which semantic and modal notions apply – via suitable coding. In addition to this, we usually require the possibility of coding finite sequences of objects and, therefore, a modicum of number theory to represent functions such as projections. A useful notion capturing in one go all these desiderata is the notion of *sequential* theory (Pudlák 1983; Visser 2013), which is formally defined as a theory that relatively interprets *Adjunctive Set Theory*

$$\exists y \, \forall u (u \notin y) \tag{AS1}$$

$$\exists v \, \forall u (u \in v \leftrightarrow u \in x \vee u = y) \tag{AS2}$$

without relativizing quantifiers and preserving identity. In what follows we will therefore stipulate that all base theories that we employ are sequential.

A particularly nice feature of sequential theories is that they can formalize, via a suitable interpretation, the usual development of the syntax of first-order theories as it is required, for instance, by the standard proofs of Gödel's second incompleteness theorem. To see this, it suffices to notice that AS interprets theories such as Buss' S_2^1 or $I\Delta_0 + \Omega_1$, which are precisely up to the task of formalizing syntactic notions and operations. This interpretation guarantees the right syntactic structure to the objects of truth and modalities.[4] This fact, combined with the possibility of coding sequences of all objects in the domain of a sequential theory, makes it possible to formulate the usual clauses for satisfaction that are a necessary condition for extending our base sequential theory with semantic or modal predicates.

Inside the family of sequential theories, we will distinguish between schematically presented or *schematic* theories featuring a finite set of axioms together with a finite set of axiom schemata, and *finitely axiomatized* theories. More precisely, we will mostly deal with a subclass of schematic theories that we will call *inductive*, in the sense that they satisfy (possibly via translation) the full induction schema for $\mathcal{L}_B$:

$$\varphi(0) \wedge \forall x \in \omega \, (\varphi(x) \leftrightarrow \varphi(x+1)) \rightarrow \forall x \in \omega \, \varphi(x) \tag{3.1}$$

The possibility of a translation suggests, of course, that the set of natural numbers ω does not necessarily have to be available in B; in fact, on many occasions, it won't be. At any rate, inductive (sequential) base theories are capable of developing partial truth predicates satisfying Tarski's inductive clauses for each of their finite

---

[4]For details on the subexponential theories S_2^1, $I\Delta_0 + \Omega_1$ the standard reference is still (Hájek and Pudlák 1993, §V).

subtheories B_0. This, in turn, entails that they can prove a canonical consistency statement for B_0. By the second incompleteness theorem, therefore, the class of inductive and the class of finite sequential theories are disjoint. Examples of inductive base theories are Peano arithmetic PA, Zermelo-Fraenkel set theory ZF, full second-order arithmetic Z_2; examples of finitely axiomatized sequential theories are elementary arithmetic EA, the extension of PA with arithmetical comprehension and class variables ACA$_0$, and Von Neumann-Gödel-Bernays set theory NBG.

We assume a fixed Gödel coding and a formalization of the syntax for our base theory B as it can be standardly carried out in S_2^1. In particular, we write $\ulcorner e \urcorner$ for the term of $\mathscr{L}_B$ representing the Gödel code of the $\mathscr{L}_B$-expression e. We assume that $\mathscr{L}_B$ contains, besides its characterizing arithmetical function symbols, also finitely many additional function symbols for syntactic operations: for instance, $\dot{\neg}$ is a function symbol expressing the syntactic operation of negation. It is also convenient to have in our language a function symbol for the operation of replacing a term z for a free variable v in a formula φ.

We will mostly deal with extensions of B with principles governing semantic and modal notions. For simplicity, I will refer to these extensions as *modal theories*. For our purposes it suffice to consider the language $\mathscr{L}_B \cup \{\Box\}$, where $\Box$ is a fresh unary predicate. We abbreviate $\Box\ulcorner \varphi \urcorner$ with $\Box\varphi$ and, similarly $\Box\ulcorner \varphi(\dot{x}) \urcorner$ – that is the result of formally substituting in $\varphi(v)$ the variable v with the numeral for x, usually expressed via the substitution function and the numeral function sub$(\ulcorner \varphi(v) \urcorner, \text{num}(x))$ – with $\Box\varphi(x)$.

With these little preliminaries at hand, we can introduce the extensions of our base theories we will be mainly interested in. The first set of principles that we consider is encompassed in the schema:

$$\forall x (\Box\varphi(x) \leftrightarrow \varphi(x)) \tag{M}$$

for $\varphi \in \mathscr{L}_B$. Several theories of truth, but also some of the theories of necessity in Stern (2016), feature this principle. (M) is manifestly plausible for alethic modalities such as truth and necessity, it essentially tells us that the extension of $\Box$ restricted to standard sentences of $\mathscr{L}_B$ agrees with the $\mathscr{L}_B$-truths of the modal theory. We will call M[B] the result of adding (M) to B.

The second group of principles we are interested in reflects the possibility a uniform distribution of the semantic or modal predicate over the logical connectives starting with truths of B and building up inductively the extension of $\Box$. In this case, we treat negation on a *case by case* manner:

$$\forall x \left((\Box R(x) \leftrightarrow R(x)) \wedge (\Box\neg R(x) \leftrightarrow \neg R(x)) \right) \quad \text{for any relation } R \in \mathscr{L}_B \tag{G1}$$

$$\forall x, y \left(\mathsf{Sent}_{\mathscr{L}_B}(x \wedge y) \rightarrow (\Box(x \wedge y) \leftrightarrow \Box x \wedge \Box y) \right) \tag{G2}$$

$$\forall x, y \left(\mathsf{Sent}_{\mathscr{L}_B}(x \wedge y) \rightarrow (\Box\dot{\neg}(x \wedge y) \leftrightarrow \Box\dot{\neg}x \vee \Box\dot{\neg}y) \right) \tag{G3}$$

$$\forall v, x \left(\mathsf{Sent}_{\mathscr{L}_B}(\dot{\forall}v x) \rightarrow (\Box(\dot{\forall}v x) \leftrightarrow \forall y \, \Box x(y/v)) \right) \tag{G4}$$

$$\forall v, x \, (\text{Sent}_{\mathscr{L}_B}(\forall v x) \rightarrow (\Box(\neg\forall v x) \leftrightarrow \exists y \, \Box\neg x(y/v))) \tag{G5}$$

$$\forall x \, (\text{Sent}_{\mathscr{L}_B}(x) \rightarrow (\Box\neg\neg x \leftrightarrow \Box x)) \tag{G6}$$

Since the predicate $\Box$ occurs only positively in the clauses G1–G5, the resulting cluster of theories G[B], although only dealing with typed predicates, may be taken to capture the *grounded* development of $\Box$. We will discuss shortly the choice of a typed predicate, that is a predicate that only applies to $\mathscr{L}_B$ sentences and not to sentence containing $\Box$.

Finally, we will consider the theories C[B] obtained by extending B with the axioms G1, G2, G4 and the axioms stipulating the full commutativity of $\Box$ with negation:

$$\forall x (\text{Sent}_{\mathscr{L}_B}(x) \rightarrow (\Box\neg x \leftrightarrow \neg\Box x)) \tag{$\neg$}$$

It is clear that, in C[B], $\Box$ is not treated positively. However, it will be treated *compositionally*. This property is clearly desirable for truth and arguably for other alethic modalities such as necessity.

The theories M[B], G[B], and C[B] will be the three case studies we will be occupied with in the rest of the paper. Two remarks are in order. On the one hand, since we will vary the base theory B while keeping the principles for $\Box$ fixed, we need to be clear about the role of the nonlogical axiom schemata of B, if present. Unless otherwise specified, we do *not* extend nonlogical axiom schemata of B to $\Box$. On the other, as it is clear from the principles above, we will impose a restriction on the applicability of the predicate $\Box$ in the theories M[B], G[B], and C[B]: in particular, these theories will force no sentence containing $\Box$ into the extension of $\Box$ itself. In other words, the theories M[B], G[B], and C[B] are *typed* treatments of $\Box$. Both the non extension of nonlogical axiom schemata of B to $\Box$ and the typed nature of our theories are motivated by our interest in the thin line separating reducible and non-reducible modal theories (to the base theory B): extending schemata will in fact result in non-reducible theories, whereas considering type-free notions will often only complicate the study of the reductions we are interested in without affecting the overall conceptual point, since the theories M[B], G[B], and C[B] are obviously contained in their type-free versions and in virtually all known semantic and modal theories.

In the next three sections we consider three well-known notions of reduction of the modal theory to the base theory B: conservativeness, relative interpretability, non elementary speed-up.

3.3 Deductive Strength

The proof-theoretic notion of conservativeness provides one with a precise sense in which a semantic or modal extension of a reasonable base theory B may involve *insubstantial* concepts, namely concepts that do not play a significant role in the explanation of non-modal or non semantic facts. This reading is reminiscent of certain formal renderings of truth-theoretic deflationism (see for instance Horwich's 1998, Shapiro 1998, Ketland 1999, and Horsten 2012) according to which the truth predicate should not play a substantial role in the explanation of non-semantic facts. A modal theory T is *conservative* over B if any theorem φ in the language of B that is provable in T is already provable in B alone. Of course there is no consensus on whether notions such as truth, necessity, possibility and the like ought to be insubstantial in the sense just hinted at. For our concerns, in fact, it suffices that such interpretations exist.

For B sequential, the theory $M[B]$ is a conservative extension of B. This can be established by a well-known argument dating back to Tarski's (1956). In a proof of a sentence φ of $\mathscr{L}_B$, in fact, only finitely many occurrences of the schema (M) can occur. This means that there is a finite set of formulas $\varphi_1, \ldots, \varphi_n$ that occur in instances of (M). Then one can simply define in B a predicate

$$P(x, y) :\leftrightarrow (x = \ulcorner \varphi_1(\dot{y}) \urcorner \land \varphi_1(y)) \lor \ldots \lor (x = \ulcorner \varphi_n(\dot{y}) \urcorner \land \varphi_n(y)) \qquad (3.2)$$

and replace occurrences of $\Box\mathsf{sub}(\ulcorner \varphi_i \urcorner, \mathsf{num}(y))$ in the proof of φ with $P(\ulcorner \varphi_i \urcorner, y)$. The resulting proof is a legitimate proof of φ in B, witnessing the conservativity of $M[B]$ over B.

Also the conservativity of $G[B]$ over sequential B can be easily established, although via standard semantic considerations. By slightly extending an argument contained in Cantini (1989), in fact, any model of $\mathscr{M} \vDash B$ can be expanded to a model of $(\mathscr{M}, S)$ of $G[M]$, where S is the extension of the predicate $\Box$. The fundamental reason for this is that a set of $\mathscr{L}_B$ sentences satisfying the $G[M]$ axioms can be characterized as a fixed point of a positive inductive definition. In what follows $f^{\mathscr{M}}$ stands for the denotation in M of the function symbol f of $\mathscr{L}_B$ and we assume an expanded language $\mathscr{L}_B^+$ featuring names for all objects in the domain $|\mathscr{M}|$ of $\mathscr{M}$:

$a \in X \Leftrightarrow a$ is a sentence of $\mathscr{L}_B^+$, and

$\Big(a$ is a true atomic formula or negated atomic formula, or

a is $b \land^{\mathscr{M}} c$ and $b \in X$ and $c \in X$ for some sentences b, c, or

a is $\neg^{\mathscr{M}} (b \land^{\mathscr{M}} c)$ and $\neg^{\mathscr{M}} b \in X$ or $\neg c^{\mathscr{M}} \in X$ for some sentences b, c, or

a is $\forall bc$ and for all $d \in |\mathscr{M}|$, $\mathsf{sub}^{\mathscr{M}} (c, \mathsf{num}^{\mathscr{M}} (d)) \in X$

for c a formula with one free variable and b a variable, or

a is $\neg^{\mathscr{M}} \forall bc$ and for some $d \in |\mathscr{M}|$, $\neg^{\mathscr{M}} \mathsf{sub}^{\mathscr{M}} (c, \mathsf{num}^{\mathscr{M}} (d)) \in X$

for c a formula with one free variable and b a variable $\Big)$

In a fixed point of the positive inductive definition associated with the right-hand side of the biconditional above, that is a set S such that $(\mathscr{M}, S)$ satisfies the above

Table 3.1 Conservativeness

Modal theory over a reasonable B	Inductive	Finite
M[B]	✓	✓
G[B]	✓	✓
C[B] ($\supseteq$ IΔ_0(exp))	✓	✓

equivalence, we will have that, for instance, a sentence $\varphi \wedge \psi$ is in S if and only if φ is in S and ψ is in S, and similarly for all other axioms of G[B].

The conservativeness of the theory C[B] over suitable B cannot be achieved so easily. For a substantial chunk of sequential theories, that is the ones extending (either directly or via interpretation), a specific subsystem of first-order arithmetic called IΔ_0(exp) or EA, Leigh (2015) has produced an argument showing how to eliminate cuts on formulas of the form $\square\varphi$ in proofs of $\mathscr{L}_B$-formulas. The role of EA, a system designed to capture exactly a proper subclass of primitive recursive functions, the so-called Kalmar's elementary functions, is roughly the one of controlling the complexity of boxed formulas in derivations so that the usual induction on the length of the derivation of $\mathscr{L}_B$-theorems to push $\square$-cuts upwards could be carried out.[5] It is still unknown whether the theory M[B] is conservative over B for *all* sequential B.[6]

Table 3.1 summarizes the results sketched in the last few paragraphs.

My aim in this paper is to highlight how conservativeness and other notions of proof-theoretic reduction are highly sensitive to the choice of the underlying, sequential base theory. Conservativeness, however, appears to behave in a very stable way. The overall analysis that we are going to propose in the concluding paragraph will be that this stability is only due to the coarse-grained nature, if compared with other notions, of the relation of conservativeness.

If we move to a setting in which schemata of B may be extended to $\square$, however, non-uniformity phenomena appear also in the context of the conservativeness of the modal theories considered over B. We consider one simple example: the theory C[PA]$^+$, for instance, where the $^+$ denotes the extension of the induction schema of PA to the $\square$, is strong enough to formalize the soundness of PA by reading $\square$ as truth. Therefore, C[PA]$^+$ will prove Con(PA) and will not be conservative over PA.[7] However, if we leave arithmetic and move to base theories that are able to prove strong forms of reflection, such as ZF or Z$_2$, the situation changes. It is well-known, in fact, that ZF proves, for any sentence φ of the language of set theory, its equivalence to its relativization φ^{V_α} for some limit ordinal α (see for instance Jech

---

[5]It should be noticed that Halbach's (1999) contains a cut-elimination argument that was later shown to contain a gap. The problem there was exactly this absence of a suitable machinery to control the complexity of boxed formulas in proofs of $\mathscr{L}_B$-formulas.

[6]For the reader not familiar with subsystems of first-order arithmetic, it may be useful to know that it's easy to find theories that are proper sub-theories of EA but still sequential. One example is Buss' theory S$_2^1$ mentioned above.

[7]The argument is folklore, see for instance Halbach (2014, §8).

2008, Thm. 12.14); similarly, in Z_2 or, equivalently, Π^1_ω-CA, we can prove that any sentence φ of the language of second-order arithmetic has a countable ω-model (see Simpson 2009, Lem. VIII.5.2).

Now consider, seeking a contradiction, a sentence φ of the language $\mathscr{L}_2$ of Z_2 such that, for instance, $C[Z_2]^+$ proves φ and Z_2 does not prove φ. It's important to notice that $C[Z_2]^+$ features the schema of $\mathscr{L}_2$-induction extended to the $\square$ but *not* the extended full-comprehension schema. Let A be the finite subsystem of $C[Z_2]^+$ proving φ – which is given by the compactness theorem – and the set B of the axioms of Z_2 employed in the proof of φ. By the ω-reflection principle Z_2 proves that there is an ω-model of B. The truth predicate associated with this model suffices to interpret A in Z_2, including the instances of the extended induction. Therefore, Z_2 proves φ after all. The argument transfers to ZF without essential modifications except for the use of the Lévy-Montague reflection principle instead of the ω-reflection principle: again we highlight that $C[ZF]^+$ is formulated by means of a schema of ω-induction extended to $\square$ but without extending the schemata of separation and replacement of ZF.

It should be now clear how the results mentioned in the last paragraphs amount to a case of non-uniformity for the notion of conservativeness relative to the choice of base theories, at least in relation to one of our modal theories: if $C[PA]^+$ is non conservative over PA, the theories $C[ZF]^+$ and $C[Z_2]^+$ are indeed conservative over their respective base theories. One objection is in order: how can one justify the extension of the *ω-induction* schema of, say, ZF to $\square$ and the simultaneous non-extension of the other non-logical schemata of ZF? One straightforward answer may be that, since $\square$ has to be interpreted as truth, necessity, possibility, knowledge, and the like, its range of application are *syntactic* objects: as a consequence, since it is well-known that the structure of syntactic objects – strings of a finite alphabet in particular – is isomorphic to that of natural numbers, the extended principle of induction can be easily justified as a syntactic principle (see also Fujimoto (2017) on this point). By contrast, surely we would *not* consider the full replacement or separation schemata syntactic principles.

3.4 Conceptual Reducibility

The connection between the conceptual reducibility of a modal notion to the arithmetical resources and the *relative interpretability* of a modal theory to the base theory has been recently highlighted by a number of authors (see for instance Horsten 2012, Fischer and Horsten 2015, and Nicolai 2015). Informally, a theory U is relatively interpretable in a theory V if there is a translation of the primitive concepts of the language of U into the language of V that commutes with propositional connectives, possibly relativizes quantifiers, and preserves theoremhood.[8] U is locally interpretable in V if every finite subtheory of U is interpretable in V.

---

[8] For a formally precise definition, see for instance (Visser 2013).

As we just did for conservativeness, we ask ourselves whether $M[B]$, $G[B]$, and $C[B]$ are interpretable in B. To do so, however, we need to introduce a bit of terminology. A formula $\varphi(v)$ is said to be *progressive* in U if U proves $\varphi(0)$ and $\forall x\,(\varphi(x) \to \varphi(x+1))$; $\varphi(v)$ is a *cut* in U if it is progressive in U and downwards closed under $\leq$, that is if U proves $\forall y, x(y \leq x \wedge \varphi(x) \to \varphi(y))$. A remarkable fact due to Robert Solovay is that, in extensions V of Robinson's arithmetic Q, for every inductive formula $\varphi(v)$ one can find a V-cut ψ such that V proves $\forall x(\psi(x) \to \varphi(x))$ (see Hájek and Pudlák 1993, Lem. 5.9).

It turns out that, when relative interpretability is involved, the distinction between inductive and finitely axiomatized sequential theories matters. If B is inductive, in fact, it can prove the consistency of any of its finite subtheories. By a well-known result often called Orey's compactness theorem, if U is locally interpretable in V and V is inductive, then U is also relatively interpretable in it.[9] Therefore, if B is inductive, the proof of the conservativeness of $M[B]$ over B, which can be easily seen to amount to a proof of local interpretability, is indeed a proof of the interpretability of $M[B]$ in B. If, by contrast, B is finite, a recent unpublished argument by Albert Visser shows that $M[B]$ is *not* interpretable in it.[10] For B finite, in fact, $M[B]$ can define a cut $\mathscr{I}$ that is the intersection of all B-definable cuts. At the same time, by a result of Pudlák (1985), for *each* $n \in \omega$, B can define a cut $\mathscr{J}$ such that $B \vdash \mathrm{Con}_n^{\mathscr{J}}(B)$ – where $\mathrm{Con}_n^{\mathscr{J}}(B)$ says that there is no proof in $\mathscr{J}$ of contradiction from axioms of Gödel number smaller than n and with a proof whose elements contain at most n quantifiers. The two claims just made entail that $M[B]$ can define the intersection of all such $\mathscr{J}$'s. Therefore B can interpret, by relativizing all quantifiers to the cut $\mathscr{I}$, the theory $S_2^1 + \{n \in \omega \mid \mathrm{Con}_n(B)\}$. But, by another result of Pudlák (see Pudlák 1985, Cor. 3.5), no finitely axiomatized sequential theory T can interpret $S_2^1 + \{n \in \omega \mid \mathrm{Con}_n(T)\}$; so if $M[B]$ was interpretable in B, we would get a contradiction by the transitivity of interpretability.

We have therefore just seen a first non-uniformity result: if B is inductive, $M[B]$ is interpretable in B. If it is finite, $M[B]$ is not interpretable in B. We find a similar scenario when we move to the theories $G[B]$ and $C[B]$. For our purposes it is better to start with considering $C[B]$.

If B is inductive, it will prove the consistency of any of its finite subtheories. By Leigh's proof of the conservativity of $C[B]$ over B considered above, therefore, which is formalizable in EA, we have for B containing EA,

$$B \vdash \mathrm{Con}(B_0) \to \mathrm{Con}(C[B_0]) \tag{3.3}$$

---

[9]Here's a proof: Let's assume that U is locally interpretable in V. In V, either $\mathrm{Con}(U)$ or $\neg\mathrm{Con}(U)$. If the former, an interpretation can be found via the Henkin-Feferman arithmetized completeness theorem. If the latter, then there is a finite $V_0 \subseteq V$ such that $V \vdash \mathrm{Con}(V_0) \to \mathrm{Con}(U_0)$ for all finite $U_0 \subset U$. Therefore V proves $\mathrm{Con}(U_0)$. But then, by the well-known argument due to Feferman (1960), one can find an intensional consistency statement $\mathrm{Con}^*(U)$ such that $V \vdash \mathrm{Con}^*(U)$. We can then employ the Henkin-Feferman construction again.

[10]Personal communication.

where B_0 is a finite subsystem of B. Therefore $\mathsf{C}[B]$ will prove $\mathsf{Con}(\mathsf{C}[B_0])$ and so it will also be reflexive. It is well-known, however, that the relative interpretability of T_0 in T_1, with T_0, T_1 reflexive, follows from the arithmetical Π_1-conservativity of T_0 over T_1. This therefore yields the interpretability of $\mathsf{C}[B]$ and $\mathsf{G}[B]$ in B for B inductive.

By contrast, if B is a finitely axiomatized sequential theory, $\mathsf{C}[B]$ and $\mathsf{G}[B]$ are not interpretable in B. We first consider the idea of the argument for $\mathsf{C}[B]$. By the inclusion of $\mathsf{M}[B]$ in $\mathsf{C}[B]$, the axioms of B are all in the extension of $\Box$. Assuming that B is formulated in a calculus in which Modus Ponens is the only rule of inference, we move to a $\mathsf{C}[B]$-definable cut $\mathscr{K}$ in which all (codes) of logical axioms of B are in the extension of $\Box$. We close $\mathscr{K}$ under provability in $\mathsf{C}[B]$, by moving to another cut $\mathscr{N} \subset \mathscr{K}$: this is possible by Solovay's method because the property 'the last element of a B-proof x in $\mathscr{K}$ is in the extension of $\Box$' is provably progressive in $\mathsf{C}[B]$. All theorems of $\mathsf{C}[B]$ that belong to $\mathscr{N}$, therefore, are in the extension of $\Box$. If a proof of a falsity $\bot$ was in $\mathscr{N}$, then, $\Box\bot$ would be provable in $\mathsf{C}[B]$ and so would $\bot$. In other words, we can prove the consistency of B relative to $\mathscr{N}$ in $\mathsf{C}[B]$. By Pudlák's result, therefore, B cannot interpret $\mathsf{C}[B]$.[11]

Let's now turn to $\mathsf{G}[B]$ for finitely axiomatized B: it also contains $\mathsf{M}[B]$, and so it proves that all nonlogical axioms of B are in the extension of $\Box$. To replicate the argument just given for the non-interpretability of $\mathsf{C}[B]$ in B, we would need a workable axiom of full commutation of $\Box$ with negation.[12] Since we cannot have it in full, we notice that the 'formula'

'for every sentence φ of $\mathscr{L}_B$ of logical complexity $\leq x : \neg\Box\varphi$ if and only if $\Box\neg\varphi$'

is provably progressive in $\mathsf{G}[B]$. Therefore we can find a $\mathsf{G}[B]$-cut $\mathscr{H}$ such that the formulas belonging to it enjoy full commutation of negation. We intersect this cut with the analogue of the cut $\mathscr{K}$ capturing a portion of the true logical axioms of B. We can then close $\mathscr{H} \cap \mathscr{K}$ under provability and prove the consistency of B on a cut in $\mathsf{G}[B]$ as well, which yields its non-interpretability in B (for more details we refer to Nicolai 2016b).

Relative interpretability, that we have associated with the conceptual reducibility of the modal notion to the syntactic resources of the underlying base theory, displays therefore a deeply non-uniform behaviour. Table 3.2 summarizes the situation: For B inductive and 'reasonable' – that is interpreting EA – $\mathsf{M}[B]$, $\mathsf{C}[B]$ and $\mathsf{G}[B]$ are all interpretable in B. For all sequential and finitely axiomatized B, $\mathsf{M}[B]$, $\mathsf{G}[B]$ and $\mathsf{C}[B]$ are not interpretable in B.

---

[11] For a full argument, see Nicolai (2016a).

[12] Notice for instance, that a form of full commutation with negation is needed even for the principle

$$\forall\varphi, \psi(\Box(\varphi \to \psi) \wedge \Box\varphi \to \Box\psi) \tag{K}$$

that is required to close provability under $\Box$.

Table 3.2 Relative interpretability

Modal theory over a reasonable B	Inductive	Finite
M[B]	✓	✗
G[B]	✓	✗
C[B]	✓	✗

The message to extract from these results is clear: if one associates the conceptual reducibility of a modal notion characterized via a modal theory to the relative interpretability of the latter into its base theory, as Horsten (2012) and Fischer and Horsten (2015) suggest, one has to face the fact that the *very same* principles for $\Box$ may be associated to a conceptually reducible or irreducible notion depending on facts that have nothing to do with the characterization of $\Box$. Inductive and finitely axiomatized sequential theories, for all we know, satisfy the desiderata imposed to the characterization of the bearers of modal ascriptions that can be found in the literature (see again Horsten 2012, Halbach 2014, and Stern 2016). We will see in the concluding section that perhaps something is indeed missing in these accounts.

3.5 Instrumentalism

If conservativeness behaves uniformly, relative interpretability exhibits a discouraging discontinuity. As a possible tie-breaker, I consider the capability of the modal theories of shortening proofs of theorems of B, the so-called *speed-up* phenomenon. As Fischer suggests in Fischer (2014) in relation to truth, a notion that possesses this property would enable us to significantly enhance our expressive resources becoming an indispensable *instrument* to express in a more concise and tractable way some facts concerning an underlying ontology: in our case the structure of the bearers of modal ascriptions. We will soon consider some concrete examples. The leading question of this section is, therefore: when moving from an inductive to a finite base theory B, is the capability of M[B], G[B], and C[B] of shortening B-proofs affected?

As before, we briefly introduce some technical tools and terminology from Pudlák (1998) and Fischer (2014). Given a Hilbert-style formulation of a theory U, we let $|\varphi|_U$ be the shortest (code of a) U-proof of φ, if that proof exists at all. Given theories U, V with $U \subseteq V$, we say that V has *at most polynomial speed up* over U if there is a polynomial p such that, for φ provable in U, if $|\varphi|_V \leq n$, then $|\varphi|_U \leq p(n)$. We are encouraged from complexity theory to set aside polynomial speed up. By contrast, *non-elementary speed up* is not to be overlooked: V has non-elementary speed up if the are $\mathscr{L}_U$-formulas $\varphi_1, \ldots, \varphi_k$ and no function F of elementary growth rate such that, if $|\varphi_i|_V \leq n$, then $|\varphi_i|_U < F(n)$ for $1 \leq i \leq k$. By function with an elementary growth rate, I mean a function $f(x)$ that can be majorized by a superexponential function 2_m^x.[13]

---

[13] Where $2_0^x = x$ and $2_{y+1}^x = 2^{2_y^x}$.

As in the case of conservativeness, the main question of this section can be addressed uniformly for M[B] for sequential B, regardless of their being finite or inductive. A version of the argument witnessing the conservativeness of M[B] over B sketched on page 44 can be given so that when we transform a M[B] proof, say of length n, to a B-proof, the size of the latter only grows by a polynomial in n. The essential ingredient of the argument is the observation that for any formula $\varphi(x)$ of $\mathscr{L}_B$ we can find a partial truth predicate for it such that the proof of its 'disquotational' property

$$\mathsf{Tr}_\varphi(\varphi(x)) \leftrightarrow \varphi(x)$$

can be proved in B with a proof of size polynomial in the code of φ (see Pudlák 1998, Thm. 3.3.1). Therefore, for B sequential, M[B] *has at most polynomial speed up over B*.

As before, the situation for G[B] and C[B] is more complex. When B is finite and sequential, we have seen that they prove the consistency of B on a cut definable in G[B] or C[B]. By contrast, by a fundamental result of Friedman and Pudlák, the consistency statements $\mathsf{Con}_{2_n}(B)$ are, for each n and some constant c, only provable in B with a proof of size greater than 2_n^c (see Pudlák 1998, Thm. 6.2.3). However, for theories that prove the consistency of B on a cut, proofs of $\mathsf{Con}_{2_n}(B)$ become *linear* in n.[14] Therefore C[B] and M[B] have non-elementary speed up over B for B finitely axiomatized and sequential.

What happens when B is inductive? Unfortunately we don't have a clear answer, but only the strong conjecture that the proofs of the interpretability of C[B] in B, for B inductive, may give a polynomially bounded reduction of C[B]-proofs to B-proofs. This would in turn yield the lack of non-elementary speed up of G[B] over B for inductive B.

Table 3.3 summarizes the situation for speed up, which is more open than the ones for relative interpretability and conservativeness, although the proofs of the interpretability of C[B] in B for B inductive strongly support that idea that also speed up has a non-uniform behaviour. From the conceptual point of view, again if our conjectures are true, we would have that the capability of the predicate □ of shortening proofs and therefore the usefulness of the modal notion in question to increase the expressive capabilities of our language crucially depends on the presentation of the axiom system that shapes our syntactic/arithmetical world. I will elaborate more on this point in the next, final section.

---

[14]See Fischer (2014, 4.11) for a proof in the case of PA. The method can however be generalized to all B.

Table 3.3 Non-elementary speed up over B	Modal theory over a reasonable B	Inductive	Finite
	M[B]	✗	✗
	G[B]	?✗?	✓
	C[B]	?✗?	✓

3.6   A Dilemma?

Let's pause for a moment and reconstruct in some more detail the main line of argumentation that may have been blurred by the necessary technicalities introduced in the previous sections. We started with arithmetical theories in their role of theories of the bearers of modal ascriptions. We then asked ourselves what are the sufficient conditions for characterizing the objects to which intensional notions apply: our answer led us to identify a class of arithmetical systems, the *sequential* theories, as satisfactory candidates for such a role (Sect. 3.2). This is not an unconventional choice. Vann McGee, for instance, in his classic on truth writes that we require from a theory of the objects of truth

> ... the ability to describe expression types and their syntactic properties, the ability to talk about natural numbers and their mathematical properties, and the ability to talk about the coding relations. (McGee 1991, p. 18)

Via the interpretation of suitable syntactic theories, such as S_2^1, we can even require the provability of some universally quantified statements concerning the structure of syntactic objects.[15]

Once the properties of a suitable arithmetical/syntactical theory are identified, it is possible to formulate extensions of our base theory with principles characterizing alethic modalities. We have considered principles that can be satisfied by a cluster of intensional notions, broadly conceived as to include truth and propositional attitudes: for a sequential base theory B, they amount to the *minimal theory* M[B], the theory of *grounded* modalities M[B], the theory of *compositional* modalities C[B]. Admittedly, the principles characterizing the three theories are not aimed at pinpointing a single modal notion, but at being sufficiently general to capture the fine boundary between intensional notions that are *reducible* to the resources of the object theory, and intensional notions that are not.

---

[15]This desideratum was clearly emphasized in Halbach's influential monograph:

> ... a base theory must contain at least a theory about the objects to which truth can be ascribed. [...] The axioms of the truth theory can serve their purpose only if the base theory allows one to express certain facts about syntactic operations; [...] it should be provable in the base theory that a conjunction is different from its conjuncts and different from any disjunction.[...] The decisive advantage of using Peano Arithmetic is that I do not have to develop a formal theory that can be used as base theory...(Halbach 2014, §2)

What's the sense of *reducible* that is relevant here? We considered three possible choices: the *non conservativeness* (conservativeness) of the modal theory over the base theory, witnessing – following the tradition initiated by Shapiro (1998) and Ketland (1999) – the property of the modal theory of explaining (not being able to explain) non-modal facts concerning B not already explained by B; the relative interpretability (non-interpretability) of the modal theory in the base theory B, witnessing – as envisaged by Horsten (2012), Fischer (2014), and Fischer and Horsten (2015) – the conceptual reducibility (irreducibility) of the modal notion to the inferential resources of B; the modal theory's having (not having) non-elementary speed up with respect to the base theory B, witnessing the expressive extra-power (or the lack of it) with respect to B (Fischer 2014).

The scenario that resulted from considering the reducibility of the three modal theories to a sequential theory B turned out to be at least puzzling. If conservativeness behaves uniformly for sequential theories extending a weak arithmetical system (i.e. EA), relative interpretability is highly dependent on the presentation of the theory B, and speed up is likely to behave much more similarly to relative interpretability than to conservativeness.

A natural thought would then be the following:

1. One should disregard notions of reduction that display a non-uniform conduct. In the case at hand, we should only take seriously the conservativeness (or non-conservativeness) of a modal theory over a sufficiently strong sequential theory. Given their idiosyncrasies, relative interpretability and speed up should not be trusted.

Let's reflect a bit on 1. Looking back at the philosophical theses associated with the three reducibility relations we have taken into account, a consequence of 1. would be to consider as legitimate the equation 'conservativeness = lack of explanatory power', whereas the conceptual reducibility of an intensional notion could not be coherently matched with the relative interpretability of the modal theory in the base theory, and similarly for the attempts of combining expressive power and non-elementary speed up – under the assumption of the truth of our conjectures above. The reason should now be clear but it's worth repeating in a direct and informal way: in constructing a modal theory we identified two main components. In the first place a satisfactory theory of the objects of modal ascriptions, that we identified with the class of sequential theories. Secondly, some principles governing some modal notion, that do not depend in any way on the choice of the underlying sequential theory. However, we have seen that the very same principles characterizing a modal notion may be reducible or not reducible, expressively stronger or not, depending on which sequential theory we choose.

However, despite its initial plausibility, thesis 1. cannot be accepted. It is in fact clear that, in the study of the semantic or modal notions characterized by the axioms of M[B], G[B] and C[B], conservativeness is a coarser grained notion if compared to, for instance, relative interpretability. On the one hand, in fact, we have seen that there are cases in which M[B], G[B] and C[B] are conservative over B but not interpretable in B. On the other, let us consider *natural and essential* extensions of, say, $G[B]$ and $C[B]$, namely the theories $G[B]^+$ and $C[B]^+$ that will all contain the results of extending the nonlogical schemata of $B \supseteq S_2^1$ to the predicate $\square$. Both theories will prove $Con(B)$ and therefore will not be conservative over B by Gödel's second incompleteness theorem. However, this also entails, both in the case in which B is inductive and finite, that neither $G[B]^+$ nor $C[B]^+$ can be interpreted in B. In other words, for natural modal theories, relative interpretability implies conservativeness but not vice versa: *as a consequence, the study of the relative interpretability of the modal theory in the base theory adds substantially more information about the notion in question than what is obtained investigating its conservativeness.* Although this is not a direct rejection of 1., I take it as a sign that a significant amount of important data would be lost once endorsing it.

A second reaction to the situation depicted in Tables 3.1, 3.2, and 3.3 could be the following:

2. Although uniformity is lost if we consider *both* classes of finite and inductive theories, it is clearly regained if we focus on *each* of them. We must therefore choose between finite and inductive base theories. In particular, inductive theories feature an open-ended schema for the natural numbers that, given the close connection between the structure of syntactic and natural numbers, enables us to capture without arbitrary restrictions the intended domain of bearers of modal ascriptions. Therefore we must disregard finite theories.[16]

Also this option should be resisted, for reasons that are similar to the ones contained in my reaction to 1. If one is interested, for instance, in the conceptual reducibility of the notion captured by $\square$, one would clearly realize that, for the very specific behaviour of relative interpretability in the context of inductive theories witnessed by Orey's theorem above and similar results, the relative interpretability of the modal theory in the inductive base theory cannot be a real test. It's only in the context of finite theories that these general collapsing phenomena between relations of interpretability are blocked, and the question of the conceptual reducibility of the modal predicate becomes relevant. It's there, and not elsewhere, that one can really see how the modal predicate helps in structuring the ontology of numbers (such as when we defined cuts on with the consistency of the base theory can be proved) in ways that are not otherwise available in the base theory. And it's also there that, as we have seen, the modal predicate *is* doing some conceptually irreducible work, and it's precisely this that would be hidden when we restrict our attention to inductive base theories only.

Similarly, under the assumption that the modal theories $G[B]$ and $C[B]$ have no significant speed up over an inductive B – a fact that would likely be extracted from the relative interpretability of $C[B]$ in B for B inductive – one should conclude that the modal predicate in question has no expressive power or cannot be a useful tool to express facts concerning the syntactic base theory in a more concise and shorter

---

[16]A suggestion along these lines has been recently made by Fujimoto in Deflationism beyond arithmetic. Unpublished manuscript.

way. However, by focusing on finite base theories, one suddenly realizes that there is a clear sense in which the modal notion in question *can* shorten proofs in the base theory: although it cannot cope with an infinite presentation of *B*, when the starting point is a finite set of axiom the modal predicate is indeed expressively stronger than any of the resources of *B*. In both cases, therefore, endorsing 2. and focusing only on inductive base theories would lead to a severe loss of significant information.

But are 1. and 2. two horns of a dilemma? In a sense, if one aims at clear-cut solutions that could result in a concise and effective philosophical message concerning all reductions at once, the answer seems to be yes. The results mentioned above, however, also point at an irreducible level complexity that cannot be so easily resolved as it happens in both horns of the dilemma. One way to react to the deadlock is therefore to add an additional parameter to the evaluation of reduction of semantic and modal notions to an underlying base theory: the presentation of the base theory. In the light of what we said above, it is not sufficient to consider a modal theory as compounded by (i) a theory of bearers of modal ascriptions satisfying some minimal sufficient conditions and (ii) a set of principles characterizing a modal notion. We should also consider how the theory specified in (i) is presented to us: schematic, inductive, infinitely axiomatized, finitely axiomatized.

This conclusion, however, does not hold for *any* uses of modal theories: when only paradoxes and consistency are at stake, there is no need to consider subtle issues like the specific axiomatization of the truth bearers. But even outside the analysis of reductions of a modal theory to the base theory, for instance when one compares different principles belonging to a single solution to paradoxes, as Halbach and Nicolai do in Nicolai and Halbach (2017) for Kripke's theory of truth, the presentation of the base theory is an non-eliminable parameter in measuring the proof-theoretic strength of such solutions.

Acknowledgements This research has been supported by the European Commission, grant no. 658285 FOREMOTIONS. I would like to thank Martin Fischer, Albert Visser, and an anonymous referee for useful comments and insights.

References

Cantini, A. 1989. Notes on formal theories of truth. *Zeitschrift für Logik un Grundlagen der Mathematik* 35: 97–130.

Feferman, S. 1960. Arithmetization of metamathematics in a general setting. *Fundamenta Mathematicae* 49(1): 35–92.

Field, H. 1999. Deflating the conservativeness argument. *Journal of Philosophy* 96(10): 533–540.

Fischer, M. 2014. Truth and speed-up. *Review of Symbolic Logic* 7(2): 319–340.

Fischer, M., and L. Horsten. 2015. The expressive power of truth. *Review of Symbolic Logic* 8(2): 345–369.

Fujimoto, K. 2017. Deflationism beyond arithmetic. *Synthese* 1–25. Online-first version.

Halbach, V. 1999. Conservative theories of classical truth. *Studia Logica* 62(3): 353–370.

Halbach, V. 2014. *Axiomatic theories of truth*, Revised ed. Cambridge: Cambridge University Press.

Horsten, L. 2012. *The Tarskian turn*. Oxford: MIT University Press.

Horwich, P. 1998. *Truth*. Oxford: Clarendon Press.

Hájek, P., and P. Pudlák. 1993. *Metamathematics of first-order arithmetic*. Perspectives in mathematical logic. Berlin, New York: Springer.

Jech, T. 2008. *Set theory*, The 3rd millennium ed. Berlin: Springer.

Ketland, J. 1999. Deflationism and Tarski's paradise. *Mind* 108(429): 69–94.

Leigh, G. 2015. Conservativity for theories of compositional truth via cut elimination. *The Journal of Symbolic Logic* 80(3): 845-865

McGee, V. 1991. *Truth, vagueness, and paradox*. Indianapolis/Cambridge: Hackett.

Nicolai, C. 2015. Deflationary truth and the ontology of expressions. *Synthese* 192(12): 4031–4055.

Nicolai, C. 2016a. A note on typed truth and consistency assertions. *Journal of Philosophical Logic* 45(1): 89–119.

Nicolai, C. 2016b. More on systems of truth and predicative comprehension. In *Objectivity, realism, and proof. FilMat studies in the philosophy of mathematics*, ed. F. Boccuni, and A. Sereni. Cham: Springer.

Nicolai, C., and V. Halbach. 2017. On the costs of nonclassical logic. *Journal of Philosophical Logic*, Online First: https:doi.org/10.1007/s10992-017-9424-3

Pudlák, P. 1983. Some prime elements in the lattice of interpretability types. *Transactions of the American Mathematical Society* 280: 255–275.

Pudlák, P. 1985. Cuts, consistency statements, and interpretations. *Journal of Symbolic Logic* 50: 423–441.

Pudlák, P. 1998. The lengths of proofs. In *Handbook of proof theory*, vol. 137 of *Studies in logic and the foundations of mathematics*, ed. Samuel R. Buss, 547–637. Amsterdam/Lausanne/New York: Elsevier.

Shapiro, S. 1998. Proof and truth: Through thick and thin. *Journal of Philosophy* 95(10): 493–521.

Simpson, S.G. 2009. *Subsystems of second order arithmetic*. Perspectives in logic. New York: Association for symbolic logic.

Stern, J. 2016. *Towards predicate approaches to modality*. New York: Springer.

Tarski, A. 1956. Der Wahrhetisbegriff in den formalisierten Sprachen. In *Logic, semantics, metamathematics*, 152–278. Oxford: Clarendon Press.

Visser, A. 2013. What is the right notion of sequentiality. In *New studies in weak arithmetics*, CSLI lecture notes, vol. 211, ed. C. Charampolas, P. Cegielski, and C. Dimitracopoulos, 229–272. Stanford: Publications and Presses Universitaires du Pole de Recherche et d'Enseingement Superieur Paris-est.

Chapter 4
Intensionality in Mathematics

Jaroslav Peregrin

Abstract Do mathematical expressions have intensions, or merely extensions? If we accept the standard account of intensions based on possible worlds, it would seem that the latter is the case – there is no room for nontrivial intensions in the case of non-empirical expressions. However, some vexing mathematical problems, notably Gödel's Second Incompleteness Theorem, seem to presuppose an intensional construal of some mathematical expressions. Hence, can we make room for intensions in mathematics? In this paper we argue that this is possible, provided we give up the standard approach to intensionality based on possible worlds.

4.1 Which Sentence of PA *Says* that PA Is Inconsistent?

Suppose that my friend Charles is bald. Hence the sentence

(1) Charles is bald

is true. Suppose, moreover, that Charles happens to be the tallest bald man (in the whole world, say). Hence the sentence

(2) The tallest bald man is bald

and (1) appear to "say the same thing"; namely, they ascribe baldness to Charles. Yet while (1) appears to convey something nontrivial, (2) is trivial: it says that a bald

J. Peregrin (✉)
Institute of Philosophy, Czech Academy of Sciences, Prague, Czech Republic

Faculty of Philosophy, University of Hradec Králové, Hradec Králové, Czech Republic
e-mail: peregrin@flu.cas.cz

© Springer International Publishing AG, part of Springer Nature 2018   57
M. Piazza, G. Pulcini (eds.), *Truth, Existence and Explanation*,
Boston Studies in the Philosophy and History of Science 334,
https://doi.org/10.1007/978-3-319-93342-9_4

man is bald, hence a truism.[1] How does this fact square with the intuition that the
two sentences "say the same"?

Quite easily, of course. The two sentences say the same only given the fact that
Charles is the tallest bald man, and this is an *empirical* fact which need not obtain.
Hence (1) and (2) say the same only on the assumption that

(3) Charles is the tallest bald man

It follows that the sense of "saying the same" just considered cannot be identical
with "meaning the same", i.e. with synonymy. The point is that the meaning of (1)
and (2), and hence the identity or non-identity of their meanings, cannot depend on
empirical facts of the kind spelled out by (3). And if (1) and (2) were to mean the
same thing, then in a case where (3) were false they could not cease to mean the
same thing (though there would be a sense in which they would no longer say the
same).

Now consider a *prima facie* similar case. Suppose that some formal theory, say
Peano arithmetic (PA), is consistent, hence the sentence

(1*) PA is consistent

is true. It follows that PA is the biggest consistent theory included in PA, hence that

(2*) The biggest consistent theory included in PA is consistent

Just as in the previous case, the two sentences appear, on the one hand, to "say the
same"; namely, they ascribe consistency to PA while not being, on the other hand,
really synonymous. But why are they not really synonymous? We might be tempted
to give an answer wholly analogous to the previous case, namely that (1*) and (2*)
"saying the same" is conditioned by

(3*) PA is the biggest consistent theory included in PA,

which is an empirical fact that need not obtain. But this would not work here: (3*) is
not an empirical fact and it can*not* fail to obtain. If PA is consistent, there cannot be
a situation in which it would fail to be identical with the biggest consistent theory
included in PA.

Is it then the case that (1*) and (2*), unlike (1) and (2), *do* have the same
meaning?

We may tend to say that this perhaps is not a question worth dwelling on. In a
sense, (1*) and (2*) do mean the same thing, while perhaps in another sense they do
not. But this cavalier approach would lead us into some difficulties, not the smallest
of them being that the answer to the synonymy question is presupposed by one
of the most significant findings in logic of the twentieth century, Gödel's Second
Incompleteness Theorem.

As pointed out by Auerbach (1985), while the First Incompleteness Theorem is
a plain statement regarding the structure of certain formal systems (if the axioms of

---

[1] Well, perhaps it says, over and above this, that there is one and only one tallest bald man; however,
this again does not seem to be too nontrivial.

such a system include a modicum of arithmetic, then there is bound to be a sentence which is independent of them), the nature of the Second Theorem is different – it presupposes that certain arithmetical sentences have certain meanings. (Therefore, Feferman 1960 takes it to be an *intensional* problem.[2]) The Second Incompleteness Theorem says that no system of the above kind proves a theorem that says of the system that it is consistent. Hence, the very formulation of the Theorem assumes that we know what it takes for a sentence of the system to say (mean?) that the system is consistent.

At first sight, it might seem that this is not a problem and that we do not need to enter the muddy waters of the considerations displayed above. To say that a system is consistent is to say that it does not prove the absurdity (often, the sentence "$0 = 1$" is used as its embodiment), or that there is something which it does not prove. Hence everything we need is the provability predicate, which, in turn, is based on the binary predicate Prf that relates two numbers iff the first is the Gödel number of a proof of a theorem whose Gödel number is the second argument.

Thus it might seem that a sentence's saying that the system is consistent can be reduced to it being of the form $\neg \exists x \mathrm{Prf}(x, \ulcorner 0 = 1 \urcorner)$, where $\mathrm{Prf}(x, y)$ says that x is (the Gödel number of) a proof of (the Gödel number of) y. And $\mathrm{Prf}(x, y)$'s saying this in turn reduces to its relating all and only numbers of proofs with numbers of theorems they are proofs of. But this unfortunately is not the case, and we are led directly into the muddy waters that we hoped to avoid.

Consider the binary predicate $\mathrm{Prf_R}$ defined in the following way (it was, in effect, proposed by Rosser 1936):

> $\mathrm{Prf_R}(x, y) \equiv_{Def.} x$ is a proof of y and x is not greater than the number of any proof of any negation of y.

Given PA is consistent, the existence of a proof of y excludes the existence of a proof of its negation, and hence the number of a proof of y can never be greater than the number of a proof of its negation; hence $\mathrm{Prf_R}(x, y)$ iff $\mathrm{Prf}(x, y)$. It follows that in so far as saying that x is (the Gödel number of) a proof of (the Gödel number of) y amounts to nothing more than relating all and only numbers of proofs with numbers of theorems they are proofs of, $\mathrm{Prf_R}(x, y)$ says this just as $\mathrm{Prf}(x, y)$ does. But the trouble is that if we substitute $\mathrm{Prf_R}$ for Prf in the consistency sentence, we get a sentence that is *provable*.[3] Hence, either Gödel's Second Incompleteness Theorem does not hold, or $\mathrm{Prf_R}(x, y)$ does not say that x is a proof of y after all. And, as the

---

[2]A thorough discussion of the ways in which the Second Incompleteness Theorem is intensional is given by Halbach and Visser (2014a,b).

[3]This is quite obvious if, instead of $\mathrm{Prf_R}$, we use still another predicate, namely $\mathrm{Prf_C}(x, y) \equiv_{Def.}$ $\mathrm{Prf}(x, y) \wedge \neg \mathrm{Prf}(x, \ulcorner 0=1 \urcorner)$. Given $\mathrm{Prf}(x, \ulcorner 0=1 \urcorner)$ for no x, $\mathrm{Prf_C}(x, y)$ holds for the same x's and y's as Prf (x, y); while $\neg \exists x \mathrm{Prf_C}(x, \ulcorner 0=1 \urcorner)$ is $\neg \exists x (\mathrm{Prf}(x, \ulcorner 0=1 \urcorner) \wedge \neg \mathrm{Prf}(x, \ulcorner 0=1 \urcorner))$, which is obviously provable. In fact, if we consider a further modified version of $\mathrm{Prf_C}(x, y)$, $\mathrm{Prf_{C'}}(x, y)$ $\equiv_{Def.}$ $\mathrm{Prf}(x, y) \wedge \forall x \neg \mathrm{Prf}(x, \ulcorner 0=1 \urcorner)$, we may return to our example (1*) vs. (2*), for we may think of (1*) as roughly capturing the sense of the standard consistency sentence based on the predicate Prf, and of (2*) as capturing that of its variant based on $\mathrm{Prf_{C'}}$ – for $\forall x \neg \mathrm{Prf}(x, \ulcorner 0=1 \urcorner)$ clearly amounts to the consistency of PA. (Cf. Auerbach 1992).

Second Theorem would be taken as a well-established fact, we seem to be back to pondering what does it take for an expression of a formal system to say something.[4]

4.2 Possible Worlds

Let us return to sentences (1) and (2). The fact that there is a sense in which they do "say the same" and a sense in which they do not is clear, and by now it has a standard explanation: they do "say the same" on the level of reference, while they do not "say the same" on the level of meaning. The fact that semantics must consist of two levels, such as meaning and reference, now an established commonplace, was most clearly articulated by Frege (1892a); and since his seminal analyses, almost no theory of semantics has avoided a stratification of this kind.

Why do we need the two levels? An answer follows from a consideration of what it is that language is good for. On the one hand, we certainly need language for talking about objects that are around us: the chair on which I am sitting, my computer, my friend Charles, the current president of the USA, etc. On the other hand, we need language to be very general and universal – its semantics should not depend on there being my computer or my friend Charles. Doing justice to both these requirements works in such a way that linguistic expressions have general *meanings*, which conspire together with the empirical facts to produce *referents*. Thus, the expression *the current president of the USA* has a general meaning, independent of the actual possessor of this office, but given the current state of the USA, it is able to refer to a concrete person.

Frege famously called the two levels of semantics the level of *Sinn* (sense) and that of *Bedeutung* (meaning) – a not very happy choice, for it is his *Sinn*, rather than his *Bedeutung*, that corresponds to meaning in the intuitive sense of the word. From this viewpoint, it was helpful when Carnap (1947) proposed altering the terminology and rechristened Frege's *Sinne* as *intensions* and his *Bedeutungen* as *extensions*. This is the terminology that has become standard and which lets us say that sentences (1) and (2), and especially their subjects, *Charles* and *the tallest bald man*, share the same extension while they differ in intension.

How best to get an explicitly semantic grip on the difference between intension and extension was foreshadowed already by Carnap, but came to full fruition only after Kripke (1963a,b, 1965) introduced his semantics for modal logic based on the notion of possible world. The basic idea of so-called intensional semantics[5] that grew from the Carnapian considerations regarding the nature of intensions and the

---

[4]We may think about a "non-intensional" content of the Second Incompleteness Theorem: for example we may take it as saying that we cannot prove $Pr(\ulcorner 0 = 1 \urcorner)$ for any predicate Pr fulfilling Löb's derivability conditions (Löb 1955; cf. also the thorough discussion of Boolos 1995). But then the question is why this result should be very interesting; unless we show that there is a reason to think that fulfilling Löb's conditions amounts to being the provability predicate, it is a far cry from what is usually taken as the Second Incompleteness Theorem (see Detlefsen 1986).

[5]Put forward by Montague (1974) and others.

apparatus of possible worlds was that the intension may be seen as something like an extension relativized to possible worlds; more technically, it can be considered as a function mapping possible worlds on corresponding extensions.[6] Hence, while the extension of *the tallest bald man* is (presumably) a concrete person (my friend Charles), its intension is a function that maps any possible world on its tallest bald man (if any).

Let me call the result of this development the *possible-worlds account for intensionality*, or PWAI, for short. To many philosophers it appeared to be the ultimate solution to the problem of intensionality. However, it is usually taken to have no impact on mathematics. The facts of mathematics, it is usually assumed, are independent of the way the world is, hence they are constant across possible worlds and if we therefore assign intensions to mathematical sentences they will be constant functions mapping every possible world on one of the truth values.

Hence, given PWAI, there is no space for intensionality in mathematics. According to PWAI, all true mathematical sentences have the same intension and so do all the false ones. (This was also behind the most spectacular failures of the accounts for the so called propositional attitude reports provided by PWAI – see Bigelow 1978 or Partee 1982.) Some semanticists proposed that semantics must go "hyperintensional" (see, e.g., Lewis 1972 or Cresswell 1975).

Most working mathematicians do not seem to care – after all, when doing mathematics and using its language there is no need to poke into the details of its semantics. It is clear that for many mathematical expressions, such as e.g. "1+3" and "2+2", though they have the same extension (in this case the number 4) and from the viewpoint of PWAI also the same intension (namely the function mapping every possible world on the number 4), they do not have, intuitively, the same meaning; but it is not clear that mathematicians (in contrast to philosophers of language) need to rack their brains over this apparent clash with everyday intuition. However, we have seen that some problems, such as the Second Incompleteness Theorems, which are often understood as eminently mathematical, do presuppose a clarification of meaning far beyond simple intuition.

4.3 Consequence and Meaning

From the viewpoint of PWAI, mathematical discourse is anomalous in that it lacks the extension/intension distinction, which appears to be vital for non-mathematical discourse. Let us consider the ways in which this anomaly influences the concept of consequence and then the very concept of meaning.

What is consequence? We can say that

(4) Fido is an animal

---

[6]This is an oversimplification; technically, the systems of intensional logic tend to be somewhat more complicated. See Peregrin (2006).

is a consequence of

(5) Fido is a dog

and similarly we can say that

(4*) L has the smallest element

is a consequence of

(5*) L is a lattice.

What, in general, does it take for a sentence to be a consequence of other sentences?

Intuitively, A is a consequence of $A_1, \ldots, A_n$ iff the truth of $A_1, \ldots, A_n$ "forces" the truth of A, i.e. if the latter is true whenever the former are true. The "whenever" just employed indicates that this definition contains a tacit quantification: A is true in all cases when $A_1, \ldots, A_n$ are – but what are the cases over which we quantify? For (4) and (5), the cases would seem to be the possible states of the world (which may be captured by Kripkean possible worlds): all states in which Fido is a dog are bound to be states in which Fido is an animal (while in other states this need not be the case). However, this does not work for (4*) and (5*), for if L is a lattice, it remains a lattice whichever state of the world obtains.

Let me stress that just like the case of *Fido* in (4) and (5), L in (4*) and (5*) is supposed to be the name of a specific object (not just a kind of variable), for (4*) and (5*), like (4) and (5), are supposed to be sentences, not just schemata. But could we not use the Tarskian definition of consequence, which, in effect, reduces the consequence among sentences to consequence among the corresponding logical forms (where the latter obtains iff everything that is the model of the premises is also a model of the conclusion)?

Not really. The fact that (5*) follows from (4*), obviously, is not a matter of their logical forms, it is a matter of the fact that lattices are bound to have smallest elements. But could we not say that (4*)'s being a consequence of (5*) amounts to the quantificational statement

(6*) for every x, if x is a lattice, then x has the smallest element?

This is also problematic. For consider the analogue for the case of (4) and (5).

(6) for every x, if x is a dog, then x is an animal.

It is true; but it would not be generally held that its truth would guarantee the consequence. Consider the sentence

(7) for every x, if x is a featherless biped, then x is a human.

This is also true, but it certainly does not follow that *Charles is a human* is a consequence of *Charles is a featherless biped*. The reason is that the truth of (7) is *contingent*; in terms of PWAI we can say that though it is true in the actual world, it is not true in *every* possible world.

Hence we can see that a conditional of this kind guarantees consequence again only if it is true not only about the actual world, but across possible worlds. Does this hold of (6*)? Yes, but somewhat vacuously, for a mathematical sentence is true in every possible world as soon as it is true in the actual one. It follows that consequence in mathematics, if considered in this standard way, boils down to material implication; hence it also suffers from all the well-known maladies of this kind of implication.[7]

Maybe this is something we should simply accept. But what about meaning? Similar arguments indicate that furnishing an expression with an extension is insufficient for making it genuinely meaningful.

Suppose we stipulate that a word refers to the set of current humans. What have we thus made the word *mean*? *Human*? Or *featherless biped*? These are definitely two different meanings, hence our new term has to mean one or the other (or still something else). The answer is that in this way we do not make it mean anything, for conferring extension is notoriously short of conferring meaning. To make a word meaningful, we need to furnish it with an intension. Now how is it with meanings of mathematical expressions?

Mathematical expressions, of course, *may* be thought of as having intensions, but then the only intensions available are those that are constant over possible worlds and hence are uniquely determined by extensions (and hence are somewhat superfluous). Thus to furnish a mathematical expression with an extension *is* to furnish it with an intension, albeit trivially. However, this would mean that, e.g., the expressions *PA* and *the biggest consistent part of PA* mean the same thing, which does not seem to be viable.

One possible approach to this situation is to revise the concept of possible world on which we base the PWAI. A proposal, going back to Hintikka (1975), is to consider not only worlds that are "ontologically" possible, but also worlds that are "epistemologically" possible (though they are ontologically impossible). This means that, for example, in a situation where there is no proof of the consistency of PA, there is an (impossible) possible world in which PA is not consistent.

It seems to me that this proposal is on the right track – as far as the semantics of possible-worlds goes. An obvious objection, however, is that there is no way of determining what is epistemologically possible. Is it what any speaker of the language in question can hold for true? But would the ignorance and/or folly of some speakers stop short of anything – would there be anything that would be at all impossible?

Perhaps we can resort to a whole society as a measure of epistemic possibility. Perhaps something is epistemically possible iff the contrary has not been "ascertained" (proved, in the case of mathematics) by the community. According to this proposal, our community no longer allows for an (epistemologically) possible world in which Fermat's Last Theorem would not hold, but it does allow for one in which the Goldbach conjecture does not hold. But this would mean that there is also no world in which PA is inconsistent (at least if we do not want to challenge the existing proofs of its consistency) and *PA* would come out again as co-intensional with *the biggest consistent part of PA*.

---

[7] See Peregrin (2014, Chapter 7).

4.4   Back to Frege

Frege's approach to intensionality differs from that of his continuator Carnap in at least two respects. Frege's *Sinn*, "intension", is a *way of givenness* (*Art des Gegebensein*; see Frege 1892a) (and this is pretty much everything Frege tells us about it). And, intuitively, it would seem that specifying PA by way of *PA* and specifying it by way of *the biggest consistent part of PA* are two *different ways*.

But, of course, we must not assume that two nominally different phrases always mean different *ways of givenness* – not all different expressions need to mean different things (independently of which particular theory of meaning we subscribe to). And possible worlds appear to offer us precisely a criterion of determining when two individual descriptions are semantically different and when not: they are different if there is a possible world in which they do not describe the same entity. Is it possible that *the biggest consistent part of PA* does not describe PA? Not according to PWAI.

Frege's opinion was arguably different. For what he gives as one of his pivot examples of different ways of giving the same object is the following:

> Let *a*, *b*, *c* be the lines connecting the vertices of a triangle with the midpoints of the opposite sides. The point of intersection of *a* and *b* is then the same as the point of intersection of *b* and *c*. So we have different designations for the same point, and these names ('point of intersection of *a* and *b*,' 'point of intersection of *b* and *c*') likewise indicate the mode of presentation; and hence the statement contains actual knowledge. (Frege 1892b, p. 26)

As it is clear that the point of intersection remains the same however we vary the world, the two descriptions have the same intension (on the standard construal of possible worlds) – from this viewpoint they therefore do *not* appear as different descriptions. Hence Frege thought that two descriptions can be semantically different even if they are (according to PWAI) co-intensional.

Let us consider the idea underlying PWAI. To have, aside an extension, also an intension is to have an extension not only now, but also in various counterfactual situations – to be, as it were, "counterfactually robust". Perhaps we need not be sure what the extension of the term might be in some very exotic circumstances (what would be the extension of *human* in a world with many more or less humanoid species?), but to understand the word is to know at least how its extension moves when the circumstances shift from the current ones into their neighborhood.

Of course, if we consider changes of circumstances, we must concentrate on the way the changes affect the object or objects that constitute the extension in question – to know, that is, the rules that regulate the movements. Unless the object is affected, the change is not really relevant. Hence considering counterfactual circumstances is, in effect, considering variants of the object or objects in question and considering which of the variants still count as "the same" object or "the same" kind of objects. And this depends on what we are, and what we are not, willing to consider "the same".

PWAI is based on the notion of "ontological possibility", which is taken to be an objective matter, wholly independent of us. (In this way, the standard treatment is usually not only supposed to *account for* meaning but also to *explain it*, to reduce it to something non-semantic.) The above considerations indicate that this is doubtful. Why is it that the sentence *Fido is a dog and is not a dog* cannot be true? Because there is no possible world in which it is true? But how do we know, have we checked all the possible worlds, one by one? We know it because we know (explicitly or implicitly) the rules of our language (in this particular case the rules governing *not*), and we know the truth is excluded by these rules. Thus we do not know that the sentence is necessarily false because it is false in every possible world; rather, we know that it is necessarily false, because this follows from the rules of our language that we mastered on learning it, and we therefore know that there cannot be a world in which the sentence is true. Hence, possible worlds are a certain (useful) means of envisaging certain rules of our language, not something that would explain why the rules are such as they are.[8]

Take (1). It would be taken as a prototype of a contingent, empirical sentence – hence a sentence that is *not* true in every possible world. But why is this? Clearly it depends on what kind of entity we consider the individual Charles to be. When I point at Charles, I point at an intricately complex configuration of molecules making up an entity with a lot of physical features, including a bald head – and precisely this particular configuration cannot lack the bald head, for it would be a *different* entity if it did. The fact that we take (1) to be contingent is because we individuate individuals more loosely than the individual configurations of molecules – many more or less different configurations of this kind may count as the same individual. (Though what is and what is not the same individual is not always clear; and it is much more unclear in cases of things other than animals – *viz.* the celebrated case of the ship of Theseus.) Hence, saying that what is and is not contingent is clearly determined by "ontology" is a huge exaggeration.

The situation is not very different in the case of extension for general terms, such as, e.g., *human*. How much can the extension of the class of current humans vary while still remaining the class of humans? In contrast to the extension of an individual term, such as Charles, here we must take into account not only the variations of the extension itself, but also those of the world around it. It is clear that the extension may gain a lot of kinds of individuals in many ways differing from the current ones; and it can lose any of the current ones (if they are no longer part of the world). It is also clear that gaining individuals differing in more substantial ways (e.g. not having brains) might compromise the status of the set as the set of humans; as well as losing some current individuals (if they still are part of the world). Thus, though the boundaries of the species *human* are far from clear-cut, there are many features that obviously do not move an individual outside the boundaries and there are many such that do. And to understand the term *human* is to understand these.

Note also that we need not be able to see all the conditions that shape the boundaries of the extensions. Imagine a kind of unobvious bodily feature X without which one cannot have a brain. Thus, insofar as we would hardly take a brain-less

---

[8]I discussed this in great detail in Peregrin (1995).

organism to be human, the lack of X would prevent an individual from being a member of the corresponding extension, though we need not know about this lack. In general, this is a matter of the fact that we need not see all the consequences of the rules of our language. It follows that we may consider a class of organisms containing some X-less ones as admissible classes of humans.

In fact, the claim that humans have X would belong to what Kripke (1972) calls the "necessary a posteriori". (While Kant famously urged that there are not only analytic a priori and synthetic a posteriori judgments, but also synthetic a priori ones, Kripke argues that we can have, the other way around, analytic a posteriori judgments.) The criteria constitutive of a meaning of a term may have consequences we are not able to see, and hence there are properties which anything falling under the term must have in force of the meaning the term (and hence the claim that it has the property is analytic); though to find out about them might be a discovery (thus the claim is a posteriori).

The important point is that this holds both for empirical and mathematical terms. Our normal usage of the name "PA" is such that their bearer gets individuated in terms of the axioms constitutive of it. We certainly intend to refer to the unique system determined by the axioms with all its properties, but in the case of us not knowing all the properties, or in the case of not all the properties being obvious, the meaning of "PA", as embodied in its use, does not exclude some variants of the intended reference. (Unlike reference, the meaning is something we know when we understand an expression. Thus, what is contingent – in the sense of Kripkean a posteriori – cannot be part of the meaning.) Note that if this were not the case, the *meaning* of "PA" would be unknown also to many of those who would know the axioms of the system perfectly well, which would seem to be strange.

Perhaps, then, we can see (1*) as contingent, for it is a posteriori in the Kripkean sense. If PA were slightly different from what it is, it would not need to be consistent; and the slight difference would not make us say it is no longer PA (because we need not be able to see all its consequences). Then we could say that just as the fact that *Charles is bald* is contingent because we can imagine something non-bald and close to the current configuration of molecules making up Charles (enough so that we are content to call it "the same individual"), the fact that *PA is consistent* is contingent because we can imagine something inconsistent and close to PA in a strict enough sense that we are content with calling it "the same system".

This opens up a novel view of intensionality. Entities of the actual world may have "neighborhoods" consisting of entities more or less similar to them that are considered to be "identical with them". The configuration of molecules called at this moment *Charles* continues to be the same Charles even if it undergoes various kinds of changes. The set of humans of the actual world, which is the extension of *human*, remains a set of humans even after some of its members are removed and new members – of the appropriate kind – are added. And PA remains the same system providing we tamper with it only in ways which we see as not threatening its identity.

4.5   Intension as "Robustness"

What is important is that to know the intension of a term is to know in which
directions its extension can be moved without ceasing to be the extension, and
in which directions this is not possible. Consider the conglomerate of molecules
constituting the current Charles. There are many changes that can happen in this
entity that do not cause it to no longer be Charles. Indeed, almost any minor change
is insubstantial; only an entity substantially different from the current one may not
count as Charles. (Though even an entity very radically different from the current
Charles may still count as Charles if it results from certain transformations of the
current Charles – the criteria of what still counts as Charles and what no longer does
are quite complicated and not always unequivocal.)

Thus, we may exploit the wisdom behind PWAI in a slightly different way than
PWAI itself does: we can say that an expression is furnished not only with extension,
but also with intension if it is determined to what extent it is "robust" – how various
kinds of changes of the *status quo* tamper with its extension, especially which kinds
of variants of the current extension are prone to take over its role. And, in contrast
to PWAI, this alternative view of intensionality may also be usable in the realm of
mathematics.

Consider the binary relation that links $\ulcorner x \urcorner$ with $\ulcorner y \urcorner$ iff *x is a proof of y*. It is
clear that if a (number of a) proof plus a (number of a) sentence are in this relation
w.r.t. a given axiomatic system, this cannot be changed by adding more axioms to
the system (nor, for that matter, to another one). The only thing that can result from
the extension of the list of axioms is that more numbers will be in this relation.
Hence if we consider the relation links $\ulcorner x \urcorner$ with $\ulcorner y \urcorner$ iff *x is a proof of y and PA
is consistent*, we can see that it behaves differently – adding axioms to PA so that it
becomes inconsistent exempts any sentences from this relation. (The same holds for
the more complicated version *x is a proof of y and there is no proof z of a negation
of y such that z < x*).

Now given an axiomatic system, the claim *there is something that is not derivable
from the empty set of premises*, if false, can be made true by subtracting axioms, but
not by adding more of them to the system (and certainly not to PA, if PA is not
the system under consideration); whereas if we used *x is a proof of y and PA is
consistent* instead of *x is a proof of y*, this would not be the case. Hence if we
defined *is consistent* as *there is something that is not derivable from an empty set of
premises*, it is ok for the former derivability predicate, but not for the latter one.

If we thus accept that a necessary condition for a term having an intension is its
"determinateness" in the vicinity of its actual extension (in the sense that it is clear
which variants of the current extension still are its potential extensions, and which
not), we have a criterion for deciding which predicates are reasonable candidates
for expressing the property of consistency (and similarly for other mathematical
terms). The criterion is not sharp and its applicability is limited; however, it brings
forth a possible way of explicating the intuition that mathematical terms also have
intensions.

Now let us return to the statement

(1*) PA is consistent

Does it hold necessarily? Consider the statement (where X is still a feature an organism must have to have a brain and such that a normal speaker is not likely to know about it):

(1**) Charles has X.

Does *it* hold necessarily? Yes, it does, in the sense that without X Charles would not be what we are normally willing to call a human being, and hence *a fortiori* would not be Charles. However, this is a piece of empirical knowledge (based on the assumption that the biological theory conditioning having a brain, and hence being human, by having X holds) – it is not a matter of the rules of our language. To fail to know that Charles has X (which, we assumed, a normal speaker does) is not to fail to understand the term *human* or the name *Charles*, it is to fail to know something about humans and especially about Charles.

Similarly (1*) does hold necessarily, for we know that the lack of consistency would be possible only when PA were a different theory. Unlike the case of X and Charles, this is not empirical knowledge, but just like in this case it is conditioned by the fact that there are no errors in the proofs of the consistency of PA. (What if all the proofs of the consistency of PA presented so far are in fact erroneous, it is only that we have missed the errors? This is, needless to say, very improbable, but certainly neither impossible in some metaphysical way, nor excluded by the rules of our language.) Hence we can say though (1*), in contrast to (1**), does not amount to empirical knowledge, it does amount, in accordance with (1**), to knowledge that is a posteriori in the Kripkean sense.

In contrast to this, there is no such possibility for (2*). However we vary PA, it not only cannot be the case that its biggest consistent part would not be consistent, but also it cannot be the case that we would have to discover that it is consistent. This, I think, accounts for both the intuitive distinction between (1*) and (2*) and the intuitive parallel between (1) and (2), on the one hand, and (1*) and (2*), on the other – both pairs get equated only via an additional nontrivial fact that is a posteriori, though in the first case in the ordinary synthetic a posteriori sense (the fact is simply empirical), while in the second case in the sense of the Kripkean necessary a posteriori.

4.6 Conclusion

Where does this bring us w.r.t. intensionality and meaning in mathematics? I think that what is clear is that the fact that *to have a genuine meaning* is *to have an intension* holds for mathematical expressions just as for empirical ones. However, to make sense of the concept of intension as applied to mathematical expressions we must disregard the almost universal acceptance of PWAI and slightly alter our understanding of it. I think that, roughly, an expression has an intension iff it does not only have an extension,[9] but if there are rules determining which changes of the extension are admissible so that it still counts as the same extension, and which changes would move us beyond this border.

Take (1*), i.e. the sentence *PA is consistent*: I do not think that anyone accepting it would be taken as not knowing what PA *means*. If you know the axioms of PA, but do not know about the consistency proofs, you would hardly be considered as not knowing what PA is, you would most probably be considered as merely missing something *about PA*. This seems to me to indicate that the rules of our language, by themselves, do not exclude the inconsistency of PA – at least they do not exclude it in the way in which the rules governing *not* exclude the truth of *Fido is a dog and Fido is not a dog*.

One of the morals to be drawn from this is that though the idea of epistemologically possible, but ontologically impossible, possible worlds is not wrong, we should not take possible worlds as the ultimate unexplained explainers. The ultimate level of explanation is the level of the rules of language. The sentence *Fido is a dog and is not a dog* is necessarily true because whoever would accept it would be deemed an incompetent speaker of English, in particular not knowing what *not* means.

Does this attitude towards meaning not lead us to an excessively fuzzy notion of meaning? If meaning is a matter of the rules of usage, the boundary of which can never be clearly specified, does it not mean that our theory of meaning goes awry? Not really – it is the meaning itself, not our theory of it that is to blame. We should become reconciled to the fact that meaning is not an object with sharp boundaries, but rather a role conferred on an expression by an often vague cluster of rules (as Wittgenstein, Quine, and Sellars all taught us.[10])

This, however, does not imply a humptydumptyism according to which expressions can mean whatever anybody wants them to mean. There are clear arguments distinguishing predicates that do mean *consistent* from co-extensional predicates that do not. They probably do not allow us to delimit the former ones quite sharply; hence they may not clear all the puzzles of the 'intensional' problems of mathematics, such as the Second Incompleteness Theorem. However, they do allow us to gain a much clearer view of their nature than when we ignore the intensionality of mathematical expressions or when we settle for its trivial theory.

Acknowledgements Work on this paper was supported by Grant No. 17-15645S of the Czech Science Foundation.

---

[9]To be sure, there may be expressions, especially definite descriptions, that have an intension but do not have an extension – in the case of mathematical expressions an example might be the greatest prime. But this is because they are composed out of meaningful subexpressions in such a way that they are themselves meaningful, though they do not pick up any extension.

[10]See Peregrin (2014).

References

Auerbach, D. 1985. Intensionality and the Gödel theorems. *Philosophical Studies* 48: 337–351.
Auerbach, D. 1992. How to say things with formalisms. In *Proof, logic and formalization*, ed. M. Detlefsen, 77–93. London: Routledge.
Bigelow, J.C. 1978. Believing in semantics. *Linguistics and Philosophy* 2: 101–144.
Boolos, G. 1995. *The logic of provability*. Cambridge: Cambridge University Press.
Carnap, R. 1947. *Meaning and necessity*. Chicago: University of Chicago Press.
Cresswell, M.J. 1975. Hyperintensional logic. *Studia Logica* 35: 25–38.
Detlefsen, M. 1986. *Hilbert's program: An essay on mathematical instrumentalism*. Dordrecht: Reidel.
Feferman, S. 1960. Arithmetization of metamathematics in a general setting. *Fundamenta Mathematicae* 49: 35–92.
Frege, G. 1892a. Über Sinn und bedeutung. *Zeitschrift für Philosophie und philosophische Kritik* 100: 25–50.
Frege, G. 1892b. Über Begriff und Gegenstand. *Vierteljahrschrift für wissentschaftliche Philosophie* 16: 192–205.
Halbach, V., and A. Visser. 2014a. Self-reference in arithmetic I. *The Review of Symbolic Logic* 7: 671–691.
Halbach, V., and A. Visser. 2014b. Self-reference in arithmetic II. *The Review of Symbolic Logic* 7: 692–712.
Hintikka, J. 1975. Impossible possible worlds vindicated. *Journal of Philosophical Logic* 4: 475–484.
Kripke, S. 1963a. Semantical considerations on modal logic. *Acta Philosophica Fennica* 16: 83–94.
Kripke, S. 1963b. Semantical analysis of modal logic I (normal modal propositional calculi). *Zeitschift für mathematische Logik und Grundlagen der Mathematik* 9: 67–96.
Kripke, S. 1965. Semantical analysis of modal logic II (non-normal modal propositional calculi). In *The theory of models*, eds. L. Henkin, J.W. Addison, and A. Tarski, 206–220. Amsterdam: North-Holland.
Kripke, S. 1972. Naming and necessity. In *Semantics of natural language*, eds. D. Davidson, and G. Harman, 253–355. Dordrecht: Reidel. Later published as a book.
Lewis, D. 1972. General semantics. In *Semantics of natural language*, ed. D. Davidson, and G. Harman, 169–218. Dordrecht: Reidel.
Löb, M. 1955. Solution of a problem of Leon Henkin. *The Journal of Symbolic Logic* 20: 115–118.
Montague, R. 1974. *Formal philosophy: Selected papers of R. Montague*. New Haven: Yale University Press.
Partee, B. 1982. Belief sentences and limits of semantics. In *Processes, beliefs and questions*, ed. S. Peters, and E. Saarinen, 87–106. Dordrecht: Reidel.
Peregrin, J. 1995. *Doing worlds with words*. Dordrecht: Kluwer.
Peregrin, J. 2006. Extensional vs. intensional logic. In *Philosophy of logic*, Handbook of the philosophy of science, vol. 5, ed. D. Jacquette, 831–860. Amsterdam: Elsevier.
Peregrin, J. 2014. *Inferentialism: Why rules matter*. Basingstoke: Palgrave.
Rosser, J.B. 1936. Extensions of some theorems of Gödel and Church. *Journal of Symbolic Logic* 1: 87–91.

Chapter 5
Deflationary Truth Is a Logical Notion

D. Bonnay and H. Galinon

Abstract The thesis that truth is a logical notion has been stated repeatedly by deflationists in philosophical discussions on the nature of truth. However, to prove the point, one would need to show that the truth predicate does classify as logical according to some reasonable criterion of logicality. Following Tarski, invariance criteria have been considered to provide an adequate rendering of the generality and formality of logic. In this article, we show how the deflationist can use invariance criteria in support of her claim that deflationary truth is a logical notion.

5.1 Deflationism and the Logicality Thesis

The claim that the deflationist's notion of truth is a logical notion is found already in Quine (1970) and later in the work of contemporary deflationists such as Horwich (1998), or Field (1999).[1] These authors have typically meant the claim to be suggestive rather than a key formulation of their philosophical stance, and they did not provide arguments for it. Of course, some quick remarks naturally come to mind. For instance that the truth predicate[2] enjoys a form of topic-neutrality that is typical of logical expressions; or that deflationary truth works more or less

---

[1] Among others. See Horwich (1998) pp. 2–5, Field (1999), p. 76. McGinn (2000) takes truth as one of the logical properties. See e.g. Wyatt (2016) Footnote 28 for further references.

[2] We allow ourselves to move freely from talk about predicates and expressions to talk about notions. We do not think that much turns on this: logical expressions express logical notions and conversely if a notion is logical then an expression that expresses it is a logical expression.

D. Bonnay (✉)
Département de Philosophie, Université Paris Ouest Nanterre La Défense, Nanterre, France
e-mail: dbonnay@parisnanterre.fr

H. Galinon
Département de Philosophie, Université Blaise Pascal, Clermont-Ferrand, France
e-mail: henri.galinon@univ-bpclermont.fr

© Springer International Publishing AG, part of Springer Nature 2018 71
M. Piazza, G. Pulcini (eds.), *Truth, Existence and Explanation*,
Boston Studies in the Philosophy and History of Science 334,
https://doi.org/10.1007/978-3-319-93342-9_5

as a device for expressing infinite conjunctions; or that it can be systematically eliminated in context with the help of substitutional quantification. These remarks are typically unconclusive. As an informal take on logicality, topic-neutrality is not precise enough (how shall we know that something is topic neutral? are mathematical notions topic-neutral ? etc.), and the claims that infinite conjunction and substitutional quantification are logical are by no means self-evident. On the other hand, the logicality thesis is certainly an interesting thesis for the deflationist to build upon, as it articulates the idea of an expressive but non explanatory property, in a way which is compatible with various logical facts that philosophers and logicians have drawn our attention to.[3] We would make some progress in the discussion of deflationism if we could deepen our understanding of the relationships between the notion of deflationary truth and logical notions.

A favorable circumstance is that, despite *prima facie* elusiveness, the boundary between logical and non logical notions is a well-trodden philosophical subject. Among various frameworks developed in the literature in order to test expressions for logicality, the major contender is possibly the 'invariance' framework, which makes it a reasonable option to inquire about the logicality of truth. Since some philosophers have doubted that the deflationist's thesis about the logical nature of truth could survive a test based on a precise criterion of logicality such as invariance,[4] we hope that the results reported in this essay will be of interest to the sceptics as well as to professed deflationists. Indeed what we show in the following is that the notion of truth, as the deflationist understands it, *is* a logical notion.

To achieve our goal, we will proceed as follows. The remainder of this section will set the stage: we first introduce the basics of the invariance approach to logicality and then decide on what it would mean for truth – "truth as understood by deflationists", as we said – to be logical in that setting. The main uptake is that truth is logical in so far as adding a truth predicate adds to the expressive power of a language *only when the underlying logic is not as expressive as the bounds of logic allow*. This is a thesis about deflationary truth, because the focus is on the expressive power of a language equipped with the truth predicate, and the approach is invariance based because the bounds of logic are to be understood as given by some invariance criterion. The second section defines what it means for a logic (not) to be as expressive as invariance allows, by putting forward the concept of a *generated logic*. Some sets of logical expressions are, as we shall argue, somehow "incomplete", and being generated is the property of a logic which does not suffer this special kind of expressive defectiveness. In the third section, we develop a small formal apparatus that allows us to get a grip on the expressive power of the deflationist's notion of truth, in particular regarding the increase in expressiveness that results from adding a truth predicate. This leads to defining the concept of a

---

[3] In particular Shapiro (1998) and Ketland (1999). See also the essays in Achourioti et al. (2015). On the philosophical significance of the thesis of logicality of truth in this context, we may refer the reader to Galinon (2015).

[4] See e.g. Wyatt (2016), pp. 15–17.

truth-complete logic. In a nutshell, a truth-complete logic is a logic in which what can be expressed by means of a deflationary truth predicate can be expressed without it – more precisely, whatever class of structures can be defined using a deflationary truth predicate should be definable without it. The crucial step in Sect. 5.4, consists in showing that the generated logics are exactly the truth-complete ones. We argue that the truth-completeness of generated logics supports the claim that deflationary truth is a logical notion. In the last section, we briefly compare our methodology to the proof-theoretic approach by Shapiro (1998) and Ketland (1999).

5.1.1 Logicality as Invariance

What does it mean to say that an expression is logical? In the sense we are interested in, what makes an expression logical is its denotation – that is, a semantical feature of a linguistic expression, not a feature of the way it is used, or of the specific discursive purposes it serves, though there is of course no denying that the denotation of an expression, the ways that are appropriate to use it and its role in discourse are somehow tied together. With this preamble in mind, the philosophical starting point for the invariance approach to logicality is a shared intuition that logic is a maximally general science, a science which is not about any specific domain of the world in particular. Then given this basic idea, one tries to characterize the logical expressions as those whose semantic value comes out *invariant* under an appropriate class of transformations. As Tarski (1986) puts it, this is a natural extension of Klein's Erlangen program. Klein had showed that different geometries can be characterized as studying those notions that are invariant under different classes of transformations – Euclidian geometry deals with notions which are invariant under isometric transformations, topology with those invariant under continuous transformations, etc. In the same spirit, one may try to capture the generality of logic by thinking of it as the science studying those notions that are invariant under the most general, or *biggest*, class of transformations. And indeed Tarski has argued along these lines that logical notions were those notions which were invariant under all permutations (over the fixed universal domain).

To get a better grasp of what is going on, let us see how the extension of a logical expression, for example the existential quantifier, and the extension of a non-logical expression, say "red", fare with respect to permutations of objects. The extension of "red" shall simply be the set of red objects, but what about the extension of the existential quantifier? In keeping with the literature on logicality, we shall handle quantifiers in a Fregean way, as second-order predicates. Thus, a unary quantifier such as the existential quantifier will be interpreted by a class of structures of the form $\langle M, P \rangle$, $P \subseteq M$. Writing $Q_\exists$ for the interpretation of the existential quantifier we get that $Q_\exists$ is the class of structures $\langle M, P \rangle$ with $P \neq \emptyset$.

Given a domain M, a subset P of M is *permutation invariant* if and only if, for all permutations π on M and for all $a \in M$, $a \in P$ if and only if $\pi(a) \in P$. This definition lifts naturally to the type of quantifier extensions. The interpretation Q of a unary quantifier, which is a class of structures of the form $\langle M, P \rangle$, is *permutation invariant* if and only if, for all M and all permutations π on M, for all $P \subseteq M$,

$$\langle M, P \rangle \in Q \text{ if and only if } \langle M, \pi(P) \rangle \in Q$$

Consider now a toy universe with domain $M = \{a, b, c\}$, and assume that two objects in that universe are red, say $\text{Red}_M = \{a, b\}$. There are 6 permutations on M:

M	π_1	π_2	π_3	π_4	π_5	π_6
a	a	a	b	b	c	c
b	b	c	a	c	b	a
c	c	b	c	a	a	b

Compare, under Tarski's invariance criterion, the behavior of the intuitively logical notion expressed by the existential quantifier and the behavior of the notion expressed by "red". Since a is in $\text{Red}_M(a)$ but $\pi_4(a)$ is not, the extension of "red" in M in not invariant under permutation. On the other hand $\exists_M = \{\langle M, \{a\}\rangle, \langle M, \{b\}\rangle, \langle M, \{c\}\rangle, \langle M, \{a, b\}\rangle, \langle M, \{b, c\}\rangle, \langle M, \{a, c\}\rangle, \langle M, \{a, b, c\}\rangle \}$, and it is easy to check that for any subset P of M, $\langle M, P \rangle$ is in $\exists_M$ if and only if $\langle M, \pi_4(P) \rangle$ is, and similarly for all the other π_is. Permutations may take us out of redness, but they may not take us out of non-empty sets. To put it another way, we may say that "red" allows one to make some distinctions between structures ($\langle M, Red \rangle$ and $\langle M, (\pi_4(Red)) \rangle$ for instance) that a purely logical notion like the existential quantifier cannot distinguish.[5]

The generality of Tarski's specific approach, however, is questionable: paying attention to *permutations on a fixed domain* may be seen as an unacceptable restriction of the generality intuition – why don't we consider more general class of transformations that allow shrinking and expanding the domain for instance? And indeed one of the outcomes of Tarski's approach was an expectable overgeneration problem.[6] To generalize Tarski's fixed-domain approach and escape its framing effects, one is led to consider all possible structures of interpretation of the language and seek out to define an appropriate relation of "logical similarity" between them: the logical notions then come out as the notions that are invariant under the chosen similarity relation. The question "what is a logical notion?" then amounts to asking "what does it mean for two structures to be logically similar?", or what is the similarity relation between structures, whose matching notion of invariance yields logical extensions.

---

[5]One might worry that *Red* and $\exists_M$ are not of the same type, but nothing important hinges on this. We invite the reader to check that the first-order relation $=_M$ (the identity on M) comes out as logical under the proposed criterion, and that the second-order predicate *tobeacolor* does not.

[6]Intuitively: the smaller the transformation class, the bigger the resulting class of logical operators; with Tarski's criterion, the scope of logic is identified with the scope of mathematics.

Various answers have been proposed to this question. Taking the existence of an isomorphism as our target similarity relation between structures, we get Tarski-Sher's criterion,[7] that is to say Tarski's criterion extended so that isomorphic structures not sharing their domain may now count as "similar".[8] However, entirely new possibilities open up in this setting. Feferman (1999), for instance, has considered the possibility of taking the existence of a surjective homomorphism between two structures for the first one to be similar to the other; and Bonnay (2008) has argued at length that potential isomorphism should be considered a better candidate, achieving a good balance between the absence of mathematical and empirical content on the one hand, and triviality on the other hand.[9]

Regarding the constraints that are appropriate to characterize a relation of 'similarity' between structures which would be universally accepted as purely *logical* similarity, various sets of conceptually motivated conditions are possible, each package giving rise to different proposals such as those just mentioned. It is fair to say that there is still room for disagreement when it comes to implementing the invariance idea. Fortunately, and this an important point to make in regard to our project here, the fine-tuning of the invariance criteria does not matter for our purpose. What we will show below is that it is sufficient that one natural constraint, namely closure under definability, be in the package for the corresponding logic to make the addition of an interpreted truth predicate idle. Thus at the end of the day, the validity of our conclusion regarding the logicality of deflationary truth will certainly depend (among other things) upon the validity of the invariance approach to logicality, but it will *not* depend on the validity of the choice of any specific 'logical similarity' relation between structures.

5.1.2 How Can Truth Be Logical?

How then should we handle truth in the invariance framework? Obviously, we would run into problems if we were to apply *directly* an invariance criterion of logicality to the truth predicate. Logical notions are assumed to be maximally general in the sense that they do not distinguish among objects. But truth applies only to a very special kind of objects (sentences, say), and it does make differences among them.

However, we think that it would be superficial to take this point as a refutation of our project right from the start. The reason is that the truth predicate, as understood by the deflationist, is not really meant as a way to talk about sentences and the

---

[7]Sher (1991).

[8]This in line with the idea that logical notions should not be sensitive to the identity of objects.

[9]Note that just considering the biggest similarity relation over structures (the universal relation) would lead nowhere and that consequently the class of candidates for a good "similarity relation", that is a good notion of what it is for two structures to be logically indistinguishable, has to be somehow constrained.

language – it is rather employed to cancel talk about the language. Thus, when it is stated that all theorems of arithmetics are true, the statement is not really about sentences with such and such properties, but is really about numbers that are the subjects of the theorems themselves. Recall Quine's deflationist insight that by attributing truth to the sentence "snow is white" we basically just attribute whiteness to snow. The point of the truth predicate, Quine insists, is that it allows one to talk about the world even when, for some technical reasons (epistemic reasons), one has been forced to semantic ascent:

> Where the truth predicate has its utility is in those places where, though still concerned with reality, we are impelled by certain technical complications to mention sentences. Here the truth predicate serves, as it were, to point through the sentence to the reality; it serves as a reminder that though sentences are mentioned, reality is still the whole point. (Quine 1970, p. 11)

Thus, to vindicate the logicality of the deflationist's notion of truth, one should analyze the behavior of the truth predicate in combination with devices expressing generality and naming sets of sentences. The question is what happens when they are used together as tools to talk about the (possibly non-linguistic) world, and one should thus abstract away from the possible role of the truth predicate as a tool to talk about sentences. From that perspective, it becomes clearer that applying invariance criteria to the denotation of the truth predicate would be at odd with the deflationist's semantic stance.

We then have to adopt an *indirect* strategy. The question we wish to ask is whether what the truth predicate enables us to say about the world is nothing more than what logic contributes to what we say about the world. If this is so, deflationary truth will rightly count as logical: its specific expressive power remains within the bounds of logic. If this is not so, deflationary truth should not be regarded as logical: we may say with a truth predicate things about the world which do not possess the hallmark of logical generality. In order get a formal grip on this question, we shall proceed in the following way. First, we will assume that a particular invariance criterion for logicality is given, in the form of a relation of logical similarity between structures which logical notions should be invariant under (e.g., Tarski's criterion corresponds to picking up invariance under automorphisms). Then consider the set of logical constants that are invariant for the given similarity relation – maybe these are the logical constants of an extension of first-order logic.[10] Then again, add to the logical constants the expressive possibilities made available by the use of a deflationary truth predicate and its companion devices. Finally, ask whether the classes of structures that are now definable with these extended alethic means were already definable in the logic (this is our property of truth-completeness). Again, if the answer is yes, it follows that deflationary truth talk just is logical talk by the light of invariance. We shall show that the answer is indeed positive, not just for one specific choice of invariance criterion, but for any such criterion which enjoys

---

[10]Remember, quantifiers are interpreted as class of structures, and conversely any class of structures can be seen as the interpretation of a putative quantifier.

a natural property of closure under definability (quantifiers, as class of structures, that can be defined by sentences built from invariant quantifiers are still invariant). This will show up as our main result to the effect that a logic is truth-complete if and only if it is exactly generated, because a logic is exactly generated when it may be viewed as the logic one gets on account of an invariance condition closed under definability.

5.2 From Invariance to Logic and Back

In Sect. 5.1.1, we introduced invariance under permutation as a logicality criterion. The present Section generalizes invariance under permutation to invariance under an arbitrary similarity relation, to allow for stronger invariance conditions such as those favored by Feferman or Bonnay. We then make implicit the connection between invariance criteria and first-order logics with generalized quantifiers. But not every similarity relation yields such a logic: sometimes, adding invariant quantifiers may allow one to define on a purely logical basis operations which are not invariant. This does not seem right if invariance is meant to capture what it means for an operation to be logical, so we accordingly restrict our attention to invariance criteria for which this does not happen (Principle of closure of definability).

5.2.1 Similarity Relations, Invariance and Logics

A **similarity relation** S is a relation between structures respecting signatures (i.e. S is a family of relations S_σ between σ-structures for all signatures σ), the notation is $\mathcal{M} \, S \, \mathcal{M}'$. S is meant to capture what it means for two structures to be indistinguishable from a logical point of view (e.g.: being identical up to an automorphism).

We are then interested in the **invariance** of operators under similarity relations. An operator is just a class of structures. Recall the example in Sect. 5.1: the operator associated with $\exists$ is the class $Q_\exists$ of structures $\langle M, P \rangle$ such that P is not empty. We write $Q_\exists(\langle M, P \rangle)$ for $\langle M, P \rangle \in Q_\exists$. The connection with the standard satisfaction clause for $\exists x$ is the following (where $\mathcal{M}$ is an arbitrary model and τ an arbitrary assignment over $\mathcal{M}$):

$\mathcal{M} \vDash \exists x \, \phi(x) \, \tau$
iff
there is an $a \in M$ such that $\mathcal{M} \vDash \phi(x) \, \tau[x := a]$
iff
$Q_\exists(\langle M, \{a \in M \, / \, \mathcal{M} \vDash \phi(x) \, \tau[x := a]\} \rangle)$

We say that an operator Q is S-**invariant** iff, for any structures $\mathcal{M}$, $\mathcal{M}'$, if $\mathcal{M}$ S $\mathcal{M}'$, then $Q(\mathcal{M})$ iff $Q(\mathcal{M}')$. We denote by $Inv(S)$ the class of operators which are S-invariant.

By a *logic L*, we shall mean whatever class of interpreted logical symbols of first-order type (quantifiers of type level at most 2 and propositional connectives), plus the required syntactic sugar (first-order variables x, y, ... and parentheses).[11]

Let K be a class of operators. The **logic L_K associated with K** consists in first-order variables and logical constants interpreted by operators in K. For any signature σ, we thus obtain a language $L_K(\sigma)$ with extra-logical symbols corresponding to σ, whose interpretation varies freely, and logical symbols whose interpretation is taken from K and is kept fixed. For example, let $Q \in K$ be a class of structures of the form $\langle M, R \rangle$ where $R \subseteq M \times M$, L_K contains a logical symbol $\overline{Q}$ which is interpreted by Q. This means that, in the recursive definition of satisfaction for L_K, the clause for Q is the following one:

$$\mathcal{M} \models \overline{Q}x, y \; \phi(x, y) \; [\sigma] \text{ iff } Q(M, ||\phi(x, y)||_{\mathcal{M},\sigma})$$

where $||\phi(x, y)||_{\mathcal{M},\sigma}$ is the interpretation of ϕ over $\mathcal{M}$ according to σ, that is the set of pairs $\langle a, b \rangle$ of elements of $\mathcal{M}$ such that $\mathcal{M} \models \phi(x, y) \; [\sigma][x := a][y := b]$. We just gave the example for $Q_\exists$ of type $\langle 1 \rangle$. As an example for a quantifier of type $\langle 2 \rangle$, consider Q_{WF} the class of relational structures $\langle M, R \rangle$ where R is a well-ordering. $\overline{Q_{WF}}$ is the well-foundedness quantifier, $\overline{Q_{WF}}x, y \; \phi(x, y)$ being true iff $\phi(x, y)$ defines a well-ordered relation.

Conversely, given a logic L and an operator Q – for simplicity, we assume the type of Q is the same as before – we shall say that Q **is definable in** L iff there is a sentence ϕ_Q of $L(\overline{R})$ such that:

$$Q(\langle M, R \rangle) \text{ iff } \langle M, R \rangle \models_L \phi_Q$$

5.2.2   Closure Under Definability and Generated Logics

We shall now look at the conditions under which a logic may be construed as the logic generated by an invariance criterion. The invariants of a similarity relation should be *closed under definability*, this is the constraint we have alluded to in the introductory section. Let us now explain what it is about. Given a similarity relation S, $Inv(S)$ is the class of S-invariant operators. We are interested in the class $Inv(S)$ as the putative class of logical operators. This means that we want to use the operators in $Inv(S)$ as building blocks for the logical part of a language. Given a language, it is possible to define in it certain operators in a purely logical way.

---

[11]The definition is very (maximally !) liberal about what logic should be: at this point we do not impose any special semantic properties that 'truly logical' symbol should have.

For example, let us consider the class K of operators containing just the existential and universal quantifiers, the operator for equality and the boolean operators. The logic associated with K is just FOL. Now, in FOL, it is possible to define new logical operators. For example, the purely logical formula "$\exists x, y, z \, ((Px \wedge Py \wedge Pz) \wedge (x \neq y \wedge x \neq z \wedge y \neq z))$" defines the operator $Q_{\geq 3}$ ("there are at least three"), which is the class of all structures of the form $\langle M, P \rangle$ where P is a subset of M containing at least three elements.[12] Even if Q was not in K, it was "implicitly" there, because it is definable in a language based on K.

We claim that operators which are definable in a purely logical manner are logical. We just do not see how a non-logical element could creep in the logical elements of the definition and make the defined operator non logical. This is what we might call the principle of closure under definability:

Principle of closure under definability *An interpreted symbol definable only by means of logical constants is a logical constant.*

A similarity relation S is **closed under definability** if and only if every operator which is definable in a language L whose logical constants are interpreted by S-invariant operators is S-invariant. If we accept the principle of closure under definability, every similarity relation which can be used to characterize logical operations should be closed under definability.

Keeping in mind this closure principle, we shall say that a logic L is **exactly generated by a similarity relation** S when an operator Q is in $Inv(S)$ iff Q is definable in L. A logic is exactly generated *tout court* iff it is exactly generated by some S.[13]

The similarity relations which can be used to generate a logical language are precisely those that are closed under definability. If it were not the case, there would be a discrepancy between what the logic can express and the invariants of the similarity relations. This intuition is made precise by the following fact:

Fact 1 *S is closed under definability iff there is a logic L which is exactly generated by S.*

Proof This straightforwardly follows from the definition of closure under definability. If S is closed under definability, taking as logical operators all the S-invariants yields a logic L whose elementary classes are S-invariant,[14] since S is closed under definability. Conversely, If S generates a logic L, $El_L \subseteq Inv(S)$, hence S is closed under definability by definition.

---

[12]The formula is purely logical since the so-called *non logical constants* are left uninterpreted here

[13]Writing El_L for the class of elementary classes of L (following the usual terminology, an elementary class of structure is a class of structures which is definable by a single sentence), we have that L is exactly generated iff there is an S such that $El_L = Inv(S)$.

[14]Of course, in real life, we are interested in finding simpler logics generated by S, that is, the game consists in finding some relevant subset of invariant operators that can be used to define all the other invariants.

Finally, note that not all familiar logics are exactly generated and not all familiar similarity relations are closed under definability. As a case in point, $S = Iso_\omega$, partial isomorphisms of finite length is not closed under definability (see the proof in the Appendix). However, Iso_ω was the natural candidate to generate first-order logic, since there is a partial isomorphism of finite length between two structures $\mathscr{A}$ and $\mathscr{B}$, $\mathscr{A}$ Iso_ω $\mathscr{B}$, iff $\mathscr{A}$ and $\mathscr{B}$ are elementary equivalent structures in ordinary first-order logic, but one may prove that actually no similarity relation exactly generates standard first-order logic. On the other hand, some of the natural similarity relations in this context happen to be closed under definability. This is the case for the relations chosen as a basis for the distinction between logical and non logical constants by Tarski and Sher ('being isomorphic'), by Feferman ('being strictly homomorphic') and by Bonnay ('being potentially isomorphic').

5.3   A Framework for Deflationary Truth

We shall now turn to the formal setting for the truth predicate, which we will use to test whether the extra-expressive power it provides remains within the bounds of logic, as given by means of an invariance criterion. We are thus interested in the *expressive power* of the truth-predicate. The truth predicate, as noted above, does not come alone. It applies to naming devices describing set of sentences and those devices are called for to display the expressive abilities of truth. A framework aimed at the mathematical study of the deflationary concept of truth must thus provide a way to deal with sentences, sets of sentences or whatever truth applies to. Recall also that, under deflationist interpretation, talk of sentences in truth ascriptions is primarily understood as a mere technical detour and is essentially parasitic on the intended domain of discourse. To meet those requirements, we will consider some *sorted languages* and related *sorted structures*, which in some sense mimic a familiar distinction between object-language and metalanguage where the metalanguage is understood as a tool to state in a 'formal mode' (as a Carnapian would have it) what could be said (or not!) in a material mode.

Grammatically, we associate to any langage L in signature σ a sorted language which is built out of it by adding a vocabulary of a distinguished sort, including a truth predicate, new distinguished variables, and perhaps other predicates and relation symbols of the new sort.[15] The set of sentences of this extended langage is simply the union of two sets: the set of sentences made out of the vocabulary of the first sort[16] and the set of sentences made out of vocabulary of the second sort.[17] As for the semantics of such a language, the domain of discourse over which the

---

[15]The metalanguage sort of vocabulary will thereafter be printed in bold face.

[16]Using the usual building rules on the vocabulary of the first sort.

[17]We conventionally take the logical constants as being available for constructing sentences in each of the two langages – nothing hinges on that.

variables of the metalinguistic sort range and the predicates of the metalinguistic sort
– including the truth predicate – are interpreted is simply a set assumed to contain
at least the sentences of the object language. We call such a particular choice of a
metalangage an "alethic extension" (of a logic), it comes with fixed and interpreted
metalinguistic vocabulary.

Thus, let L_σ be a language of signature σ in an arbitrary logic L. An *alethic
extension* $L_{\sigma,\mathscr{A}}$ for L_σ consists in:

1. adding to the basic vocabulary of L_σ variables of a new sort (printed in boldface)
   $\mathbf{x}, \mathbf{y}, \ldots$, new sort predicates $\mathbf{P_1}, \ldots, \mathbf{P_\alpha}$ and a truth-predicate $\mathbf{Tr}$. The set of
   formula and sentences in the sorted language are obtained as the union of the
   standardly defined sets of formula and sentences in the respective vocabularies
   of the two sorts.[18]
2. selecting a structure $\mathscr{A} = \langle A, \overline{\mathbf{P_1}}, \ldots, \overline{\mathbf{P_\alpha}} \rangle$ where $\mathscr{A}$ is a set containing all the
   sentences of L_σ,
3. associating with each σ-structure $\mathscr{M}$ a *sorted* expansion $\mathscr{M}^{\mathscr{A}}$ of the form
   $\langle \mathscr{M}, \mathscr{A}, \overline{\mathbf{Tr}} \rangle$ where $\mathscr{A}$ is as above and $\phi \in \overline{\mathbf{Tr}}$ iff ϕ is a σ-sentence such that
   $\mathscr{M} \vDash \phi$.

Note that the perspective on truth we adopt is entirely semantic. In alethic
extensions, the metalinguistic predicates and the truth predicate are added as
interpreted predicates. Their extension is assumed to be uniquely defined: the
metalinguistic predicates are interpreted on a fixed model $\mathscr{A}$ of the metalanguage
and the extension of the truth predicate is uniquely determined in $\mathscr{M}^{\mathscr{A}}$ for each
σ-structure $\mathscr{M}$.

Given a logic and a signature, defining an alethic extension yields a standard
notion of validity. A formula ϕ in the extended language $L_{\sigma,\mathscr{A}}$ is valid (notation:
$\vDash \phi$) iff for all σ-structures $\mathscr{M}$, $\mathscr{M}^{\mathscr{A}} \vDash \phi$. Our extensions are *alethic* extensions
in the sense that the T-equivalences are valid, no matter how we defined our alethic
extensions. First, $\vDash \phi \leftrightarrow \mathbf{Tr}(\mathbf{x}) [\phi]$ by clause 3. In case there is a name 'ϕ' for ϕ in
$L_{\sigma,\mathscr{A}}$, we get as a valid formulas the full-fledged T-equivalence: $\vDash \phi \leftrightarrow \mathbf{Tr}('\phi')$.

We leave much freedom regarding what is in an alethic extension and what is
not. In a given alethic extension, we might get much more than the T-equivalences.
For example, assume $L_{\sigma,\mathscr{A}}$ has a predicate $Sent(\mathbf{x})$ which is such that for all $a \in A$,
$a \in \overline{Sent}$ iff there is a sentence ϕ of L_σ with $a = \ulcorner \phi \urcorner$, and a function $not(\mathbf{x})$ such
that for all $a, b \in A$, $\overline{not(a)} = b$ iff b represents a sentence which is the negation
of a sentence represented by a, then $\vDash \forall \mathbf{x} (\mathbf{Sent}(\mathbf{x}) \rightarrow \mathbf{Tr}(not(\mathbf{x})) \leftrightarrow \neg \mathbf{Tr}(\mathbf{x}))$.
Similarly, assume that the syntactic operation of conjoining two sentences is
encoded by a syntactic function $\mathbf{and}(\mathbf{x}, \mathbf{y})$ such that for all $a, b, c \in A$, $\overline{\mathbf{and}(a, b)} =$
c iff c represents a sentence which is the conjunction of two sentences represented
by a and b. We will get $\vDash \forall \mathbf{x}, \mathbf{y} [\mathbf{Tr}(\mathbf{and}(\mathbf{x}, \mathbf{y})) \leftrightarrow \mathbf{Tr}(\mathbf{x}) \wedge \mathbf{Tr}(\mathbf{y})]$. So, as we

---

[18]Remark that the definition does not allow for 'mixed' formula containing vocabulary of the two
sorts. This is a simplification, and we could have allowed for a richer syntax. For instance, boolean
combinations of sentences of the two sorts would not harm. See Footnote 19.

leave the notion of alethic extension underspecified, the theory of truth associated with alethic extensions is underspecified too. In the minimal case, we get just the T-equivalences as truth-theoretic truths. Using richer alethic extensions, we can get more, e.g. the previous recursive clause for $\wedge$. As it will turn out, leaving the notion of alethic extension underspecified is fine, because the result we eventually prove will hold for all alethic extensions, that is, no matter how weak or strong the induced theory of truth is.[19]

Here is an important fact which holds no matter the alethic extensions we choose to consider:

Fact 2 *For any L and any alethic extension $L_{\sigma,\mathscr{A}}$ for L, if $\mathscr{M} \equiv \mathscr{M}'$ then $\mathscr{M}^{\mathscr{A}} \equiv \mathscr{M}'^{\mathscr{A}}$.*

Fact 2 seems a reasonable property[20] for extensions by an interpreted truth-predicate: since the truth predicate only speaks of the truth or the falsity of the sentences of L_σ, one should not get non-elementary equivalent models from elementary equivalent ones. Note however that Fact 2 does *not* tell us that adding a truth predicate for L_σ does not add some expressive power. To see this, take L to be standard first-order logic with equality and σ the empty signature. Consider the sentences ϕ_n of the form $\exists x_1, \ldots, x_n \forall y \, (y = x_1 \vee \ldots \vee y = x_n)$ and pick an alethic extension $L_{\sigma,\mathscr{A}}$ such that there is a formula $\boldsymbol{\phi}(\mathbf{x})$ satisfying, for every $\mathscr{M}^{\mathscr{A}}$, $\mathscr{M}^{\mathscr{A}} \vDash \boldsymbol{\phi}(\mathbf{x}) \, [a]$ iff $a = \ulcorner \phi_n \urcorner$ for some n. If needed, such an alethic extension can be created by fiat: take a predicate $\mathbf{P_1}$ interpreted by $\{a \in A \, / \, a = \phi_n \text{ for some } n\}$. Then $\exists \mathbf{x} \, (\boldsymbol{\phi}(\mathbf{x}) \wedge \mathbf{Tr}(\mathbf{x}))$ is a sentence of $L_{\sigma,\mathscr{A}}$ which defines the class of finite structures. By compactness, there is not even an infinite set of sentences of L_σ defining that class. This is perfectly compatible with Fact 2 being true: a finite structure and an infinite structure are not elementary equivalent in $L_{\sigma,\mathscr{A}}$, but they are not elementary equivalent in pure first order logic either.[21]

---

[19]It has to be kept in mind, however, that in any case our theory of truth is strong in the sense that we use an interpreted truth-predicate.

[20]The fact is obvious since $\mathscr{M} \equiv \mathscr{M}'$ just means that the same sentences of L_σ are true in $\mathscr{M}$ and $\mathscr{M}'$ and by our definition of alethic extensions.

[21]Note that for Fact 2 to hold it is crucial that a sorted language is used. Assume we are dealing with a very simple language which has only one sentence which says that there are no more than n objects. Consider two structures $\mathscr{M}$ and $\mathscr{M}'$ with $|M| = n - 1$ and $|M'| = n$. In our toy language, these two structures are elementary equivalent. Now consider an alethic expansion in which the syntactic domain A has only one object representing the only sentence in the language. If one were to use a non-sorted language, $\mathscr{M}^{\mathscr{A}}$ and $\mathscr{M}'^{\mathscr{A}}$ would not be elementary equivalent anymore, just because now $|\mathscr{M}'^{\mathscr{A}}| > n$ whereas $|\mathscr{M}^{\mathscr{A}}| = n$. However, the constraints we have imposed on the syntax of $L_{\sigma,\mathscr{A}}$ could be somewhat relaxed so as to allow sentences combining variables and predicates of the two sorts. If the combinations allowed are reasonable, then expect Fact 2 to remain true.

5.4 Truth as a Logical Notion

5.4.1 Truth Completeness and Closure Under Definability

Recall our main purpose: to prove that *if* our underlying logic is well behaved, adding a truth predicate will not gain us any expressive power. For such a 'well behaved logic', what does it mean to say that adding an interpreted truth predicate to it does not increase its expressive power? We understand this as saying that truth extensions would provide us with new tools to express things which were already expressible in principle by purely logical means. In an abstract model-theoretic fashion, we take the expressive power of a logic to be given by the elementary classes associated with it. The following notion of truth-completeness of a logic then captures what we are after:

Definition 1 A logic L is truth-complete iff for every signature σ, for every alethic extension $L_{\sigma,\mathscr{A}}$, for every sentence ϕ of $L_{\sigma,\mathscr{A}}$, there is a sentence $\hat{\phi}$ of L_σ such that for all L_σ-structures $\mathscr{M}$:

$$\mathscr{M}^{\mathscr{A}} \vDash \phi \text{ iff } \mathscr{M} \vDash \hat{\phi}.$$

By quantifying universally over alethic extensions, we require this no matter how powerful the alethic extensions (intuitively, an alethic extension is more powerful if it has more definable sets of sentences to which the truth predicate can be applied).

We can now state the following:

Theorem 1 *L is exactly generated iff it is truth-complete*

When L is exactly generated, it will be truth-complete hence truth will be implicitly definable[22] *no matter how strong the underlying alethic extension is.* Therefore the condition that L is exactly generated is quite powerful. Even if we hardwire a lot of definable sets over alethic extensions through interpreted predicates over the set of sentences of L_σ, so that the truth predicate has a lot to speak about, L will still be powerful enough to say all that can be said with that truth predicate. On the other hand, note that the implicit definability of truth we get for a truth-complete logic is not uniform: for every sentence containing the truth predicate, there is an equivalent sentence which does not contain it, but this does not imply that a translation procedure exists.

The right-to-left part of the theorem hinges on the fact that our notion of alethic extension is very liberal. All invariant operators we need to define can be defined with the truth predicate only because we make the very strong assumption that all what we might need to make definition through the truth predicate can be made available by the syntactic predicates. We remained silent on the strength of the metatheory we need to fix the interpretation of all those interpreted syntactic predicates, which of course will depend itself on how fine-grained the underlying similarity relation is.

---

[22]By *implicitly definable* we mean that any class of structures which is definable (in the sense given on page 2) with the truth predicate, is definable without it.

5.4.2   Interpreting the Result

The Theorem establishes an equivalence between two properties of logics: being truth-complete and being generated by a similarity relation. The connection with closure under definability, which we take to be well-motivated constraint on similarity relations, is provided by Fact 1.

Now consider a particular logic L. Is L truth-complete? If it is, there's no reason not to count the deflationist truth-predicate as logical. But what if it is not? By Theorem 2, L is not exactly generated. Whatever the notion of logical similarity S we think is appropriate, if the logical operations of L are S-invariant, then there are some S-invariant operations which are not definable in L. So the failure of truth-completeness for L should not count as an argument against the logicality of the truth-predicate, because it may result from L's unduly restricted range of available logical operations. And indeed, it does, by Theorem 2, if we were to move from L to an extension L' strong enough to define all S-invariants, then L' would indeed be truth-complete. To repeat, it might well be the case that a given logic is not truth complete. But if we consider that this logic only partially expresses the S-invariants for some S which is closed under definability, then we know that the extra expressive power provided by the truth predicate should not count as substantial. It stays within the realm of logic, in the sense that if the logic had been powerful enough to express all the invariants, then the truth predicate would have been implicitly definable.

We now turn to examples of generated logics. Here are some examples, with the similarity relations that generate them:

- propositional calculus over finite signatures
- on a fixed domain, $L_{\infty,\omega}$ and potential isomorphisms.
- on a fixed domain, $L_{\infty,\infty}$ and isomorphisms.
- on a fixed domain, $L_{\infty,\infty}^-$ and strong homomorphisms

Note that the last three logics are infinitary. In those cases, it was quite expected that adding an (interpred) truth predicate will not add any expressive power: the infinitary combinations of sentences that can be expressed thanks to the truth predicate can already be expressed in the logic.

Finally, it should be emphasized that only one direction of our theorem is required to support our claim that deflationary truth is a logical notion. For the logicality of the deflationary truth predicate is understood as a consequence of the fact that closure under definability implies truth-completeness. And this part of the result holds no matter the class of alethic extension one may find conceptually acceptable. If one is willing to allow only sets of sentences to interpret the metalinguistic predicates in alethic extensions, or only those sets of sentences that are computable, or those that are definable in such and such theory, or whatever, this direction of the result still holds. Although it would certainly be interesting to find a natural characterization of the accordingly modified notions of truth-completeness in terms of invariance conditions, our not doing so does not count against our main conclusion concerning the concept of truth.

5.5   Expressive Power of the Truth Predicate or Proof-Theoretic Strength of Truth Theories?

Our result supports the deflationist's claim that truth can be thought of as logical and, in this sense, a "non-substantial" notion. It might be of interest to compare our methodology to the one driving the so-called "conservativity argument" against deflationism.[23] Very briefly, the idea behind the conservativity argument is the following: were "true" not to have any explanatory power, the Tarskian truth theoretic extension $T(A)$ for a theory A should be a conservative extension of A. But, the argument continues, this is not the case: for many theories A their Tarskian truth-theoretic extension explains (in the sense of proving) facts statable in non-alethic A-terms which the theory A does not explain. The classical example of this phenomenon is the provability of the consistency of PA in Tarski's theory of truth for PA. Hence deflationism is false, or so it argued.

Conservativity-based approaches to the truth-predicate focus on the strength of *theories supposed to somehow define the truth predicate*. They ask: how strong must such a theory be to adequately constrain the interpretation of the truth predicate over a given base theory B ? Our standpoint is different. To study the logicality of an expression we first take its interpretation as *given* and look out whether it is logical or not. The logicality of an expression, if the invariance approach has got it right, is something about what kind of distinctions an expression can do between different structures of interpretation of the language. There are those expressions, such as "red", that distinguish between structures on account of some empirical features of them, or strong mathematical content of them, while logical expressions should distinguish among structures only on account of what could be argued to be purely non-mathematical, formal, or most general, features of them.

Both approaches to the "substantiality of truth" face difficulties when the time comes to carry out their program. Conservativity-based approaches have to face the fact that non-conservativity results are not robust, in the sense that there are base theories over which the target Tarskian truth-theoretic extensions *are* conservative. For instance, the Tarskian truth-theoretic extension A is conservative over A when A is first-order Peano arithmetic augmented with an omega-rule or, to take a quite different example, when A is some specific arithmetic theory weaker than PA.[24] The alleged philosophical conclusion to be drawn from the logical fact of non-conservativity is thus afflicted of unstability: the notion of Tarskian truth appears as substantial or not depending on the choice of the base theory. Yet the choice of a base theory over another as a suitable basis for assessing the substantiality of truth is by no mean settled.[25] On our semantic road, we can do better.

---

[23] Shapiro (1998) and Ketland (1999).

[24] See Halbach (1999) for a finitist proof of this. The theory can be taken to be Robinson Arithmetic Q with suitable "unique readability principles".

[25] For instance the following questions do not have universally accepted answers. Is the relevance of the base theory for a philosophical assessment of truth to be judged from the fact that its truth-

As an interpretation for the truth-predicate in a language L, we have taken, in each structure $\mathcal{M}$, the set of truths-in-$\mathcal{M}$ in the standard model-theoretic sense. That is, given a language L, the interpretation of truth-in-L is taken to be a function of L but is not taken to bring out just a "fixed" set of L-sentences. Rather, truth-in-L is interpreted as a set of such sets varying along the extensions of other non-semantical terms appearing in the different models. In our model, the extension of "true" in a structure $\mathcal{M}^*$ is entirely dependent on non-semantical states of affairs as given by the structure $\mathcal{M}$, a feature which fits well with the broadly received idea that truth supervenes on non-semantical facts. So our truth predicate *has* a definite extension over the class of all models, and it is, in each model $\mathcal{M}$, the set of truths-in-$\mathcal{M}$. Why should our above main result be understood as a case for the logicality of this truth predicate? At first sight our semantic approach faces a problem quite similar to the one we have just mentioned in the case of conservativity-based approaches, one of unstability. For our interpreted truth-predicate increases the expressive power of some logics while it does not increase the expressive power of other, so that the truth-predicate appears as substantial in some cases while it does not appears so in others. But, importantly, we have offered a philosophically significant necessary and sufficient condition under which it appears as "substantial": it is when the underlying logic is not generated. This condition is significant because closure under definability of a similarity relation seems to be by itself a well-motivated constraint on similarity relations. And as a consequence, in an acceptable logic, that is in a logic which is generated by an acceptable similarity relation, truth will indeed be deflated as purely logical.

5.6   Conclusion

We do not claim that our results show a certain brand of deflationism to be the best view of truth available on the market, nor that the ordinary notion of truth is a purely logical notion. Our purpose was more modest. It was to lend support to the view that *deflationary truth* is a logical notion. More precisely we have shown that, on the background of an invariance-based conception of logicality, it is consistent to claim that *true* has the meaning the deflationist says it has, and that truth so understood is a logical notion.

---

theoretic extension meets some adequacy conditions, for instance the provability of some reflection principles over the base theory, as Ketland (1999) has it? (See also Leitgeb (2007), Halbach and Horsten (2015). See also Fischer (2015) for a useful follow-up.) Or should it be assessed solely from its intrinsic virtues as a syntactical theory – assuming that this 'intrinsic' talk could be made precise? But then, regarding the notion of truth itself, what is the proper interpretation of the fact that conservativity of Tarskian axioms for truth over PA (say), obtains or does not obtain depending on whether induction axioms in the base theory are understood as a list or as a scheme? On this last issue, see e.g. Field (1999), and more recently, in connection with the logicality of truth, our remarks in Galinon (2015).

Appendix: Proofs

Fact Iso_ω is not closed under definability.

The proof is an elementary exercise in model theory.

Proof We shall consider the operator $Q_{\geq \aleph_0}$. $Q_{\geq \aleph_0}$ is Iso_ω-invariant: let $\langle M, P \rangle$ and $\langle M', P' \rangle$ be two structures such that we have $Q_{\geq \aleph_0}(\langle M, P \rangle)$ but not $Q_{\geq \aleph_0}(\langle M', P' \rangle)$ (thus, $|P| \geq \aleph_0$ where as $|P'| < \aleph_0$). There is an integer n such that $|P'| = n$, but then it is not the case that $\langle M, P \rangle Iso_{n+1} \langle M', P' \rangle$, hence it is not the case that $\langle M, P \rangle Iso_\omega \langle M', P' \rangle$. We shall now consider the operator Q' defined by the sentence "$\overline{R}$isanequivalencerelation $\wedge \ \exists x \overline{Q_{\geq \aleph_0}} y \ x \overline{R} y$" ($Q'$ picks out the relational structures $\langle M, R \rangle$ such that R is an equivalence relation with an infinite equivalence class). Since $Q_{\geq \aleph_0} \in Inv(Iso_\omega)$, Q' is definable from Iso_ω-invariant operations. Hence to show failure of closure under definability, it suffices to show that Q' itself is not Iso_ω-invariant. We construct two $L(\overline{R})$-structures $\mathcal{M} = \langle M, R \rangle$ and $\mathcal{M}' = \langle M, R' \rangle$ such that:

- The interpretation of $\overline{R}$ on both models is an equivalence relation.
- $\mathcal{M}$ contains an infinite number of R-equivalence classes of arbitrary big finite cardinality, but no infinite equivalence class.
- $\mathcal{M}'$ is just as $\mathcal{M}$ but it contains also an infinite equivalence class.

It is clear that $\mathcal{M} \ Iso_\omega \ \mathcal{M}'$. But we have that $Q'(\mathcal{M}')$, whereas $\mathcal{M}$ is not in Q'.

Theorem 2 *L is exactly generated iff it is truth-complete*

Proof **If L is exactly generated then it is truth-complete**
Let L be a logic and S a similarity relation such that $El_L = Inv(S)$. Let σ be an arbitrary signature, $L_{\sigma,Tr}$ an arbitrary alethic extension for L_σ and ϕ a sentence of $L_{\sigma,Tr}$. It is sufficient to show that $Q_\phi = \{\mathcal{M} \ / \ \mathcal{M}^* \vDash \phi\}$ is closed under S, because then $Q_\phi \in El_L$ by $El_L \supseteq Inv(S)$. Assume that $\mathcal{A} \in Q_\phi$ and $\mathcal{A} \ S \ \mathcal{B}$. We want $\mathcal{B} \in Q_\phi$.

By hypothesis, $El_l \subseteq Inv(S)$, hence $\mathcal{A} \ S \ \mathcal{B}$ implies $\mathcal{A} \equiv_{L_\sigma} \mathcal{B}$ (or there would be a sentence ψ of L_σ which is true in $\mathcal{A}$ and not in $\mathcal{B}$, which implies that the class of models of ψ is not closed under S, contradicting $El_l \subseteq Inv(S)$). By Fact 2, $\mathcal{A} \equiv_{L_\sigma} \mathcal{B}$ implies in turn $\mathcal{A}^* \equiv_{L_{\sigma,Tr}} \mathcal{B}^*$. Since $\mathcal{A}^* \vDash \phi$, $\mathcal{B}^* \vDash \phi$ as well, hence $\mathcal{B} \in Q_\phi$ as required.

If L is truth-complete, there is an S such that $El_L = Inv(S)$. We take for S the relation $\equiv_L$ of L-elementary equivalence. First, $El_L \subseteq Inv(S)$. Let Q be an operator in El_L, then $\mathcal{M} \in Q$ and $\mathcal{M}' \equiv_L \mathcal{M}$ straightforwardly implies $\mathcal{M}' \in Q$: the formula ϕ defining Q is true in $\mathcal{M}$, hence in $\mathcal{M}'$ as well.

For $El_L \supseteq Inv(S)$, let $Q \in Inv(\equiv_L)$. We choose a signature σ matching Q's similarity type and we are looking for a formula χ of L_σ such that its models are precisely the structures in Q. Let I be an indexing of the models $\mathcal{B}$ in Q. We note $\mathcal{B}_i$ the structure indexed by i for $i \in I$. We pick a particular alethic extension $L_{\sigma,Tr}$ such that there is a formula $\psi(\mathbf{x}, \mathbf{y})$ of L_σ, Tr such that for all $\mathcal{M}*$, for all

$i, j \in A$, $\mathscr{M}^* \vDash \psi(\mathbf{x}, \mathbf{y})$ $[i, j]$ iff $i \in I$ and $j = \ulcorner \phi \urcorner$ for some sentence ϕ of L_σ with $\mathscr{B}_i \vDash \phi$.[26] Let χ be the formula $\exists \mathbf{x} \forall \mathbf{y} \, (\psi(\mathbf{x}, \mathbf{y}) \rightarrow Tr(\mathbf{y}))$. By truth-completeness of L, there is a formula $\hat{\chi}$ such that for all $\mathscr{M}$, $\mathscr{M} \vDash \hat{\chi}$ iff $\mathscr{M}^* \vDash \chi$. $\hat{\chi}$ is the sentence we are after. If $\mathscr{M} \in Q$, $\mathscr{M} = \mathscr{B}_i$ for some $i \in I$. So, by definition, $\mathscr{M}^* \vDash \psi(\mathbf{x}, \mathbf{y})$ $[i, j]$ implies $j = \ulcorner \phi \urcorner$ for some ϕ with $\mathscr{M} \vDash \phi$, hence $\mathscr{M}^* \vDash Tr(\mathbf{y})$ $[j]$. Therefore, $\mathscr{M}^* \vDash \chi$, hence, $\mathscr{M} \vDash \hat{\chi}$. Conversely, assume $\mathscr{M} \vDash \hat{\chi}$. Then, $\mathscr{M}^* \vDash \chi$, so there is an $i \in I$ with $\mathscr{B}_i \in Q$ such that for all b with $b = \ulcorner \phi \urcorner$ for some ϕ, $\mathscr{M}^* \vDash \psi(\mathbf{x}, \mathbf{y})$ $[i, b]$ implies $\mathscr{M}^* \vDash Tr(\mathbf{y})$ $[b]$. By definition of ψ and Tr, we have $\mathscr{M} \vDash \phi$ for all ϕ such that $\mathscr{B}_i \vDash \phi$, hence $\mathscr{M}$ and $\mathscr{B}_i$ are L_σ-elementary equivalent. Since $Q \in Inv(\equiv_L)$ and $\mathscr{B}_i \in Q$, this implies that $\mathscr{M} \in Q$ as well.

References

Achourioti T., H. Galinon, J. Martinez, and K. Fujimoto. 2015. *Unifying the philosophy of truth*. New York: Springer.

Bonnay, D. 2008. Logicality and invariance. *The Bulletin of Symbolic Logic* 14(1): 29–68.

Feferman, S. 1999. Logic, logics, and logicism. *Notre Dame Journal of Formal Logic* 40(1): 31–54.

Field, H. 1999. Deflating the conservativness argument. *Journal of Philosophy* 96: 533–540.

Fischer, M. 2015. Deflationism and instrumentalism. In Achourioti et al. (2015), pp. 293–306.

Galinon, H. 2015. Deflationary truth: Conservativity or logicality? *Philosophical Quarterly* 65(259): 268–274.

Halbach, V. 1999. Deflationism and infinite conjonctions. *Mind* 108: 1–22.

Halbach, V., and L. Horsten. 2015. Norms for theories reflexive truth. In Achourioti et al. (2015), pp. 263–280.

Horwich, P. 1998. *Truth*, 2nd edn. Oxford: Oxford University Press.

Ketland, J. 1999. Deflationism and Tarski's paradise. *Mind* 108: 69–94.

Leitgeb, H. 2007. What theories of truth should be like but cannot be. *Philosophy Compass* 2(2): 276–290.

McGinn, C. 2000. *Logical properties: Identity, existence, predication, necessity, truth*. Oxford: Oxford University Press.

Quine, W.O. 1970. *Philosophy of logic*. Cambridge: Harvard University Press.

Shapiro, S. 1998. Truth and proof: Through thick and thin. *Journal of Philosophy* 95(10): 493–521.

Sher, G. 1991. *The bounds of logic: A generalized viewpoint*. Cambridge: MIT Press.

Tarski, A. 1986. What are logical notions. *History and Philosophy of Logic* 7: 143–154.

Wyatt, J. 2016. The many (yet few) faces of deflationism. *Philosophical Quarterly* 66(263): 362–382.

---

[26] Should we be worried whether such an alethic extension exist? As noted in Sect. 5.3, alethic extensions are so defined that we are free to use interpreted syntactic predicates, the only limit, so to speak, being that Fact 2 will always hold. So in our case, we could just add a syntactic relation $\overline{R}$ doing what we want, that is, such that for all $i, j \in A$, $\langle i, j \rangle \in R$ iff $i \in I$ and $j = \ulcorner \phi \urcorner$ for some sentence ϕ of L_σ with $\mathscr{B}_i \vDash \phi$. The only thing we have to worry about is that the alethic extension has to be big enough to encode the indexing I. Note that if Q is a proper class, this will have the somewhat awkward consequence that our alethic extension is itself a proper class. In the examples given in the next paragraph, restriction to a fixed domain ensures that proper classes are not needed.

Chapter 6
Making Sense of Deflationism from a Formal Perspective: Conservativity and Relative Interpretability

Andrea Strollo

Abstract The contemporary study of the notion of truth divides into two main traditions: a philosophical tradition concerned with the nature of truth and a logical one focused on formal solutions to truth-theoretic paradoxes. The logical results obtained in the latter are rich and profound but often hard to connect with philosophical debates. In this paper I propose some strategy to connect the mathematics and the metaphysics of truth. In particular, I focus on two main formal notions, conservativity and relative interpretability, and show how they can be taken to provide a natural way to read formally the simplicity of the property and the simplicity of the concept of truth respectively. In particular, I show that, this way, we obtain a philosophically interesting taxonomy of axiomatic truth theories.

6.1 The Simplicity of Truth

Axiomatic theories of truth can be compared with respect to different mathematical properties. In particular, current debates pay attention to those mathematical aspects that reveal some kind of simplicity. The simplicity of truth, in fact, apart from being mathematically interesting, is relevant also for its general and philosophical role. For instance, according to deflationism, one of the most prominent philosophical positions in current debates, truth is a simple and unsubstantial notion.[1] How to

---

[1] Deflationary conceptions of truth are philosophical views generally based on three main tenets: (1) that truth is fully accounted for by means of disquotational principles such as Tarskian biconditionals; (2) that a truth predicate provides a tool for expressing infinite conjunctions and disjunctions through truth generalizations; (3) that truth is unsubstantial.

A. Strollo (✉)
Department of Philosophy, Nanjing University, Nanjing, China
e-mail: andrea_strollo@libero.it

© Springer International Publishing AG, part of Springer Nature 2018   89
M. Piazza, G. Pulcini (eds.), *Truth, Existence and Explanation*,
Boston Studies in the Philosophy and History of Science 334,
https://doi.org/10.1007/978-3-319-93342-9_6

understand the simplicity of truth from a philosophical perspective, however, is not clear. In particular, it is not clear whether the unsubstantiality at stake should be exhibited by the *concept* or by the *property* of truth (or both). Thus, even though the mathematical study of formal truth theories provides a rich pool of data that are potentially philosophically relevant, we do not really know how to relate the mathematical and the metaphysical side in a clear way. This is unpleasant both from a mathematical perspective, since we cannot shed philosophical light on many subtle and tricky results, and from a philosophical point of view, since we do not know how to evaluate different axiomatic theories and are unable to test metaphysical thesis with the precision provided by formal frameworks.

In this paper I contribute to fix this situation and propose some strategy to connect the mathematics and the metaphysics of truth. In a nutshell, I focus on two main formal notions, conservativity and relative interpretability, and show how they can be taken to provide a natural way to read formally the simplicity of the property and the simplicity of the concept of truth respectively. In particular, I show that, this way, we obtain a philosophically interesting taxonomy of axiomatic truth theories.[2]

The paper is structured as follows. In the next part I discuss the distinction between concepts and properties with respect to the issue of the unsubstantiality of truth. In the third section I introduce the notion of conservativity and two different definitions of it: a model-theoretic and a proof-theoretic one. In the fourth part I connect model-theoretic conservativity with the unsubstantiality of the property of truth and in the next I argue that proof-theoretic conservativity is not apt to make sense of the unsubstantiality of the concept. In the sixth section I connect the unsubstantiality of the concept to a different formal notion, that of relative interpretation, while in the seventh part I exclude that other notions, such as proof-theoretic reduction and speed up, have any relevance in the context of the unsubstantiality of truth. In the final section I apply the connections elaborated in the paper to yield a taxonomy of axiomatic truth theories, showing how to extract philosophical information from it.

6.2   The Concept and the Property of Truth

Deflationism holds that truth is unsubstantial. Apart from the overt vagueness of such a claim, a problem that is often neglected is that it is not clear whether

---

[2]The general picture pursued in this paper is very close to the one advocated for by Fischer and Horsten (2015). The main differences are two. First, I motivate the adoption of a model-theoretic criterion on the basis of metaphysical considerations, while they rely on reflections about the role of truth in model and proof theory. Secondly, they pay much attention to the question of the expressive strength but have no interest for the problem of the substantiality of the concept of truth. Also the notion of relative interpretation is barely mentioned and not defended in detail. Although I have strong sympathy for their approach, in this paper I prefer to focus on different, yet complementary issues.

deflationists intend to characterize the concept or the property of truth. Although how to trace a distinction between concepts and properties is not an easy matter, that there should be such a distinction should be familiar. It follows that, if concepts and properties are different sorts of entities, also their alleged simplicity and unsubstantiality should amount to different things. Thus, explaining what the unsubstantiality of the former is provides no straightforward explanation of the unsubstantiality of the latter. This is important because, depending on where we locate the simplicity of truth, we gain different philosophical positions. Apparently, the natural interpretation would suggest that deflationists intend to characterize the property of truth. This interpretation is in accordance with the natural reading of different claims of the same form. If I say 'red is a simple colour' or 'courage is a virtue', what I arguably refer to is the property of redness and the property of being courageous. Of course, one might propose a paraphrase according to which talk of courage should actually be interpreted as a talk of the concept of courage,[3] but such interpretations clearly go beyond the natural and apparent literal reading. Insofar as the distinction between properties and concepts is correct, we should say that we usually speak of objects and of the properties they instantiate, not of the concepts that represent them. Also the claim that truth is unsubstantial, then, is naturally read as a thesis about the property of truth. Despite such simple observations, it is easy to find deflationists who intend to describe the concept instead of the property of truth[4] and authors, such as Horwich, that stress that the simplicity of deflationary truth covers all the dimensions associated with truth.[5] The general situation however is complicated. Hartry Field, another champion of deflationism, explicitly mentions the property of truth by writing that a truth predicate expresses a logical property.[6] This messy situation is probably one of the reasons why the debate about the unsubstantiality of truth is hard to clarify.[7]

As a straightforward consequence of such distinctions and the fact that, in principle, the nature of the concept of truth is not forced to parallel the nature of the property, a deflationary stance can take different forms. In particular, a deflationary position can be held at two different levels. A full fledged deflationist will contend that truth is unsubstantial (in a manner to be clarified) both at the concept and at the property level. A more moderate deflationist will claim that the unsubstantiality of truth holds only at one level, either claiming that the concept, but not the property of truth, is unsubstantial or claiming, vice versa, that it is the property, but not the concept that is unsubstantial. These two positions would be hybrid views,

---

[3]For instance, one could read the above claim like: 'the concept of courage is a virtue concept'.

[4]I give just a couple of disparate references for illustration purposes: Dummett (1959), Burgess and Burgess (2011) and Horsten and Leigh (forthcoming).

[5]See Fischer and Horsten (2015).

[6]Field (1999). Another example is provided by Edwards (2013), who addresses the question of the unsubstantiality of truth by taking mostly for granted that it has to do with the property of truth.

[7]In recent years, however, an increasing number of authors has become more sensitive to the differences between talk of the concept and talk of the property of truth. See, for instance, Asay (2013).

with a non completely deflationary pedigree.[8] Clearly, a view holding that neither
the concept nor the property of truth is unsubstantial is, instead, a full-fledged
inflationary position about truth. Although I am inclined to question the legitimacy
of calling *deflationary* a position holding the unsubstantiality of truth along only
one line, I put problems of classification mostly aside, since the main purpose
of drawing such distinctions is just that of having a more accurate articulation
of a deflationary stance. In particular, by distinguishing the unsubstantiality of
the concept from the unsubstantiality of the property, we automatically need to
provide two different, independent explanations of the unsubstantiality of truth.
Accordingly, it comes as no surprise that the two ideas should be correlated with
different formal requirements.

6.3 Conservativities

In a paper recently appeared on Mind, Leon Horsten and Graham Leigh contrast
their approach with the one pursued by John Burgess in the article 'The truth is
never simple'.[9] In his work, Burgess shows that the extension of the truth predicate
(with respect to an arithmetical base theory) is complicated, in the sense that the set
corresponding to such an extension has an high degree of set theoretic complexity.
By contrast, Horsten and Leigh aim at showing that, from a proof-theoretic point
of view, there is room to argue that the intension of the truth predicate is simple,
since truth can be characterized by a simple collection of axioms. In this paper,
I am neither interested in the particular strategy promoted by Horsten and Leigh
nor in the specific complexity measurements applied by Burgess, but want, instead,
develop their background motivations. In particular, I take a step back and focus on
the discrepancy between their approaches. What we have, in fact, is an example of
the classical rivalry of a model theoretic perspective with a proof theoretic one. In
order to further illustrate this tension with respect to formal theories of truth, let me
introduce a notion that proved relevant for the philosophical debate on truth, namely
conservativity.

Conservativity is an important tool to compare different formal theories and is
based on a simple idea: if we have two theories, T1 and T2, a natural question to
ask is whether the addition of T2 to T1 is able to give us more information on what
T1 is about. If this is not the case, from the point of view of T1, the addition of T2
is somehow irrelevant. We thus say that T2 is conservative over T1. Such an idea
can be specified in different ways. It is here that the rivalry between model-theoretic
and proof-theoretic approaches becomes salient since, depending on what approach
is adopted, different definitions of conservativity are obtained.

---

[8] Asay has argued in different places in favour of an hybrid view according to which the concept of
truth is substantial but the property of truth unsubstantial. Such a position is thus a combination of
metaphysical deflationism and conceptual inflationism (Asay 2013a, b, 2014).

[9] Burgess (1986) and Horsten and Leigh (forthcoming).

Proof-theoretic conservativity: a theory T2, formulated in the language $L2$, is proof-theoretically conservative over the theory T1, in the language $L1$, iff for every sentence A in $L1$, if $T1 \cup T2 \vdash A$ then $T1 \vdash A$.

In other words, if T2 is proof-theoretically conservative over T1, then every sentence in the language of T1 that can be proved by $T1 \cup T2$ can already be proved by T1 alone.

Model-theoretic conservativity: a theory T2, formulated in the language $L2$, is model-theoretically conservative over the theory T1, in the language $L1$, iff every model $M \models T1$ can be expanded to a model $M' \models T1 \cup T2$.

In other words, if T2 is model-theoretically conservative over T1, the addition of T2 leaves the models of T1 untouched. Notably, the two definitions have different extensions and properties. In particular, model-theoretic conservativity is a stronger requirement than a proof-theoretic one, so that theories that fail to be model-theoretically conservative over a certain base theory can still be conservative in the proof-theoretic sense.[10] Choosing one approach instead of the other can thus give different and non trivial results. In general, from a mathematical point of view, both definitions are on a par, since they are fully legitimate and can be exploited to obtain different information. Indeed, it is also important to have both, since the fact that a theory meets one but not the other constraint can be an important and interesting result on its own. From a philosophical perspective, however, things are sloppier. Both definitions, in fact, can be (and have been) invoked to capture the metaphysical simplicity of truth.[11] The general motivation to employ conservativity with respect to the unsubstantiality of truth is probably that conservativity seems to reflect the apparent redundancy of deflationary truth. Deflationism has its roots in the apparent equivalence between the claim that p and the claim that p is true. This equivalence seems to reveal some kind of redundancy of the truth predicate, as philosophers, e.g. Frege and Ramsey, have pointed out in several occasions. Such a redundancy, showing that the ascription of truth is somehow vacuous prompts the conclusion that truth is unsubstantial. Given that such a kind of redundancy is echoed, at a theory level, by conservativity, some deflationists have proposed to identify the two notions.[12] Indeed, since conservativity is formally clear, we could employ it to understand the philosophical idea of the unsubstantiality of truth, at least in a formal context.

Even granted the legitimacy of an explanation of the unsubstantiality of truth in terms of conservativity, it remains to determine which definition of conservativity should be employed. The discrepancy between the two definitions, in fact, has not been adequately addressed until recent times, and their differences passed, if not

---

[10] A notable example is provided by the truth theory CT-, which is, roughly, the theory obtained by turning into axioms the usual Tarskian clauses while allowing for arithmetical induction only. In other words, CT- is a Compositional Theory of truth where the truth predicate does not enter the induction scheme.

[11] See Cieslinski et al. (To appear) for a critical discussion.

[12] Horsten (1995), Shapiro (1998) and Ketland (1999).

unnoticed, at least often unappreciated. This is important for at least two reasons. First of all, since the two definitions have different extensions, using one instead of the other gives different results. Different axiomatic theories are deemed as deflationary depending on which definition is preferred. The philosophical choice between the two definitions thus can not be neglected. Secondly, given their extensional divergence, we gain two different notions of unsubstantiality. These two notions are close and share, to some extent, the same motivation, but they still give form to different ideas. Accordingly, if we appeal to conservativity to clarify the unsubstantiality of truth, we eventually get two different explanations of what the unsubstantiality of truth is. The two definitions, in fact, go in different directions. If the proof-theoretic view is centred on the explanatory strength of a truth predicate, the model-theoretic approach is focused on the availability of an extension associated with a truth predicate. What, if any, of these two dimensions is relevant to make sense of the deflationary claim about the simplicity of truth is unclear. By exploiting the distinction between the concept and the property of truth, however, I argue that the issue can be clarified.

6.4   Semantic Conservativity and Metaphysical Robustness

The distinction between the concept and the property of truth seems to parallel the distinction between proof-theoretic and model-theoretic approaches in a rather natural way. Take the model theoretic perspective first. When we focus on the semantics of the truth predicate we are clearly interested in its semantic value, namely in the entity it expresses. Given that the expression is a predicate, the associated entity must be a property. Notice that this conclusion is neutral with respect to the metaphysical pedigree of such a property. By claiming that a truth predicate expresses a property we are just putting in other words the claim that there is an extension (namely a set of entities satisfying the predicate) associated with the predicate, in the way dictated by its syntactical nature. This way of speaking is rather harmless and it is common even among deflationists.[13] Indeed, if one took the semantic clauses seriously, she could also identify the property directly with the set corresponding to the extension of the truth predicate. This would be a crude view, but one that a philosopher with a nominalistic attitude would be happy to take. Once we have given an extension to the truth predicate, we can of course wonder what kind of property is exhibited by the items in such an extension. It is at this point that the path of the deflationist and that of the inflationist divide, with the former claiming that truth is a thin property and the latter that it is a thick one. Model-theoretic conservativity fits naturally with such considerations. When we wonder

---

[13]To resist this view, one should tinker with the syntax by adopting a view where a truth predicate is only apparently a predicate. The superficial grammar of the language would then be misleading. This idea has been defended for instance, by prosententialists.

whether a truth theory is model-theoretically conservative over a base theory, we are immediately dealing with the extension of a predicate, namely with the associated truth property.

If model-theoretic conservativity is a notion found at the right level of analysis, related to the extension of the truth predicate, the next issue is deciding whether it can give us some useful information with respect to metaphysical (un)substantiality. Also in this case, I contend, the answer is positive.[14] If a truth theory is model-theoretically conservative over a base theory, then, by definition, every model of the base can be expanded to a model of the base theory plus the truth theory. This means that we can find, in any base model, an extension for a truth predicate that leaves everything else in the model unaltered. Accordingly, to give an extension to the truth predicate we only need to operate at the level of the language, not at the level of the things the language is about. Irrespective of what and how these things are, we manage to find a suitable way to gather them and talk of truth. If the theory is model-theoretically conservative, the property that is theorized is not substantial in a clear sense: it is unable to shape the items instantiating it in any (new) way, over and beyond the base theory. In other words, the truth property does not restrict the range of states of affairs that are possible from the point of view of the base theory. Remarkably, this kind of story can be related in a natural way to an independent metaphysical interpretation of the unsubstantiality of truth.[15] The fact that we can find a way to gather the items in any domain, in fact, means that we can find an extension in the model by carving it irrespective of its natural joints. This approach easily aligns with the philosophical idea that, at an informal level, the unsubstantiality of truth could be interpreted in terms of abundance, exploiting the metaphysical distinction between sparse (joints carving) and abundant (not necessarily joints carving) properties.

6.5 Proof Theoretic Conservativity and the Concept of Truth

If semantic conservativity fits the unsubstantiality of the property of truth very well, what notion could be employed to make sense of the unsubstantiality of the concept of truth? Probably the first option that springs to mind is that of resorting again to conservativity, just passing from a model-theoretic to a proof-theoretic approach. If the passage from a model-theoretic to a proof-theoretic perspective is arguably correct, the resort to conservativity, however, is questionable. Let me start with a discussion of the proof-theoretical perspective.

---

[14] See Strollo (2013) for a more extensive discussion of the metaphysical significance of expandability of models.

[15] See Edwards (2013) and Asay (2014). This shows that model-theoretic conservativity can be given a strong philosophical motivation that seems to be lacking in the original debate.

If we are not interested in the property expressed by a predicate, but in the concept associated with it, we need to shift our attention from the level of reference to a different level. We must stop focusing on the extension of a certain expression and start considering the very principles that govern that expression.[16] The subject of our investigation, in fact, is not provided by the entities the theory describes, but by the particular kind of resources that the theory employs to describe such entities.[17] Thus, when we are interested in the concept corresponding to a certain notion, we should stop looking at the models selected by a certain theory, considering, instead, the axioms of such a theory. To investigate different concepts we need to compare the theories, not the models of those theories. This motivates the passage from a model-theoretic to a proof-theoretic perspective.[18]

If the proof-theoretic level is the correct one, however, conservativity does not seem the right notion for our purposes. Even in a proof-theoretical version, in fact, conservativity is still related, although in an indirect way, to what a truth theory is about, namely to the property of truth, not specifically to the concept of truth underwritten by a truth theory. This can be shown by considering the two main motivations that have been put forward to defend the adoption of proof-theoretic conservativity. On the one hand, there is the idea that a proof-theoretically conservative truth theory would not be explanatory of anything regarding the base theory, on the other hand, there is the idea that it would also lack justificatory power.[19] Both aspects are arguably connected to the unsubstantiality of truth and both are certainly exhibited by a proof-theoretically conservative truth theory.[20] However, even if such suggestions were developed into explicit arguments, the obtained criterium would

---

[16] An example of this approach is provided by the usual reflections on provability predicates. Rosser's provability predicate, for instance, can be distinguished from other provability predicates for the peculiar axioms governing it, even if it delivers the expected results. Indeed, we can also criticize Rosser's predicate for not expressing the right concept and having the wrong intension.

[17] Notice that we are just considering formal theories with an extensional semantics. If we had a modal semantics at disposal, e.g., we could focus on the modal intension of a certain predicate. Given that our base theory is an arithmetical theory, however, a modal approach is not very suitable. Arithmetical propositions, in fact, are either necessarily true or necessarily false.

[18] It might be wondered whether the study of the concept should be related to a language instead of a theory. The problem is that we want to consider a language with a truth predicate, namely with at least a determinate class of interpretations. At the same time, we want to compare the truth concepts embodied in different truth theories. Thus posing some minimal constraints, such as selecting those interpretations validating Tarskian biconditionals (as in Halbach 2001) will not work. In this way we would lose the differences between different truth theories equally satisfying such constraints. Thus there is no alternative to consider different axiom systems.

[19] Of course I am assuming that explanatory and justificatory roles are considered with respect to a truth-free base theory. A truth theory can certainly be explanatory or justificatory with respect to truth-theoretical claims.

[20] Notice, however, that such strategies can also be criticised by showing that the lack of explanatory and justificatory power does not fit well with proof-theoretic conservativity. See Cieslinski (2015). Cieslinski also criticises the resort to model-theoretic conservativity, but I think that his arguments could be resisted. For matters of space, however, I must put the discussion of these points aside.

only provide a clarification for the wrong kind of substantiality at most. Speaking of explanatory or justificatory power of a notion, in fact, makes sense only if it is the property expressed by that notion what we have in mind. Here, however, we are looking for a clarification of the unsubstantiality of the truth concept, not of the associated property. Let me explain what I mean.[21] If we account for a phenomenon P by invoking a notion N, we explain P by posing a certain property or entity corresponding to N. The explanation of P has, very roughly, the following form: it is because certain items are N, that P occurs. Of course, since an explanation is an epistemic matter, it must be articulated at a conceptual level, so that, i.e., we need to employ the concept of N to yield it. After all, the explanation is not the phenomenon to be explained. It is for this reason that we often say that the concept N is explanatory of P, although what causes P is, strictly speaking, not the concept of N but N itself. To be fully precise, however, we should say that the concept is explanatory just because the property it expresses has a role in making P occur.[22] It is because certain items are N that P occurs, and it is because the property of N has this role that the concept of N is, in such a derivative sense, explanatory. In other words, the explanatory role of the concept is just the reflection of the metaphysical or causal work performed by the property of N.[23] It follows that in an explanation we consider a concept just for the role played by the corresponding property, we do not characterize it *qua* concept and its very nature is only indirectly touched by explanations. Thus, if a proof-theoretic conservativity criterium is adopted for its relation to the lack of explanatory power, we fail to gain a good clarification of the unsubstantiality of the concept as such. What we obtain is, at most, a specification of what characterizes a concept that expresses an unsubstantial property. The fact that an unsubstantial property is expressed by a certain concept, however, does not make, per se, also the concept unsubstantial. We can take this view only at the price of neglecting the differences between the two, treating a concept as a mere shadow of the expressed property.

It is certainly easy to fall victim of such confusions, since, especially in the case of truth, the expression 'the concept of truth' is often used in a mere rhetorical way to refer to truth simpliciter. Analogously, if we wonder about the nature of a certain concept we often find natural to focus on the kind of property it ascribes. Traditional conceptual analysis gives plenty of examples in this direction. This approach can be legitimate, however, only if the two issues are clearly distinguished. If we are interested in the nature of the concept of truth, we are not primarily interested in the property of truth, and vice versa. Although concepts and properties are related, they

---

[21]I focus on explanatory force but parallel considerations can be proposed also for justificatory roles.

[22]I am here assuming that the truth predicate ascribes a property in some sense.

[23]I have in mind ontological explanations, but merely epistemic explanations would exhibit the same basic features.

are different sorts of entities, with a different nature, so that the unsubstantiality of the former should be expected to be of a different sort from the unsubstantiality of the latter.

All of this is confirmed by how proof-theoretic and model-theoretic conservativity relate to each other. Although model-theoretic conservativity is, as remarked, a stronger requirement than proof-theoretic conservativity, in several cases they coincide, since the lack of the latter implies a failure also of the former. If a truth theory is not proof-theoretic conservative, in fact, it must also be non model-theoretic conservative. Thus, once we have identified the substantiality of the concept with the lack of proof-theoretic conservativity and the substantiality of the property with the lack of model-theoretic conservativity, we have that the concept of truth can be substantial only if also the corresponding property is substantial, so that the two dimensions are not free to vary. Under the light of the above considerations this does not come as a big surprise. If a concept expresses an unsubstantial (or a substantial) property it is natural that it might have, at least in some cases, some feature witnessing this. Proof-theoretic conservativity, in a sense, tracks the unsubstantiality of the expressed property at the conceptual level. Although this can be an important and useful aspect of formal truth theories, however, it is not enough to inform us of the nature of the truth concept embodied in such systems.

6.6 Relative Interpretation

If conservativity is not suitable for our purposes, the question becomes what other formal tool, if any, could be applied to make sense of the (un)substantiality of the concept of truth qua concept. To find such a tool, we must avoid treating the concept just for its ability to ascribe a property, but focus on features of a different sort. To capture the idea of the unsubstantiality and simplicity of the concept of truth we must focus not on the complexity of the extension of the concept but on the complexity of the intension, moving from the property described by a truth theory to the truth theory itself. Is there a suitable way to describe and compare the proof theoretic complexity of theories? There is, and it is called 'relative interpretation'.[24]

The basic idea of relative interpretation is that, if a theory is relatively interpretable in another, then the former theory can be considered reducible to the latter

---

[24]Fischer and Horsten (2015) pair a demand for model-theoretic conservativity with a particular constraint on the expressive power of a language including a truth predicate. Although their proposal is certainly valuable and interesting, I should stress the discrepancy with respect to my approach. The main difference is that here we are concerned with the unsubstantiality of the concept of truth, not with its expressive power. Although a deflationist might have reasons to claim that a truth predicate increases the expressive power of a base language, in principle, such a claim might also be independent from the unsubstantiality of the concept. Even if the relations between conceptual unsubstantiality and expressive power should be investigated, the two have different nature so that we should offer different accounts of such ideas.

theory (in a sense to be specified). In particular, this means that the complexity of the interpreted theory does not surpass that of the interpreting theory. Here is the formal definition. Let T be a theory formulated in the language LT and B a theory in the language LB. A relative interpretation of T in B, if possible, requires two steps: a relative translation of the language LT in the language LB, and the relative interpretation itself. A relative translation $\tau: LT \to LB$ is given by a pair $<\delta, \varphi>$, where δ is a LB-formula representing the domain of the translation, and φ is a map that associates to each LT-relation symbol R with arity n, an LB-formula (R) with variables among $\gamma 0, \ldots, \gamma n-1$. The translation is obtained by: (1) exploiting φ; (2) requiring that it commutes with logical connectives; (3) relativizing the domain of the quantifiers to δ. Namely:

$$(R(\gamma 0, \ldots, \gamma n - 1)) \tau := \varphi(R)(\gamma 0, \ldots, \gamma n - 1)$$
$(\cdot)\tau$ commutes with the propositional connectives;
$$(\forall \gamma A) \tau := \forall \gamma (\delta(y\gamma) \to A\tau)$$
$$(\exists \gamma A) \tau := \exists \gamma (\delta(\gamma) \wedge A\tau)$$

Once we have the translation, we can pass to the second step. We say that the relative translation τ supports an interpretation of T into B iff for all sentences A, if $T \vdash A$ then $B \vdash A \tau$. If T is relatively interpretable in B, we can say that, in a sense, B has enough resources to account for T. This rather metaphorical claim, however, should be handled with care. In particular, that T is relatively interpretable in B does not mean, per se, that T is already contained in B. Indeed, if this were the meaning of relative interpretation, we could not expect a truth theory be relative interpretable in a base theory such as Peano Arithmetic (PA[25]), given that, by Tarski's theorem, it is unable to define its own truth predicate. Thus, we can go, at most, in the opposite direction, claiming that if a theory B contains a theory T, then T must be relatively interpretable in B. What we can say when we have a relative interpretation is that B can mimic T, and that, in this limited sense, it has enough expressive and proof-theoretic resources. In particular, if T is relatively interpretable in B, the complexity of the resources of T does not exceed that of B. Accordingly, B contains T in the sense that it has the capacity to capture the notions employed by T.

It should not be hard to see that relative interpretability can be positively correlated with the unsubstantiality of the concept of truth (at least relatively to a base theory). Suppose that we have a truth theory T, in the language $T(x) \cup LPA$, obtained by adding axioms governing $T(x)$, namely a truth predicate, to PA. If T is relatively interpretable in PA, then T is conceptually not more complex than the notions theorized by PA. Namely, PA has enough resources to account for T, even if it does not define a truth predicate. Although PA does not contain T, it can still capture it somehow. In other words, the conceptual complexity of the truth theory is

---

[25]I take PA in its usual first order formulation. See Horwich (1998) for the relevant details.

conceptually reducible to that of PA.[26] I take this conceptual reducibility as a good clarification of the idea that a certain concept is (relatively) unsubstantial.

The fact that the unsubstantiality of the concept depends on the choice of the base theory might seem problematic. Suppose that our truth theory is interpretable in PA. Is this a proof that the truth theory is conceptually weak or just that the base theory is conceptually strong? Such an objection is serious, but it should be noted that the same problem emerges also with respect to the application of model-theoretic conservativity. For conservativity and its relation with the unsubstantiality of the property of truth, in fact, we could put forward the same kind of considerations. Actually, this is a well known problem that has already been discussed at some length in the literature.[27] The natural option to avoid this obstacle would be that of taking as a base theory the empty theory, namely pure logic. Conservativity and interpretability results over an empty base theory would clearly assure the unsubstantiality of the examined notions. However, it is usually agreed that, in order to formulate a truth theory in a sensible way, a base theory including a theory of syntax is necessary. This is what leads to the adoption of an arithmetical theory, which allows a formalization of syntax, as a base. It is certainly debatable whether PA is the best choice or whether we should favour something weaker like Robinson Arithmetic, Primitive Recursive Arithmetic, or something stronger such as a fragment of second order Arithmetic or even full ZFC. These are important questions that should be addressed eventually. However, given that they are, to some extent, independent issues, and given that they are already familiar problems, I put the discussion of these topics aside.

6.7 Proof Theoretic Reduction and Speed Up

Conservativity and relative interpretation are not the only formal notions that can be employed to compare different theories. It is thus worth asking what the place of those other notions is and whether they can be related to the unsubstantiality of truth. In particular, I briefly discuss two notions that are particular prominent in the study of formal theories: proof-theoretic reduction and speed up. The notion of proof theoretic reduction is, to some extent, similar to that of relative interpretation. Roughly, a theory B is proof theoretic reducible to another theory T if there is an effective method, i.e. a primitive recursive function, for transforming proofs in the theory B, of formulas belonging to a set S, into proofs in the theory T and, importantly, this can be established inside T. Clearly, in order to have an interesting reduction, the formulas in the set S must be in the intersection of *LB* and *LT*, so that basically, with proof-theoretic reduction, we want to preserve the provability of

---

[26]See Niebergall (2000) and Hofweber (2000) for a more extensive discussion of the conceptual significance of relative interpretation.

[27]See, for instance, Halbach (2011).

certain formulas in the common part of the languages. I skip the formal definition here, and just note that proof-theoretic reduction is, very roughly, a mix of relative interpretation and conservativity with a proof-theoretic attitude. Some differences, though, should be stressed. First, in the case of conservativity we ask that all sentences in the intersection of the languages (which coincides with the language of the base theory) must be preserved. In the case of proof theoretic reduction, instead, we usually select a smaller set of formulas. Second, for formulas in S, we do not want just an alternative proof in the reducing theory. We want two further things: (1) that there is an effective way to transform a proof in B into a proof in T, (which means that the transformation must be provided by a uniform algorithm) and (2) that we are able to prove (1) inside the reducing theory. Thus, we do not actually translate the language of S, as we do for relative interpretability, but we do something analogous. What we interpret in T are the proofs of formulas that are already in *LT*. Moreover, for proof theoretic reduction we prove such a reduction inside T, whereas for relative interpretation we just show that the preservation holds in the meta theory.

If proof-theoretic reduction is an important tool to study the mathematical properties of formal theories, and is also valuable to extract meta logical and foundational information, it does not seem to have any particular role to play with respect to the unsubstantiality of truth. I just point to two aspects that are related with our issue. On the one hand, the fact that only a certain set of formulas is considered already makes the reduction less significant, since there could be (and typically there will be) a part of the reduced theory, namely the formulas not in S, that is not reduced. Indeed, this part contains also the truth predicate, given that it is not in the common part of the languages. This is a potential problem, especially if we are interested in the reducibility of the concept of truth. On the other hand, a proof-theoretic reduction does not seem to have a direct metaphysical significance. In a sense, that a theory is able to prove the reduction can be taken as a sign that the theory *knows* that the reduction can be put forward. However, even putting aside the difficulties of unpacking such a metaphor, that a theory has enough resources to see this is something that seems superfluous. What is important is that truth is unsubstantial, not that a certain truth theory can be taken *to know* that a certain notion of truth is unsubstantial. The further constraints posed by proof-theoretic reduction, thus, seem redundant. There is, however, one case in which proof theoretic reduction could appear relevant. Suppose that we prove that a certain theory is conservative over or reducible to a certain base B. This is a pleasant result for a deflationist, but can she afford it? Apparently, the result is acceptable only if it is proved by means a deflationist can accept, so that a resort to strong mathematical tools like set theory or notions like that of a standard model might be potentially problematic.[28] In this case, requiring the reduction be carried out in a rather small system can be a justified request. However, also the assumptions on which such observations are based are questionable. It is not clear, in fact, whether

---

[28] See Cieslinski (2015) for a discussion of similar worries.

truth deflationists can only appeal to weak mathematical means. Certainly, austere mathematical frameworks are close in spirit to the attitude deflationists have towards truth, but they are not obliged to take an austere view with respect to every notion. Deflationism about truth is just a conception about truth, so that it can, in principle, be combined also with a robust position in philosophy of mathematics. Moreover, even if deflationists were not entitled to certain formal tools, they could still appeal to them adopting a mere instrumentalist attitude. For such reasons, I conclude that proof-theoretic reduction does not have any clear role to play in the discussion of the unsubstantiality of truth.

A different notion that has been invoked in the investigation of formal theories, and truth theories in particular, is that of *speed up*. Roughly, a truth theory has speed up over a certain base if it allows a simplification of proofs of theorems of the base.[29] The notion of speed up has a intuitive deflationist appeal. From a deflationary perspective, in fact, it can be argued that the only reason to have a notion of truth is provided by its logical and expressive roles. Accordingly, if a truth predicate permits a significant simplification of proofs, we have a good motivation to introduce it. Speed up would be a good explanation of why, even though a truth theory is conservative or interpretable in the base, it is still worth having a truth predicate at disposal.

I find such considerations attractive and I am inclined to think that a deflationist had better embrace a truth theory with speed up. However, it does not seem to have any particular role to play with respect to unsubstantiality issues. Speed up (or its lack), in fact, seems compatible both with the substantiality and the unsubstantiality of truth. There is nothing problematic in a position according to which truth is a metaphysically heavy notion incapable of any proof-theoretic efficacy. Analogously, an unsubstantial notion of truth could also have speed up, as deflationists would eagerly point out.

6.8   A Philosophical Classification of Formal Truth Theories

Let me sum up briefly. We introduced philosophical deflationism and its idea that truth is a simple and unsubstantial notion. We remarked that investigations on the simplicity of truth are common also in the formal study of truth theories, often because they are driven exactly by such philosophical suggestions. We then wondered whether the philosophical idea of unsubstantiality could be correlated with some formal demand. To clarify the issue a distinction proved crucial. Speaking of the notion of truth, in fact, is often ambiguous between talk of the concept and talk of the property of truth. We argued that the unsubstantiality of the property of truth can be captured, in a formal context, by model-theoretic conservativity, while the unsubstantiality of the concept of truth can be positively correlated with (proof-

---

[29] See Fischer (2014).

theoretic) relative interpretation. We also considered other formal notions (such as proof-theoretic reduction and speed-up) but put them aside. With such correlations at disposal, we are now in a position to yield an interesting and not trivial taxonomy of formal truth theories. We just need to check whether the various systems on the market are relative interpretable and model-theoretic conservative over the usually accepted base theory, namely PA. In particular, given that conservativity and interpretability are not co-extensional, the territory is supposed to be mapped in an interesting way. Once we have the formal classification, we can illuminate it by philosophical means exploiting the above correlations.

I now sketch some of the main points of the obtained taxonomy. For matters of space, I can not provide a full characterization and just focus on the most notable cases for illustration purposes. Moreover, for the sake of simplicity, I consider only simple typed theories, namely theories in which truth ascriptions are restricted to sentences in the language of PA. Let me start with the most basic truth theory, TB-, which is the truth theory yielded by Tarskian biconditionals with arithmetical induction only.[30] TB- meets both demands since it is both relatively interpretable and model-theoretically conservative over PA. It is thus a perfect candidate for a philosophical position holding that both the concept and the property of truth are unsubstantial. The same holds for UTB-, namely the uniform version of TB-. If we consider TB and UTB, namely the versions of those theories with full induction, (so that we permit the truth predicate to enter into the induction schema), however, we lose conservativity and retain interpretability in both cases. Thus, we have that TB and UTB underwrite a substantial property of truth, although they give shape to an unsubstantial concept of truth.[31] This is already an intriguing result. Things are interesting even when we consider theories based on compositional clauses: the classical compositional theory obtained by turning into axioms the usual Tarskian clauses, namely TC- (without full induction), and TC (with full induction). TC, in fact, is neither conservative nor interpretable in PA. We thus have that it is unacceptable to deflationists along both dimensions of the notion of truth. TC-, instead, is not conservative but it is interpretable, so that it is similar to TB and UTB for its general deflationary appeal. It is remarkable to note the different roles played by full induction in different contexts, given that sometimes it is needed to lose conservativity (TB-/TB, e.g.) sometimes it forces a loss of interpretability (CT-/CT).

We have just seen that that there are theories that can be taken to describe a substantial property of truth through an unsubstantial concept, besides theories holding or rejecting both kinds of unsubstantiality at the same time. Can we also find theories with the opposite combination, holding a robust concept but a thin property

---

[30]I do not offer a full characterization of these systems nor provide a proof of the claims. General references and relevant results can be found in Fischer (2009), Halbach (2011) and Strollo (2013).

[31]Notably, both theories (and, a fortiori, TB- and UTB-) have no significant speed up over PA. At the same time, although TB- and UTB- are acceptable by crude deflationists, lacking speed up, they are hard to defend with instrumental considerations.

of truth, namely theories that are model-theoretic conservative but not relative interpretable in PA? Fischer and Horsten[32] argued that an example of such a theory is provided by the theory PT*tot*,[33] namely the Tarskian theory obtained by positive inductive clauses plus an axiom of internal induction.[34] Notice that such a theory, far from being acceptable to a full fledged deflationist would be, instead, a perfect formal counterpart of a conceptual primitivist view such as the one advocated by Asay.[35] In any case, unfortunately, such remarks can not be developed, since they rest on a mistake. PT*tot*, in fact, has been very recently proved not to be model-theoretic conservative over PA.[36] Thus, whether there is a good axiomatic theory of truth where we have both model-theoretic conservativity and not interpretability is an interesting open question.

Similar considerations can be extended to cover also more complex truth theories, such as theories with partial induction, with iterated truth ascriptions or type free. How to proceed, however, should now be clear.

6.9 Conclusion

Truth, according to many, is simple. This idea can be investigated by formal means using different notions. The philosophical significance of such mathematical results is hardly clear however. In this paper I have proposed a way to clarify some philosophical confusions preventing an adequate philosophical reflection on formal truth theories. I have articulated the deflationary thesis of the unsubstantiality of truth along two dimensions, distinguishing the concept and the property of truth. I have then argued that the former can be related to the formal notion of relative interpretation while the latter can be interpreted in terms of model-theoretic conservativity. Employing such tools, I have sketched a taxonomy of some of the main formal truth theories on the market, showing how profitable this approach can be. Although much work remains to do, both on the mathematical and the metaphysical side of the study of truth, I take this approach to promote substantial progress toward a reconciliation of the two.

---

[32]See Halbach (2001).

[33]PT*tot* is often called PT-. Here I prefer to avoid this label to avoid possible confusions.

[34]See Halbach (2001) for details. PT*tot* has also non elementary speed up over PA.

[35]Asay defends a theory of truth combining deflationism about the property and inflationism about the concept (Asay 2013a, b, 2014).

[36]Cieslinski, Łelyk, Wcislo (To appear).

References

Asay, Jamin. 2013a. *The Primitivist Theory of Truth*. Cambridge: Cambridge University Press.
———. 2013b. Primitive truth. *Dialectica* 67 (4): 503–519.
———. 2014. Against truth. *Erkenntnis* 79 (1): 147–164.
Burgess, John P. 1986. The truth is never simple. *Journal of Symbolic Logic* 51 (3): 663–681.
Burgess, Alexis G., and John P. Burgess. 2011. *Truth*. Princeton: Princeton University Press.
Cieslinski, Cezary. 2015. The innocence of truth. *Dialectica* 69 (1): 61–85.
Cieslinski, C., M. Łelyk, and B. Wcislo. To appear. *Models of PT- with Internal Induction for Total Formulae*.
Dummett, Michael. 1959. Truth. *Proceedings of the Aristotelian Society* 59 (1): 141–162.
Edwards, Douglas. 2013. Truth as a substantive property. *Australasian Journal of Philosophy* 91 (2): 279–294.
Field, Hartry. 1999. Deflating the conservativeness argument. *Journal of Philosophy* 96 (10): 533–540.
Fischer, Martin. 2009. Minimal truth and interpretability. *Review of Symbolic Logic* 2 (4): 799–815.
———. 2014. Truth and speed-up. *Review of Symbolic Logic* 7 (2): 319–340.
Fischer, Martin, and Leon Horsten. 2015. The expressive power of truth. *Review of Symbolic Logic* 8 (2): 345–369.
Halbach, Volker. 2001. How innocent is deflationism? *Synthese* 126 (1–2): 167–194.
———. 2011. *Axiomatic Theories of Truth*. Cambridge: Cambridge University Press.
Hofweber, Thomas. 2000. Proof-theoretic reduction as a philosopher's tool. *Erkenntnis* 53 (1–2): 127–146.
Horsten, Leon. 1995. The semantical paradoxes, the neutrality of truth and the neutrality of the minimalist theory of truth. In *The Many Problems of Realism (Studies in the General Philosophy of Science: Volume 3)*, ed. P. Cartois. Tilburg: Tilberg University Press.
Horsten, Leon, and Graham E. Leigh. forthcoming. Truth is simple. *Mind* 184.
Horwich, Paul. 1998. *Truth*. Oxford: Clarendon Press.
Ketland, Jeffrey. 1999. Deflationism and Tarski's paradise. *Mind* 108 (429): 69–94.
Niebergall, Karl-Georg. 2000. On the logic of reducibility: Axioms and examples. *Erkenntnis* 53 (1–2): 27–61.
Shapiro, Stewart. 1998. Proof and truth: Through thick and thin. *Journal of Philosophy* 95 (10): 493–521.
Strollo, Andrea. 2013. Deflationism and the invisible power of truth. *Dialectica* 67 (4): 521–543.

Part II
Structures, Existence, and Explanation

Chapter 7
Structure and Structures

Reinhard Kahle

Abstract In this paper we critically evaluate the notion of the structure of the natural numbers with respect to the question how the internal structure of such a structure might be specified.

7.1 Structure

Let us start with two short examples of how structure comes into play in Mathematics (and its applications).

Tiling a Mutilated Board

Consider a board with a 8 × 8 grid on it, dividing it into 64 squares; now remove two opposite squares from the corners so that only 62 squares remain:

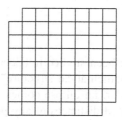

Research partially supported by the Portuguese Science Foundations FCT, projects *Hilbert's 24th Problem*, PTDC/MHC-FIL/2583/2014, and *Centro de Matemática e Aplicações*, UID/MAT/00297/2013.

R. Kahle (✉)
CMA & DM, FCT, Universidade Nova de Lisboa, Caparica, Portugal
e-mail: kahle@mat.uc.pt

© Springer International Publishing AG, part of Springer Nature 2018
M. Piazza, G. Pulcini (eds.), *Truth, Existence and Explanation*,
Boston Studies in the Philosophy and History of Science 334,
https://doi.org/10.1007/978-3-319-93342-9_7

Is it possible to tile 31 dominoes 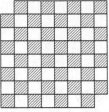 on this board so that all squares are covered?

The answer is easy to see when we put some extra structure on the board, namely the usual black and white alternation of a chess board:

A mutilated chess board

As we removed two white squares, the mutilated chess board has 32 black squares but only 30 white squares; but each domino covers exactly one white and one black square, so that we cannot tile a board which doesn't have an equal number of white and black squares.

This example is widely used to illustrate ingenious mathematical reasoning, stressing often the aspect of creativity and/or intuition which is reflected in the idea of replacing the original board by the chess board.[1]

We presented the example here as the addition of the black and white "colouring" is a typical example of how structure can be imposed on an object, structure which is useful in dealing with questions concerning the original object. The crucial point for the following discussion is that this structure, the black-white "colouring", is *not* present in the original board. It is structure we have *imposed* on the original board.

Elliptic Orbits

Ellipses as conic sections have been studied in (Euclidean) Geometry since antiquity. They exhibit a lot of "mathematical structure" which can be used to study them geometrically. In particular, an ellipse comes with two focal points which, together with the (fixed) summed distances of any point of the ellipse to the foci, define a particular ellipse. It was an observation by Kepler that the planets of the solar system move on elliptic orbits around the Sun (within astronomical accuracy).

---

[1] Historically, the example of the mutilated chessboard can be traced back to Max Black who posed it in 1946 as a problem in his book *Critical Thinking* (Black 1946, exercise 6, p. 142) (but starting off with the chess board, thus, leaving out the creative part of adding this structure as a first step). It is also reported that Emil Artin occasionally used this example in his lectures (see Zweistein 1963; Thiel 2006); it might well be that he took it from Black (or some other later source), but it was stressed in a obituary for Artin that he applied the idea of the solution within his mathematical activity, as he had "the very rare ability to detect, in seemingly highly complex issues, simple and transparent structures". ("[Er hatte d]ie so seltene Gabe, in scheinbar hochkomplexen Sachverhalten einfache, durchsichtige Strukturen aufzuspüren" (Reich 2006, p. 39).)

And the center of the Sun is one of the focal points.[2] The natural question is: what is at the other focal points (which are different for the orbits of the different planets)? Apparently, there is no specific *physical* counterpart to the distinguished mathematical (second) focal point of an orbit (while, to repeat, for the first focal point, there is one). Thus, the mathematical structure of ellipses is richer than the physical structure we encounter when we apply the mathematical notion of ellipse to an orbit in the solar system.

The lesson of this example is that (mathematical) structure should not be reduced to physical structure: we legitimately may "add" mathematical objects (here: the second focal point) to those stemming from the empirical data (here: the points of the trajectory of a planet and the Sun). Even without distinct empirical counterparts, their distinction (in a mathematical sense) is unquestionable.

7.2 Structures

Today it is common to say that mathematicians investigate structures. The term structure is, however, ambiguous: it may refer to *concrete structures*, as in the case of the structure of the natural numbers, or to *abstract structures*, which emerged from algebra and which were prominently promoted by Bourbaki. Both notions, as technical terms, are clearly different from the informal notion of structure as sketched in the previous section. Without going into a detailed discussion about the differences between concrete and abstract structures, a discussion which would lead a too far afield, we will concentrate on concrete structures.[3]

A concrete structure can be understood as a mathematical realm of discourse of semantic nature.[4] The prime example is the *structure of the natural numbers*, but there are many others, including the structure of the real numbers and the structure set theory is concerned with. The characteristic of such concrete structures is that they start from a certain universe of discourse, which is assumed to be preexistent—somehow in a platonistic manner—and that they provide "tools" (i.e., functions and relations) to investigate such a universe.

---

[2]More exactly: the center of mass of the solar system, which consists of the Sun together with all objects of the solar system; but, clearly, the Sun dominates this mass in such a way that it is reasonable to identify the center of mass with the center of the Sun.

[3]We will shortly address abstract structures in Sect. 7.4. The distinction of these two forms of structures is ubiquitous for the working mathematician: the different character of natural numbers and groups is conceptionally self-evident. For the syntactic counterpart of structures we find it explicitly discussed in (Hermes and Markwald 1958, §§ 4.2 and 4.3) as *heteronome* and *autonome Axiomensysteme*. Although the authors presuppose an empirical base for the heteronomous axiom system (what we will not do for the concrete structures), they point to the fact that these axioms systems are chosen *a posteriori* ("nachträglich gewählt"). This is in accordance with our understanding of concrete structures relating to mathematical objects which are supposed to preexist.

[4]The semantic nature implies, in particular, that it is assumed that any (properly formulated) statement about this realm has a definite truth value.

The exact shape of such concrete structure is not fully specified. Here we will deal mainly with *first-order structures*, but addressing the question of second-order structures only in Sect. 7.3.5.2. A first-order structure is a well-defined concept in Mathematics; it presupposes the definition of a *first-order language* which should be interpreted in such a structure.[5] The language dependency is not unproblematic, as it is the convictions that Mathematics—or mathematical truth—should be language independent. The following consideration can be taken, however, as an argument that the choice of the language is somehow related to structure (in the informal sense) we impose on the universe of discourse. In the following section, we will look more closely at the seemingly familiar concrete structure of the natural numbers and discuss how its internal structure is given.

7.3   Structure in Structures

> Mathematics is just the detection and investigation of structures of thinking which lie hidden in the mathematical symbols. The simplest mathematical entity, the chain of integers $1, 2, 3, \ldots$, consists of symbols which are combined according to certain rules, the arithmetical axioms. The most important of these is an internal coordination: to each integer there is one following it. These rules determine a vast number of structures; e.g. the prime numbers with their remarkable properties and complicated distribution, the reciprocity theorems of quadratic residues etc.   Max Born, (Born 1966, p. 151f.)

Let us start with the structure of the natural numbers which is often given in the form: $\mathfrak{N} = \langle \mathbb{N}, 0, +, \cdot, \ldots \rangle$. The use of the dots should make one puzzle; they are added in a sloppy presentation of a (concrete) structure, to indicate that one may add some more functions and relations which are supposed to come along naturally. For the natural numbers, for instance, it could make sense to include the less-or-equal relation $\leq$ after addition and multiplication. At second glance, however, one should realize that it is far from being obvious what should or could be part of such a structure.

7.3.1   Constitutive Structure

To start from the scratch, let us first consider the raw set $\mathbb{N}$ *without* any further structure on it. Stripped off of any structure, the set of natural numbers should appear as nothing else than a bag with infinitely many elements, and the only condition we have is that any two elements taken out of this bag are different.[6]

---

[5]The distinction of the syntax and semantics of, let say, function symbols and functions themselves is, of course, fundamental in mathematical logic, see Kahle and Keller (2015). In the present context, however, we argue essentially entirely on the semantic side and will neglect the difference as it should not give rise for confusion.

[6]The (potential) infinity of $\mathbb{N}$ could also be given by such a condition: by consecutively taking out elements from this bag, one will never completely empty the bag.

By naming the elements, let us say the first one, I pick, by 0, the next one by $S(0)$, then by $S(S(0))$ and so on, we are already *imposing* some structure on the previously completely unstructured set $\mathbb{N}$.[7] In this view, $\mathbb{N}$ "alone" would not have any particular meaning; in formal terms one would have already more than just a raw set at hand, namely the rudimentary structure $\langle \mathbb{N}, 0, S \rangle$. It is defensible that the very use of $\mathbb{N}$ presupposes this rudimentary structure. But one could also consider to put another structure on the raw infinite set, as, for instance, a (binary) word structure; but in such a case, one would probably use the designation $\mathbb{B}$; or, for a tree structure, $\mathbb{T}$.

We will say the 0 and the successor function S build the *constitutive structure* of the natural numbers; likewise, ϵ and two unary successor functions S_0 and S_1 build the constitutive structure of binary words, and a finite, non-empty set of constants together with a binary successor function provides the constitutive structure of (binary) trees. In this view the constitutive structure of a mathematical structure is nothing other than the set of constructors of a recursive data type in Computer Science.

The distinction of a constitutive structure for a universe, which is to be used as the basis for a concrete structure, is of importance as it gives a justification for both recursive definitions of functions on the underlying set, as well as the proof scheme of induction for this set.

But starting from a raw infinite set, even the specific constitutive structure is not present in the set, but imposed on it—in the very same way as the black-white colouring is not present on the original mutilated board.

7.3.2 First-Order Structure

With the constitutive structure we get naturally some more structure: for instance, for $\mathbb{N}$ (or, more exactly, $\langle \mathbb{N}, 0, S \rangle$), we obtain immediately an order structure, as the elements of $\mathbb{N}$ are now naturally ordered by the length of their names. But this description is given on the meta level. To get off the ground, in particular in a formal way, we need to "inject" some more structure on $\mathbb{N}$. In (first-order!) Peano-Arithmetic, PA, this is done by presupposing addition and multiplication as primitive functions (see below).

Assuming these functions, the less-or-equal relation can be introduced for the natural numbers by the definition: $t \leq s :\iff \exists x.t + x = s$. Formally, this can be

---

[7]Tactically, we use here 0 and S as syntactic entities to give names; if you think of their semantic interpretations, you would have always the corresponding concrete elements of the infinite set at hand; still, S would be supposed to be a semantic function telling you how to go from one element to the next one.

From another, *constructive* perspective one can also proceed the other way around: starting from 0, one *constructs* successively the elements of $\mathbb{N}$ by applying the "successor function" S. In fact, it would be far better to call this "function" in this specific context "constructor". This terminology is known from Computer Science, and the analogy is not accidental: the separation of the definition of a *datatype*, by constants and constructors, from the implementation of functions operating on this datatype corresponds to the distinction we have in mind here. The problem with the constructive perspective is that it cannot go beyond countable universes.

expressed by *definitorial extensions* (and the possibility to perform such extensions without further explanations give the justification of the dots in a sloppy presentation of a structure), see, e.g., (Rautenberg 2006, §2.6).

It is then an exercise in first-order Arithmetic to show that all first-order definable functions and relations can be given by first-order logic formulas using addition and multiplication.[8] And this leads us to the first 'result':

Structures are supposed to be closed under first-order definability!

That is why they are also called *first-order structures*.

And one can see here a reminder of the distinguished status of *logic*. Whenever one starts off with some structure, one is committed to the structure which can be build up from it by logical (first-order) definitions.

7.3.3 Primitive Structure

First-order definability, however, is not enough to get off the ground from the constitutive structure in the case of the natural numbers: as we said, we need at least addition and multiplication as introduced, for instance, in the now standard Peano Axioms. We may already note, that this extra structure would not be needed, if we had second-order definability at hand; see Sect. 7.3.5 below.

But it is also notable, that, in PA, we can already dispense with the definition of further primitive-recursive functions, like exponentiation as they are (first-order) definable from addition and multiplication. On the other hand, multiplication is not (first-order) definable from addition alone.

Primitive Recursive Arithmetic, in contrast, requires the inclusion of all primitive-recursive functions as primitive structure, as in the presence of (only) quantifier-free induction, first-order logic is not expressive enough to obtain, let say, exponentiation from multiplication and addition.

Apparently, the role of addition and multiplication in PA is very specific to the first-order framework with full induction. To our knowledge, there is no intrinsic explanation why addition and multiplication are distinguished in PA; they appear just to be the two functions which serve technically for the purpose. Conceptually, the introduction of all primitive-recursive functions seem to be a more natural choice, as they are defined by a general definition scheme over the constitutive structure of the natural numbers (and this general definition scheme can equally be applied to other constitutive structure).

Again, and somehow even more than for the constitutive structure, the primitive structure seems to be imposed by us on the set, even if it is already equipped with constitutive structure, rather than "being there".

---

[8]Intentionally, we phrased this exercise as a trivial tautology: first-order definable functions and relations are, by definition, given by first-order logic formulas over addition and multiplication. The interesting question is, of course, which are these functions; it is just an empirical observation that they include essentially all number-theoretic functions used (and defined independently) in the history of mathematics; to show this inclusion is not a trivial exercise!

7.3.4 Some Consequences

To take stock, let us briefly reflect on what we have "at hand" in PA. As already noted, addition and multiplication is enough to get all other primitive recursive functions (and even more). Also, the present structure allows one to distinguish numbers in ways which cannot be seen as directly implied by the original structure. As a prominent example, let us mention the number 1729 which, as Ramanujan observed, is the smallest number expressible as the sum of two cubes in two different ways[9]: $1729 = 1^3 + 12^3 = 9^3 + 10^3$.

Such a distinction of a number, induced by the structure, can be seen as a parallel to the physically meaningless second focal point of a planetary orbit, whose distinction is induced only by the elliptic orbit.

But there are much more far-reaching consequences of the introduction of the structure, touching even on the ontological issues. Interestingly, the first-order structure of the natural numbers provides us already with a large part of *ordinal arithmetic*. Conceptually, ordinals seem to transcend the natural numbers by continuing counting into the transfinite, starting with a new element ω as the first infinite ordinal. This is supposed, by definition, not to be an element of our "bag of natural numbers". However, by defining appropriate order relations we can, straightforwardly, encode ordinals in the natural numbers (including, of course, the intended arithmetical operations for them).[10] As consequence, we have—unintentionally—extended our mathematical realm into the transfinite.

7.3.5 More Structure?

7.3.5.1 Fast Growing Functions

Even with some primitive structure and first-order definablity there will always be some structure of the natural numbers which cannot be captured by a first-order axiomatization, due to Gödel's First Incompleteness Theorem. In fact, expressing that a function f is *provably total* by the formula $\forall x \exists y . f(x) = y$, it requires

---

[9]The story was recorded by G. H. Hardy (1921, p. lvii f): "I remember once going to see him when he was ill at Putney. I had ridden in taxi cab number 1729 and remarked that the number seemed to me rather a dull one, and that I hoped it was not an unfavorable omen. 'No,' he replied, 'it is a very interesting number; it is the smallest number expressible as the sum of two cubes in two different ways.'" One may note, however, that Bernard Frénicle de Bessy had already published this fact in 1657.

[10]For instance, by defining $\prec$ as:

$$1 \prec 2 \prec 3 \prec \cdots \prec 0,$$

where 0 is supposed to be $\prec$-greater than every other natural number, it can be taken to represent ω (and any original n represents $n - 1$ of the ordinal world).

only a short reflection on arithmetical first-order theories to see that we can always diagonalize over the provably total functions of a given theory to obtain a new function which is not provably total in this theory. As a matter of fact, PA only proves the totality of ε_0-recursive functions; and the *Hardy Hierarchy* H_α can be used to classify, quite generally, the complexity bounds of arithmetical theories (for these notions and the corresponding results, see Fairtlough and Wainer 1998).

Of course, we expect that our standard structure of the natural numbers will contain all these fast growing functions as total functions; but it remains unclear how far we can actually go in demanding that they make up part of the structure which should "automatically come with" the natural numbers. If they all should "automatically come" we have to admit that we are not able to exhaust the structure *in its very definition*.[11] If we cannot include it all, where to draw the line?

7.3.5.2   Second-Order Definability

With respect to the definition of structures, one should recall a historical fact: both, Dedekind (in *Was sind und was sollen die Zahlen?*, Dedekind 1888) as well as Peano (in *Arithmetices Principia Novo Methodo Exposita*, Peano 1889) define the structure of the natural numbers in a second-order context. In this context, addition and multiplication are logically definable (from 0 and successor). Thus, we would not need any extra primitive structure. But the reason why we no longer follow this path is well-known: second-order logic is not recursively axiomatizable. This fact became known only with Gödel's (First) Incompleteness Theorem[12]; but it discredited second-order logic as a basic framework for mathematics up to today.

From a staunch platonistic standpoint, this is almost an anthropomorphic reason to distinguish first-order structures: the restriction to recursive axiomatizability seems to be a limitation imposed by human capacity to construct formal systems, rather than anything which should be inherent to mathematical ontology. And the non-categoricity of any first-order theory for the natural numbers, expressed in Gödel's First Incompleteness Theorem, gives even more reason to challenge it: after all, we are interested in the standard structure of *the* natural numbers, aren't we?

Kreisel fought forcefully for the consideration of second-order properties when he advocated *informal rigor* (Kreisel 1967, pp. 138f and 152): "Informal rigour wants [...] not to leave undecided questions which can be decided by full use of evident properties of these intuitive notions" and "most people in the field are so accustomed to working with the restricted [first-order] language that they may simply not succeed in taking other properties seriously." He had in mind, first of all, the Continuum Hypothesis. But already for Arithmetic the consequences would be far-reaching: second-order arithmetic already provides us with a concept of real

---

[11] Remember the dots in the definition of $\mathfrak{N}$ at the beginning of Sect. 7.3.

[12] In fact, at a time where Skolem had already promoted first-order logic as the "one and only solution" for mathematics; but against the fierce opposition of Zermelo.

numbers and the very structure already incorporates Analysis. Apparently, today mathematicians are inclined to draw a line here: Analysis doesn't seem to come "automatically" with Arithmetic.[13]

7.3.6   Less Structure?

We have argued above that the first order closure of a constitutive structure should be taken for granted. But there are, at least, two contexts where this presupposition is questioned.

7.3.6.1   Intuitionisms

From an intuitionistic perspective, the very definition of a first-order structure is problematic: it fixes the meaning of the negation in the classical way and, thus, is not constructive. Therefore, intuitionism dispense with the concept of structure but reduces mathematics to the part expressible in constructive terms. This is unproblematic for arithmetic, although it even *justifies* classical arithmetic by use of the double negation interpretation. But for analysis, Brouwer wasn't able to provide the mathematical community with an alternative to the classical conception.[14] In any case, intuitionism is today—quite ironically—more a formal enterprise (but as such, very valuable) and serves the mathematical community more through its conceptual analysis of constructivity than through philosophical reflections.[15] But, as such, it approaches the natural numbers (and other concrete structures) with much less structure than the closure of the primitive structure under (classical) first-order logic provides.

7.3.6.2   Quantifier-Free Theories

We already mentioned Primitive Recursive Arithmetic which is a quantifier-free formalization for the natural numbers and, as such, exhibits less structure than Peano Arithmetic. Another prominent example for a quantifier-free theory is *Gödel's*

---

[13]This is in contrast to Hilbert who treated Arithmetic and Analysis equally in his second problem in his famous Paris problem list of 1900, as, by that time, he had no reason to restrict himself to first-order theories.

[14]"Brouwer and other constructivists were much more successful in their criticisms of classical mathematics than in their efforts to replace it with something better." (Bishop 1967, p. ix).

[15]"Intuitionism was transmuted by Heyting from something which was anti-formal to something which is formal. When one speaks today of intuitionism, one is talking of all sorts of formal systems (studied by the logicians)." Bishop in (Bishop 1975, p. 515); for the valuable results see, for instance, (Kohlenbach 2008); for constructivity (Bishop 1967).

$\mathcal{T}$. Of course, both theories should not be considered as closed under first-order logic, as one would incorporate some strength into them which is intended to be controlled. Otherwise, the consistency proof of Peano Arithmetic in terms of Gödel's $\mathcal{T}$ would be pointless. Thus, in foundational studies, one is clearly not bound to the ontological commitments of the intended standard model, as it just the task to give an independent account of it. In this sense, Shoenfield dismissed, quite correctly, any consistency proof by semantic methods (Shoenfield 2000, p. 214): "The consistency proof for P by means of the standard model [...] does not even increase our understanding of P, since nothing goes into it which we did not put into P in the first place."

7.4 Abstract Structures

Next to concrete structures, mathematics deals also with *abstract structures*. In this case, one abstracts entirely from any underlying universe but is studying structures entirely on the characteristic properties of operations and relations.[16] In our terminology, one could say that in an abstract structure the closure of the primitive structure under logical definability (i.e., normally first-order definability) is all what is available in the structure—even the constitutive structure of the concrete instances of the abstract structure is intentionally left unspecified. And one does not face the problem concerning the available structure discussed for the concrete structures; all that is there in an abstract structure is exactly that what we have put into it.

The classical example for an abstract structure is the notion of group, but today we have an endless stream of abstract structures at hand. Bourbaki (1950, p. 228) even identified some "mother structures", namely algebraic structures, order structures, and topological structures, which received a distinguished status in their "architecture of mathematics".[17] Approaching mathematics from this point of view became an incredibly successful programme so that, for instance, the term "abstract algebra" is essentially a pleonasm today.

In principle, one could also treat arithmetic as an abstract study about the consequences in PA; but this would be misleading. We are clearly not just interested in properties which are common to the standard and non-standard models of PA, excluding those which may have different truth values in different models. For

---

[16]Corry (2004, p. 259) describes this for Øystein Ore's introduction of the term *structure* in the context of his concept of lattice in the following words: "The leading idea behind this attempt was that the key for understanding the essential properties of any given algebraic system lay in overlooking not only the specific nature, but even the very existence of any elements in it, and in focussing on the properties of the lattice of certain of its distinguished subsystems."

[17]We may note that the way Bourbaki promoted the notion of abstract structure was critically evaluated by Leo Corry (2004) with the result that Bourbaki's use lacks a satisfying specification of the notion of structure.

the totality statement of a fast growing function we have a clear opinion about its truth value, as well as for the formalized consistency statement as used in Gödel's Incompleteness Theorems.

In contrast to arithmetic, in set theory the situation is not so clear. Using our terminology from above, one could think of a constitutive structure for sets which includes the power set operation among the operation to obtain new sets. It is evident that this operation is far from being clear in the way it works—the Continuum Hypothesis depends on it. Thus, all that we can do is to characterize it insofar as we have intuition about it, and that is what is done in axiomatized set theory. But, in this way, we can study it only in an abstract way—through the axioms we have formulated for it—and the possibility of different interpretations of the formal power set operation in different set-theoretic universes is possible; it is even realized, for ZFC, in Gödel's constructible hierarchy and in other "forced" set-theoretic universes. Adherents of a *multiverse* conception of set theory will probably subscribe to the abstract character of ZFC, but, of course, there are also other views.[18]

7.5 Conclusion

In this paper we have discussed how internal structure is added to a raw infinite set to obtain what we call the structure of the natural numbers. One may note that this internal structure requires some linguistic tools to express it; these tools are usually given by an underlying first-order language. As for the black-white "colouring" of the mutilated blackboard, this internal structure is not *per se* present in the raw infinite set, but was imposed by us. As a consequence, we are able to distinguish elements in this set, as the number 1729, which does not carry this distinction if they are only considered as the 1730th element in a stream of numbers (somehow in analogy to the second focal point of planetary orbits).

But we may add a word of caution: it is an everlasting controversy whether mathematics invents or discovers its concepts. By saying that one "imposes" structures, we do not advocate the former view. It should be clear, that such an imposement is far from being arbitrary.[19] Thus, there is an element of discovery when a mathematician realizes the *possibility* to put a specific structure on a specific domain. By exploiting this possibility one may speak, indeed, of a discovery.

When it comes to mathematical truth, we take the view that it makes sense only insofar as it refers to structures, see Kahle (2017). As we have seen, it is, however, far from being obvious how these structures, together with their internal structure, are given. To introduce (or discover) and investigate this internal structure of mathematical structures is, therefore, the basis of mathematical research—existence[20] and truth are notions which are, then, induced by the structural set-up.

---

[18]To give just two references: Feferman et al. (2000) and Antos (2015).

[19]One will not be able, for instance, to impose the structure of the natural numbers on a finite set; or a field structure on a set with six elements.

[20]In the sense of Bernays's "*bezogene Existenz*", (Bernays 1950).

References

Antos, Carolin, Sy-David Friedman, Radek Honzik, and Claudio Ternullo. 2015. Multiverse conceptions in set theory. *Synthese* 192(8): 2463–2488.

Bernays, Paul. 1950. Mathematische Existenz und Widerspruchsfreiheit. In *Etudes de Philosophie des Sciences*, 11–25. Neuchâtel: Éditions du Griffon. Reprinted in Bernays (1976), p. 92–106.

Bernays, Paul. 1976. *Abhandlungen zur Philosophie der Mathematik*. Darmstadt: Wissenschaftliche Buchgesellschaft.

Bishop, Errett. 1967. *Foundations of constructive analysis*. New York: McGraw-Hill.

Bishop, Errett. 1975. The crisis in contemporary mathematics. *Historia Mathematica* 2(4): 507–517.

Black, Max. 1946. *Critical thinking*. New York: Prentice-Hall.

Born, Max. 1966. Symbol and reality. *Dialectica* 20(2): 143–157. Archives de l'institut international des sciences théoriques, vol. 14: objectivité et réalité dans les différentes sciences (colloque de l'académie internationale de philosophie des sciences).

Bourbaki, Nicolas. 1950. Architecture of mathematics. *The American Mathematical Monthly* 57(4): 221–232.

Corry, Leo. 2004. *Modern algebra and the rise of mathematical structures*, 2nd revised ed. Basel: Birkhäuser.

Dedekind, Richard. 1888. *Was sind und was sollen die Zahlen?* Braunschweig: Vieweg.

Feferman, Solomon, Harvey M. Friedman, Penelope Maddy, and John R. Steel. 2000. Does mathematics need new axioms? *Bulletin of Symbolic Logic* 6(4): 401–446.

Fairtlough, Matthew V. H., and Stanley S. Wainer. 1998. Hierarchies of provably recursive functions. In *Handbook of proof theory*, ed. S. Buss, 149–207. Amsterdam: North-Holland.

Hardy, G. H. 1921. Srinivasa Ramanujan. *Proceedings of the London Mathematical Society* s2-19(1): xl–lviii.

Hermes, Hans, and Werner Markwald. 1958. Grundlagen der Mathematik. In *Grundzüge der Mathematik*, vol. I, ed. H. Behnke, K. Fladt, and W. Süss, 1–89. Göttingen: Vandenhoek & Ruprecht.

Kahle, Reinhard. 2017. Mathematical truth revisited: Mathematics as a toolbox. In *Varieties of scientific realism*, ed. Evandro Agazzi, 395–406. Cham: Springer.

Kahle, Reinhard, and Wilfried Keller. 2015. Syntax versus semantics. In *4th International Conference on Tools for Teaching Logic*, ed. M.A. Huertas, J. Marcos, M. Manzano, S. Pinchinat, and F. Schwarzentruber, 75–84. University of Rennes 1.

Kohlenbach, Ulrich. 2008. *Applied proof theory*. Berlin: Springer.

Kreisel, Georg. 1967. Informal rigour and completeness proofs. In *Problems in the philosophy of mathematics*. Studies in logic and the foundations of mathematics, vol. 47, ed. I. Lakatos, 138–186. Amsterdam: North-Holland.

Peano, Giuseppe. 1889. *Arithmetices Principia Novo Methodo Exposita*. Augustae Taurinorum: Bocca.

Rautenberg, Wolfgang. 2006. *A concise introduction to mathematical logic*, 2nd ed. New York: Springer.

Reich, Karin. 2006. Große Forschung, Große Lehre: Emil Artin. In *Zum Gedenken an Emil Artin (1898–1962). Hamburger Universitätsreden. Neue Folge*, vol 9, ed. Der Präsident der Universität Hamburg, 17–41. Hamburg:Hamburg University Press.

Shoenfield, Joseph R. 2000. *Mathematical logic*. Reading, MA: Addison-Wesley, 1967. Reprinted by ASL, AK Peters.

Thiel, Christian. 2006. Kreativität in der mathematischen Grundlagenforschung. In *Kreativität*, ed. G. Abel, 360–375. Hamburg: Mainer. Kolloquienbeträge vom XX. Deutschen Kongreß für Philosophie, 26.–30. September 2005 an der Technischen Universität Berlin.

Zweistein (alias Thomas von Randow). 1963. Logelei. *Die Zeit*, Ausgabe 31, 2. http://www.zeit.de/1963/31/logelei.

Chapter 8
Towards a Better Understanding of Mathematical Understanding

Janet Folina

8.1 Introduction

What is mathematical understanding? When we say that a mathematician has a deep understanding of her subject matter, what does this mean? I take understanding to be an epistemic property with philosophical implications, rather than a (merely) psychological property. Given this, how is mathematical understanding related to proof, explanation, and other issues regarding the epistemology of mathematics?

Consider, for example, the role of visual information in mathematics. An important role of images and diagrams is to support our understanding of abstract mathematical concepts and results. We often grasp things better once we can visualize them in some way. But what is it that drives this grasping? How do images facilitate understanding; what roles do they play? Questions like these will be explored in this programmatic paper.

The topic of understanding is in good company with other intriguing features of mathematics, such as, mathematical *explanation*, *beauty*, and *depth*, to which there has lately been increased attention. For example, are mathematical explanations a proper subset of proofs, or do they include a wider variety of justifications than proofs?[1] Can mathematical beauty be articulated; does it reduce to simplicity and unity? What is mathematical depth; does it correlate simply with the *number* of results a theorem or axiom leads to, or is depth something more elusive and harder to measure? These are several of the many interesting epistemic issues that

---

[1] For example, Steiner (1978) and Kitcher (1989); I will return to these accounts.

J. Folina (✉)
Department of Philosophy, Macalester College, St. Paul, MN, USA
e-mail: folina@macalester.edu

© Springer International Publishing AG, part of Springer Nature 2018   121
M. Piazza, G. Pulcini (eds.), *Truth, Existence and Explanation*,
Boston Studies in the Philosophy and History of Science 334,
https://doi.org/10.1007/978-3-319-93342-9_8

cannot be addressed simply in terms of the deductive properties of axiom systems. Mathematical understanding is one such issue.

Understanding may be even harder to articulate than these other issues that keep it company; it seems even more detached from simple deductive properties than issues of explanation and depth, for example.[2] Understanding is also extremely broad, encompassing a wide range of puzzles, such as: why some notation-systems seem more easily grasped than others, why examples and analogies are so helpful, what abilities we count as evidence of mathematical understanding, and of course how pictures and diagrams facilitate understanding. A further complication is that there are levels of understanding to consider. The child's understanding will be different from the graduate student, and both from the professional working mathematician. What do we expect at each level; what abilities are important for each?

Even narrowing our scope to the professional mathematician, the breadth of issues is considerable. For example, Poincaré claims that a mathematician must understand the *whole* of a proof as opposed to its parts, the *unity* of a proof as opposed to its individual steps, the *historical origins* of a definition as opposed to the mere formalism that expresses it, and the *point* of a question as opposed to its purely logical, or intra-theoretical, properties. He also emphasizes key abilities, such as the ability to *create* new mathematics, to *apply* mathematics in other areas such as physics, and to *perceive* key properties he thinks are central such as *depth, beauty, generality*, and *relationships*. (Poincaré 1900)

This is a pretty long list, which is not only broad but also somewhat vague.[3] How can we better articulate the individual features he mentions; can they be unified by a general account of understanding? I will argue that appealing to the idea of mathematical structure can help us with both of these tasks – that of clarifying the individual features associated with understanding and also with integrating them.

It is important to note that my aim here is modest. There is reason to believe that no strict *theory* that attempts to set out necessary and sufficient conditions for mathematical understanding will succeed. This is, in part, owing to the diversity of elements within mathematics that produce various sorts, or aspects, of understanding; in addition we already see from the different levels of understanding that it is a continuum concept. At best, there is a family resemblance of different features associated with understanding. My appeal to structure is thus not to provide a *theory*; rather, structure provides a unifying idea that helps articulate the family resemblance concept of mathematical understanding.

I begin in Sect. 8.2 with some preliminary criteria of understanding, differentiating several possible approaches to it. Such approaches include the various aspects of mathematics, mathematical tools that promote understanding, and the abilities we associate with mathematical understanding. These diverse approaches show

---

[2]Though perhaps not the issue of beauty, which shares with understanding both an elusive quality and a psychological connotation.

[3]For example, mathematical beauty deserves its own investigation; see Rota 1997 and Netz 2005 for some such work.

how broad a comprehensive account would have to be. In Sect. 8.3, I consider several accounts of understanding, arguing that each falls short of providing us with an adequate *theory*. They are inadequate since they each fail to capture one or more central properties commonly associated with understanding mathematics, as articulated in Sect. 8.2. Nevertheless, some insights can be gleaned from reviewing these models; these insights point to the conception I will advocate.

The ideal of unity, for example, is featured in several works connecting understanding to explanation, such as Friedman (1974), Kitcher (1989) and Tappenden (2005). Unity is also central to Poincaré's views on mathematical understanding. I will argue that the structuralist perspective makes sense of why unity – the ability to perceive unifying features of different mathematical theories – is important to understanding.

In fact, I think the structuralist view makes sense of a wide variety of the aspects, abilities and tools that are articulated in Sect. 8.2 of this paper, and thus that it provides a basis for a more comprehensive, or unified, approach to the concept.[4] To begin this more positive part of my argument, I provide a very brief explanation of what I mean by the "structuralist perspective" to which I appeal. In order to provide some evidence, I then revisit several of the features articulated in the first half of the paper, arguing that structuralism helps flesh out why they are associated with mathematical understanding. For example, I argue that the structuralist perspective helps explain, and synthesize, many of the mathematical abilities that are commonly taken to indicate understanding; it also explains why some pictures are such effective tools in producing understanding. This supports my thesis that the structuralist viewpoint provides a unifying perspective that accommodates a wide variety of tools that promote, and abilities we associate with, mathematical understanding.[5]

8.2 Understanding – Some Preliminary Properties

Even outside mathematics, understanding is difficult to articulate.[6] In the philosophy of mathematics it was perhaps even taboo for a time. Relevant here is Frege's distinction between logic and psychology,[7] which is sometimes presented as exhaustive. That is, one finds it used as a basis for disjunctive syllogism: any epistemological issue that clearly lies outside the boundaries of logic could be easily

---

[4]Though, as I have said, not a *theory* with necessary and sufficient conditions.

[5]I do not specifically address how structure relates to understanding the different *aspects* of mathematics because I think that understanding definitions vs. theorems vs. axiom systems, etc., can only be addressed in terms of abilities – though I do not argue for this here.

[6]That is, understanding of any sort is hard to explain; mathematical understanding presents special challenges given its technical nature and abstract subject matter.

[7]Frege 1884, introduction.

dismissed as "merely" psychology.[8] With Wittgenstein, and many others, I reject this assumption. Instead, understanding as I am treating it is one of many interesting epistemic issues regarding non-deductive issues in mathematics – issues that all lie *between* Frege's logic-psychology distinction.

Now, I take it that Wittgenstein showed that the *feeling* of knowing, or psychological certainty, is different from actually knowing (1972). Likewise, there is a difference between understanding and the mere *feeling*, or psychological state, associated with it. As philosophers, our interest is of course in the former, not the latter. Thus my focus will be on capacities, tools, and representations rather than the psychological states that might be associated with these. The feeling of understanding is real, and surely provides psychological guidance in our *acquisition* and *manifestations* of understanding. But it is obviously not the same as understanding itself.

"Mere" psychology thus dismissed, I think we can agree that the successful working mathematician's understanding is integrated, relational, and associated with multiple capacities. This is well expressed in the following quote:

> When a mathematician says he understands a mathematical theory, he possesses much more knowledge than that which concerns the deductive aspects of theorems and proofs. He knows about examples and heuristics and how they are related. He has a sense of what to use and when to use it, and what is worth remembering. He has an intuitive feeling for the subject, how it hangs together, and how it relates to other theories. He knows how not to be swamped by details, but also to reference them when he needs them. (Michener 1978, section 1).

Michener differentiates some "*aspects*" of mathematics, such as deductive properties and examples. Also included are *abilities*: "a sense of what to use and when to use it"; "what is worth remembering"; how it "hangs together"; the ability to work in a detailed way; but also the ability to survey the whole, or "not to be swamped by details". Though in need of clarification, these do seem important to any account of understanding. I would just add *tools* such as symbolic systems, analogies and diagrams, as also relevant. Our task can thus be approached from several perspectives: from the perspective of *aspects* or features of mathematics, from that of *tools* that mathematicians use, and from the perspective of *capacities* possessed by mathematicians who seem to have understanding.

Consider first an "aspects" approach. Aspects of mathematics are its features or properties. To articulate what we mean by mathematical understanding we could aim to catalogue the variety of *things understood* in mathematics, and explain what it means to understand each type. For example, each of the following seems quite different: mathematical concepts/ideas, axiom systems or theories, proofs, relationships between two areas of mathematics, and applications. Articulating *how* each of these is different, and thus how understanding each is different, would be an interesting way to flesh out mathematical understanding.

---

[8]This is not always an explicit assumption, but it leads to a dismissive attitude one finds in Frege and some of his defenders.

Tools provide links between mathematics and mathematicians. And different tools likely promote understanding in different ways and/or for different reasons. For example, what makes pictures so useful will likely be different from how an analogy or a good example assists with understanding. Different still may be why some representation systems seem easier to grasp than others. Articulating these individual tools in more detail and coming to grips with how they function provides a different approach to illuminating understanding.

Finally, capacities take the perspective of the mathematician, or person, rather than the subject matter or its tools. We count a wide variety of abilities as criterial for understanding, including the ability to: follow an argument, to explain, to forge new connections, to find proofs, to improve proofs, to appreciate mathematical depth, and to come up with good problems.[9] Obviously the abilities that are considered as evidence of mathematical understanding are an important component of any explanation.

Optimally, a theory of understanding would account for at least these three perspectives: the variety of *aspects* of mathematics, or things understood (theories, proofs, concepts, definitions, relations . . .); why and how certain *tools* are useful for producing understanding (pictures, notation systems, illustrative examples, informal explanations . . .); and the various capacities we count as evidence for understanding (the ability to explain, prove, illustrate, produce and follow an argument, see connections, etc.). This leads to the question of whether there are any such theories.

8.3   Existing Options for a Model of Understanding

The literature on mathematical understanding in general mostly comes from the perspective of mathematics education, whether the issue is teaching mathematics to younger children or to college students.[10] Another approach would be to use what philosophers have said about related issues, such as explanation, as a basis for a theory of understanding. In this section, I'll consider just a few of these options as representatives. Though I will argue that none of these supplies us with a *theory* of understanding, there are insights to be gained from each.

8.3.1   Understanding as Making Connections

One model of understanding comes from a popular text in the mathematics education literature. Haylock and Cockburn (2008) begins with the idea that making

---

[9]Avigad 2008 focuses on *abilities*: those behind understanding mathematical proofs.

[10]Though there is some recent philosophy of mathematics literature on the topic. In particular, see Avigad 2008 and 2010.

connections is fundamental to understanding; and this provides an appealing starting point for us too. The claim is that when children learn mathematics they must learn to connect four basic components: concrete materials, symbols, language and pictures. (p. 7, Haylock and Cockburn).

This seems a good model for teaching mathematics, especially to young children, which is Haylock's target. Furthermore, the ability to make such connections does seem necessary *for all levels* of understanding mathematics, not just that of children. For example, as Poincaré pointed out, the professional mathematician should be able to connect a theory with a potential application of it. Indeed, he argued that understanding includes many other connections as well, such as between the historical origins and the formal statement of a definition, and between different mathematical theories.

But making connections does not suffice for understanding. Consider, for example, finding proofs and improving proofs, the ability to appreciate a good problem and to perceive mathematical depth. These do not seem naturally described as, or reducible to, making connections. So although the ability to make connections does seem necessary for mathematical understanding, and (as will be clarified below) it works well with the structuralist approach I support, it is not sufficient.

8.3.2   Understanding as Knowledge

A second model to consider is from Porteous (2008), who advocates a reductionist view of understanding. Porteous considers the *phrase* "understanding mathematics", arguing that each of its uses can be construed in terms of knowledge. That is, he argues that we don't need a separate theory of understanding, since understanding is reducible to knowledge. Further, Porteous believes we have an adequate account of knowledge already – the justified true belief account.[11] His view is, thus, that an account of *propositional knowledge* suffices for an account of *mathematical understanding*.[12]

It seems obvious to me that this is far too narrow. In addition to the fact that justified true belief does not suffice for knowledge,[13] knowledge does not suffice for understanding. Someone can follow a proof or have confidence in another's argument; they may have justified true belief, thus knowledge on this account. But they can still lack understanding. As is often pointed out, the ability to follow the

---

[11] This is a paper on the *philosophy of* mathematics education (rather than mathematics education); nevertheless, I should note that its intended audience still seems different from that of mainstream philosophy of mathematics.

[12] Later in the article he helps himself to a wider use of "knowledge", as in "know-how" or "knowledge of"; however, this is a problem that I will not address.

[13] Assuming this is what Gettier examples show. Porteous agrees that these examples challenge the traditional account, but maintains that for mathematics *justified true belief* is close enough.

steps of a proof can yield a justified true belief that counts as *knowledge that* a theorem is true, when we might hesitate to attribute *understanding of* a theorem.

Importantly, the distinction between *knowledge that* and *understanding* does not depend on accepting the *justified true belief* account of knowledge, in particular. For example, consider reliabilism – a conception of knowledge favored by Giaquinto in his work on visual information and mathematical discovery. Reliabilism is the view that knowledge is true belief produced by a reliable process (which also passes some rationality constraints). (2007, p. 40) The same distinction between knowledge and understanding seems to hold here. Following the steps of a proof produces a belief based on a reliable, rational process; so it produces knowledge on this account. But we typically require more when we attribute understanding to someone.

Thus, at least on the two accounts considered here – justified true belief and reliabilism – knowledge *that something is the case* does not suffice for understanding, which is broader, perhaps also deeper, and involves multiple skills. This is because *having* understanding is more than simply carrying out a certain process or possessing a justification for a particular (true) belief.

Likewise, *showing* understanding requires more than simply *repeating* the process by which the belief was acquired or justified. We might expect the person with understanding to explain the belief *in a new way*; to show how it *relates* to other results, and so on. Thus, to understand is to *transcend* knowledge (the possession of a belief justified by an adequate warrant, or produced by a reliable process). So, there are two points so far: first (propositional) knowledge *that something is the case* does not suffice for understanding; second, the reason for this is that understanding is broader than knowledge, typically involving multiple skills.

Third, knowledge – propositional knowledge – may not even be *necessary* for understanding, at least for certain types, or degrees, of understanding. Someone might have a conviction, and see connections, in advance of finding a reliable process or traditional justification supporting the conviction. For example, I might see a diagram and have a hunch that two areas are importantly related to another, or I might see an important pattern before working out its equation. My hunch is not backed by any specific *process* or *justification*; it may not even be connected to a precise belief. So I do not (yet) have knowledge, though I might have understanding, or at least the dawning of understanding. So knowledge that a particular mathematical proposition is true may not even be *necessary* for understanding.

This suggests, fourth, that understanding comes in *degrees* as well as *types*. In contrast, knowledge that something is the case doesn't. If someone has a justified true belief, there is no more or less; they either have it or they don't. There are better and worse justifications; but provided a weaker justification is adequate, the agent has knowledge on the justified true belief account. Better justifications can produce more understanding, or an explanatory proof, or even a beautiful proof. But degrees of justification don't yield degrees of knowing in the way they yield degrees of understanding (or explanation).

Thus understanding seems different from knowledge, for at least the four reasons I have identified:

- Knowledge is not sufficient for understanding.
- This is because understanding transcends knowledge, its manifestation taking a wide variety of forms.
- Knowledge may not be necessary for understanding.
- Knowledge is by nature different from understanding, which comes in degrees, while in contrast propositional knowledge – *that* something is the case – does not.

8.3.3   Understanding and Explanation

A third potential source for a model of mathematical understanding is to borrow from a neighboring concept and try to build an analogous theory. There are two immediate reasons to consider mathematical explanation for this role. First, it is closely related to understanding: when one understands one can typically explain, and a core aim of explanation is to produce understanding. Second, literature on explanation exists in mainstream philosophy of mathematics, so we would be borrowing from a familiar type of approach.[14]

The connection between explanation and understanding is discussed in the philosophy of science literature. Michael Friedman, for example, argues that they are inextricably entwined.

> When I ask that a theory of scientific explanation tell us what it is about the explanation relation that produces understanding, I do not suppose that 'scientific understanding' is a clear notion. Nor do I suppose that it is possible to say what scientific understanding is in advance of giving a theory of explanation. It is not reasonable to require that a theory of explanation proceed by first defining 'scientific understanding' and then showing how its reconstruction of the explanation relation produces scientific understanding. We can find out what scientific understanding consists in only by finding out what scientific explanation is and vice versa. (Friedman 1974, p. 6).

Friedman's focus is *natural science* rather than mathematics.[15] He aims to find an account of scientific explanation that accords with and illuminates our intuitions about scientific understanding. He supports three criteria of explanation: *generality* (it should account for most theories taken to be explanatory), *objectivity* (it should transcend the times, or intellectual fashion), and *connection to understanding* ("it should tell us what kind of understanding scientific explanations provide and how they provide it" (p. 14)). Though he argues that all existing accounts fail on one or more of these criteria, the point for us is the strong link proposed between any adequate theory of scientific explanation and that of understanding.

---

[14]For a detailed look at some specific relationships between explanation and understanding, see Tappenden 2005. This is an interesting, useful paper; but like Kitcher, Steiner and Friedman, Tappenden focuses on *theories* or *frameworks* that promote understanding. Though more sketchy, my project here has a broader (even broader!) aim.

[15]Like Kitcher (1989) and unlike Steiner (1978), as we'll see.

We might reason, then, as follows. If scientific explanation should illuminate the idea of scientific understanding, then mathematical explanation should similarly illuminate the idea of mathematical understanding. The problem, however, is that there is no generally accepted, successful account of mathematical explanation. It is a topic still in its early stages of development.[16]

One type of approach, found in Steiner (1978) with roots in Bolzano (1810), considers mathematical explanations only as certain types of proofs. So mathematical explanations comprise a proper subset of mathematical proofs, making some proofs explanatory and some proofs not explanatory. On this view, mathematical explanation is analogous to scientific explanation. Just as the latter is typically considered to occur via a particular *use* of a theory (perhaps with some added requirement like causality), so mathematical explanation is considered as a particular *use* of a theory or axiom system to deduce another result/theorem. That is, these accounts both consider explanations to occur in the form of arguments: those with special epistemic and/or metaphysical features.

Steiner does not simply adopt Bolzano's views for his theory of "explanatory proofs", but the similarities between the two accounts are striking. Bolzano (1810) proposes a theory of ground and consequence, arguing that "proper" proofs appeal to the true grounds of a result (with additional restrictions on permitted inferences). Grounds, for Bolzano, are similar to causes – they are something like a-temporal causes. This approach to explanatory proofs thus also relates to the causal theory of explanation in science; it has an Aristotelian flavor as well.[17]

Steiner's proposal is quite similar to Bolzano's, though the central criterion added to make a proof explanatory is that of "characterizing property", rather than either Aristotelian essences or Bolzano's grounds. A characterizing property is "unique to a given entity or structure within a family or domain of such entities or structures." (143). For example, Steiner cites the number 18 as something that can be *characterized* both as the successor of 17 and as 2×3^2. (Since both of these count as "characterizing properties" an object can have more than one such property). Explanatory proofs are those that refer to, and depend on, a "characterizing property of an entity or structure mentioned in the theorem." (143)

The connection between Steiner and both Aristotle and Bolzano should be apparent.

> My view exploits the idea that to explain the behavior of an entity, one deduces the behavior from the essence or nature of the entity. Now the controversial concept of an essential property of x (a property x enjoys in all possible worlds) is of no use in mathematics, given the usual assumption that all truths of mathematics are necessary. Instead of 'essence', I shall speak of 'characterizing properties', by which I mean a property unique to a given entity or structure within a family or domain of such entities or structures. (Steiner 143).

---

[16]The volume Mancosu et al. (2005) shows that there is some progress and interest; nevertheless, most of its articles are programmatic in nature.

[17]For recent work on the concept of grounding in mathematics, start with Šebestík 2016.

Thus, Bolzano and Steiner each characterize mathematical explanations as certain sub-classes of proofs – those that make use of important identifying properties of the objects that the proofs are about.

Space does not permit me to go into the details of this interesting approach. The obstacle for our project should be obvious: even if one of these provides a good account of *explanatory proofs*, neither will help in our search for a model of *understanding*. They are both too narrow for this purpose.[18] As already argued, understanding is broader than knowledge; so it is also broader than knowledge produced by a special subset of justifications – explanatory *proofs*.

Since understanding is broader than (propositional) knowledge, for philosophical work on mathematical explanation to help, we would need to borrow from a similarly broad account of explanation. An option to consider is Kitcher's account of scientific explanation, of which mathematical explanation is a subset. (Kitcher 1989) Kitcher argues that scientific explanation should be detached from causality; the appeal to mathematics is mainly as an example, to show that there are non-causal explanations (assuming mathematical entities are abstract, and that abstract entities cannot causally interact).

Now, this approach may seem similar to that of Steiner/Bolzano, which we have just rejected as too narrow for our purpose. Indeed there is a strong connection here. Kitcher even begins his discussion of mathematics by referencing Bolzano's arguments for "a broader notion of objective dependency to which correct explanations must conform." (Kitcher 424) But there are important differences between the two approaches.

Whereas both Bolzano's concept of "proper grounds" and Steiner's "characterizing property" have a metaphysical flavor, Kitcher aims for an account of scientific explanation that avoids such connotations; he aims for an account that does *not* rest on a realist, or metaphysical, notion of causality (or, I take it, any substitution for it, like essence). Rather than causality as a ground for explanation, Kitcher advocates the reverse: what we take to be explanatory reveals the "causal structure of the world." (500) What is "causal", then, is a consequence of an epistemic or methodological inquiry rather than something more metaphysical.

How do we determine what is "explanatory" without appealing to grounds or causality? I cannot do justice to Kitcher's argument here (which spans nearly 100 pages). Two central ideas, of interest to us, are *unification* and *economy of thought*. He uses these in arguing that science provides an "explanatory store", and at any given time, science promotes "those derivations which collectively provide the best systematization of our beliefs" (430). "Best" seems to mean something like economy, or most results for fewest assumptions. (431) In appealing to more operational criteria, like stores, economy, systematization, Kitcher attempts to find a

---

[18]This is a limitation, not a criticism. Steiner is explicit about the narrowness of his project: "We have not analyzed mathematical explanation, but explanation *by proof*; there are other kinds of mathematical explanation." (147, emphasis in original)

methodological rather than a metaphysical approach to causality (or something like causality).

One may still find the influence of Bolzano here, whose theory of grounds also steps back from relying on causality for explanation. Another link between the two is that the result for both enables mathematical explanations. However, in addition to his argument-strategy, Kitcher's focus is different from that of Bolzano and Steiner. He does not simply replace causality with another property that is linked to explanation, such as ground or characterizing property. His account "sits" at a different level, concerning the explanatory power or virtues of *theories* rather than particular *uses* of theories, as in (explanatory) arguments and proofs. In addition, (scientific) unity and systematization are broad desiderata, giving his account a more open feel than those of Steiner and Bolzano.

To be clear, I do not think that Kitcher's model for *scientific explanation* provides us with a ready-made theory of *mathematical understanding*. For one thing, its aim, like Friedman's, is science in general rather than mathematics – though his theory is intended to include mathematics (Sect. 4.2). For another, despite the breadth of the ideas he uses to motivate and frame his project, in spelling out his account there is a strong emphasis on deductive relationships between scientific theories and their results. And as already clarified, understanding as I am approaching it, is broader than any set of deductive relationships. It is unclear, for example, what place could be made for explaining the value of pictures, examples and analogies via any such deductive relationships. Nevertheless, like Friedman, Kitcher proposes a strong link between explanation and understanding; and he has interesting things to say about the latter. Let us briefly see what these are.

Scientific explanation for Kitcher "advances our understanding". And scientific understanding involves "seeing connections, common patterns, in what initially appeared to be different situations." (432) Interestingly, this resonates with Haylock's emphasis on making connections for mathematical understanding. Kitcher also highlights features that bring to mind Poincaré's remarks on understanding. As stated, *systematization* and *unity* are central to Kitcher's account of explanatory theories; recall that, analogously, the ability to systematize and unify are central to mathematical understanding for Poincaré. Thus, Kitcher explicitly connects scientific explanation to understanding, and his conception of scientific explanation resonates with some of the key features of mathematical understanding that we have articulated.

However, though less narrow than the Bolzano-Steiner accounts of *explanatory mathematical proofs*, the connection between explanation and understanding that Kitcher proposes is mostly suggestive. Further, as cautioned above, the features that *are* spelled out place too much emphasis on deductive relationships for our purpose. In addition, Mancosu (2015) argues that it does not suffice as a theory of explanation.

In fact Mancosu suggests that *all* existing theories have problems, leading him to question whether a theory of mathematical explanation is really possible. "Recent work has shown that it might be more fruitful to proceed bottom-up, that is by first providing a good sample of case studies before proposing a single encompassing model of mathematical explanation. Indeed, this kind of work might also lead to the conclusion that mathematical explanations are heterogeneous and that no single theory will encompass them all." (Mancosu 2015, conclusion) I am sympathetic to the trace of pessimism lingering here from his earlier thoughts on such prospects.[19]

Whether or not a single theory of mathematical explanation is unlikely, the prospects for discovering a precise theory of understanding (whether or not based on explanation) seem worse. One strategy is to narrow the focus in some way – such as Mancosu (2000) which takes an historical approach, or Tappenden (2005) which aims to articulate particular *features* of *mathematical theories* that promote understanding. These more "bottom up" approaches can provide depth through case studies; and, for example, Tappenden focuses on just one aspect of mathematics – that of mathematical theories.[20] But any attempt to provide a general overview of mathematical understanding needs to account for a very wide range of only partially articulated aspects of mathematics and capacities, tools, and levels of understanding. The problem of mathematical understanding is thus like that of a moving target: there are multiple taxonomies to take into account (tools, aspects, capacities) each of which might also vary both with context (such as different areas of mathematics) and level of training (children, professionals, etc.).

For these reasons, I do not aim for a *theory* of understanding; the negative conclusions of this section are intended as ground-clearing work. Instead, my argument going forward is that the structuralist perspective provides a unifying framework that accommodates a wide range of the capacities and tools that we associate with – as contributing to, and exemplifying – mathematical understanding.

8.4   The Structuralist Viewpoint

By the "structuralist viewpoint" I mean something much more general than any particular philosophy of mathematics, such as is associated with work by well-known structuralists like Shapiro, Resnik, Hellman, etc.[21] I mean the core view shared by all of these particular philosophies. Emerging naturally from mathematical practice, the structuralist perspective concerns the subject matter of mathematics only; it

---

[19]For example, Hafner and Mancosu 2005 "maintain that mathematical explanations are heterogeneous" (222) – so no single theory – and that the topic is even "treacherous." (241)

[20]Avigad 2010 also recommends a "bottom up" approach in order to – at least first – better articulate the data that a theory of understanding would explain. (See conclusion.) I am grateful to an anonymous referee for pointing me to Avigad's work on this topic; although his work resonates a great deal with my views I unfortunately discovered it too late to follow his recommendation!

[21]The standard literature begins with Benacerraf (1965); see Hellman (1989), Resnik (1997) and Shapiro (1997) for more developed philosophies.

is the basic view that mathematics is *about* abstract structures rather than (just) particular objects, definitions, sets, systems, etc.[22]

For example, consider the individual number 3, the set of natural numbers, and the natural number *structure*. The structuralist view is that arithmetic is *really,* or *more importantly,* about the latter, rather than individual numbers or even the set of natural numbers (any set of particular objects). The same goes for other areas of mathematics.

The structuralist *perspective* is thus a pretty bare view, leaving most philosophical questions about structures unanswered. For example, what is the relationship between objects, systems, and structures? Do structures depend on systems exemplifying them to exist? Do they actually exist, or is reference to "structure" a mere manner of speech? How do we gain knowledge of abstract structures? These and other questions are disputed by the various philosophical views about structure, while the *perspective* to which I appeal is intentionally more general.

This perspective emerges quite early, and can be associated with the increasing abstraction of mathematics. At least by the early eighteenth century, mathematicians such as D'Alembert and Maclaurin, appealed to structuralist (or perhaps proto-structuralist) ideas to defend and explain calculus. Calculus moves mathematics away from the physical world, sense experience, etc. But in detaching itself from the concrete world, it becomes unclear what it is *about*, what determines its truths. For example, questions about the ontological status of fluxions, infinitesimals, imaginary numbers, etc. arose by this time. One way to deal with such questions, which can be seen in some defenses of calculus, was to *deflect* them by asserting a proto-structuralist view. Momentum for structuralism only increases with developments in symbolical algebra and non-Euclidean geometries during the nineteenth century.

The "proto-structuralist view" includes an increased emphasis on relations, or relation-types, rather than on specific objects or sets of objects. An emphasis on relations is a move towards structuralism; it moves away from the traditional idea that a scientific subject matter requires a determinate set of objects. One path to structure is thus by way of a focus on relations instead of the objects, or content, bearing those relations to one another.

Consider two buildings or two floral arrangements. Let's suppose that someone refers to the fact that they have the same structure. Asserting this means that though they are not made of the same *materials*, the buildings or arrangements have other properties in common. In this sense structure is the *common form* that the different pairs share. For example, they may have the same shape: both buildings might be roughly cubical; both bouquets could be conical with a particularly large flower in the center. Both buildings may have two doors and 31 windows; but they may also be quite different in specific features provided they share some architectural,

---

[22]I say "just" because one can adopt the structuralist perspective and still think mathematical objects exist. That is, one can retain the view that arithmetic is about the numbers 3, 4, 5, etc., while maintaining that *important* arithmetic results typically concern the natural number *structure* rather than individual numbers. The perspective is thus agnostic on philosophical issues about the priority or independence of structures from mathematical objects and systems.

formal or "shape" properties. When it is observed that the even numbers and the odd numbers have the same structure (when ordered "naturally"), a similar observation about "shape" is brought into focus. Just as we find common concrete structures (buildings, bouquets) or common musical structures, so mathematical structures can be unveiled by focusing not on content or materials but on the relations and/or relation-types that articulate their common form.[23] This greater focus on relations and relation-types is what I mean by "proto-structuralism", and it can be seen already in some early defenses of calculus.

Since structuralism is first and foremost a view about the subject matter of mathematics, it is partly an *ontological* view. However, it leads naturally to a *semantic* view: if mathematics need not be about particular objects, then mathematical terms need not *refer* to known objects (like quantities and triangles). Terms can also represent unknowns like infinitesimals and imaginaries if they are part of a useful, consistent theory. Which brings in new *methodological* criteria, since consistency, rather than truth and reference, is used to justify this liberation from traditional ontological and semantic restrictions. Together, these changes *lift* the subject matter of mathematics to a more abstract, general perspective.

Thus, for example, regarding calculus, one need not worry about the *nature* of fluxions or whether infinities are real if references to them are mere manners of speech.[24] The idea that mathematics is not *really* about such objects, but the relations manifested symbolically – relations that lead to useful and consistent results – shifts attention from ontological worries to methodological issues, and it does this by appealing to the basic structuralist perspective on the subject matter of mathematics.[25] So in defenses of calculus one can see *structure,* or at least *relations*, heralded as solving the problem of what mathematics is *about* as it becomes more

---

[23]The idea that it is relations and relation-types that "matter" more to mathematics is a core view of mathematical structuralism. For example, according to Shapiro, "The theme of mathematical structuralism is that what matters to a mathematical theory is not the internal nature of its objects, such as its numbers, functions, sets, or points, but how those objects relate to each other." (Shapiro undated, opening sentence.) Despite this core, the general idea of "structure" is quite hard to pin down; in addition it is ambiguous. With regard to physical objects, it can be used to refer to a particular, as in "The bank is in that big brick structure down the road," though it can also mean a type, as in "The A-frame structure is visually striking". When it comes to explaining what a mathematical structure is, the notion of abstraction is often used. For example, Shapiro explains that a mathematical system is a "collection of objects together with certain relations on those objects . . . . A *structure* is the abstract form of a system, which ignores or abstracts away from any features of the objects that do not bear on the relations. So, the natural number structure is the form common to all of the natural number systems." (Shapiro undated, section 1.) The idea is that by bracketing, or removing from consideration, certain features of two mathematical systems their underlying common form, or structure, can be unveiled. This process of ignoring some features to get at a common, usually more abstract, underlying property or form is roughly what "abstraction" means. However, the notorious problems of what philosophers mean by an abstract structure (or any kind of universal), the activity of abstraction or, especially perhaps, the relation between the two are well beyond the purview of this paper.

[24]See for example Maclaurin 1742, Introduction.

[25]See again Maclaurin 1742, Book I, Chapter I, especially opening paragraph.

abstract. This view about the subject matter emerges from mathematical practice, and in reflection by mathematicians, independent of any particular philosophical theory about the nature of such structures. It is this very general conception to which I appeal here.

8.5 The Structuralist Perspective and Mathematical Abilities

How might the structuralist viewpoint help explain the various abilities we associate with the mathematician who is deemed to have a deep understanding of her subject matter? What do we take as *evidence* when attributing such understanding? Poincaré's views on this were mentioned in Sect. 8.1 above. In fact Poincaré cites a wide variety of abilities as central to genuine understanding. These citations mostly occur in the context of his negative arguments against too much focus on formality or logic in mathematics, but they result in some positive views about understanding.

For example, Poincaré argues that mathematicians should understand proofs as *wholes* rather than as mere sums of their parts; they should thus be able to perceive what he calls the *unity* of a proof. In addition, he thinks that the *historical origins* of a definition can provide context and depth that helps one grasp a formal definition within a theory. Historical origins can also illuminate the *point* of a question, theorem, or area of mathematics – the "why" behind the work rather than just the "what". Similarly, also important are the abilities to *apply* mathematics and to *relate* different areas of mathematics to one another. Of course a mathematician with a deep understanding will be able to *create* new mathematical results. These new results will ideally be *deep*, *beautiful*, and *general*; and even apart from such proof-practices, Poincaré thinks there is such a thing as the ability to perceive, appreciate, and strive for mathematical *depth*, *beauty*, *generality*, and *unity*.

What to make of these hazy metaphors? At this time I can only remark on several, urging that the structuralist viewpoint assists in fleshing them out and connecting them. For example, consider Poincaré's remarks about seeing the *whole* and *unity* of a proof. These occur in the context of complaints against the turn towards formality in mathematics (Poincaré 1900). One argument famously compares focusing on the individual logical steps of an argument to looking at the cells of an elephant under a microscope. Poincaré argues that the cells, or atoms, of an elephant are not the "sole reality". "The way in which these cells are arranged and from which result the unity of the individual, is not it also a reality much more interesting than that of the isolated elements, and should a naturalist who had never studied the elephant except by means of the microscope think himself sufficiently acquainted with that animal?" (1900 V) Obviously not.

The cellular biologist and the zoologist possess very different forms of knowledge about elephants. Analogously, the logician and the holist can each possess different forms of knowledge about one and the same proof. Closely following the individual inferential steps, and verifying the validity of each step, is one way of knowing a proof – that of "the logician". However, Poincaré thinks that the holist perspective, the understanding of the proof as a whole, is what we really mean by such understanding: "what constitutes the unity of a piece of reasoning, of what makes, so to speak, its soul and inmost life." (Poincaré 1900 VI) To grasp "the real meaning of the demonstration", its "entire reality", one must perceive the "unity of the demonstration". (Poincaré 1900 V) Poincaré asserts, further, that the holistic understanding is necessary to both create new mathematics and "really to comprehend the inventor." (*op cit.*)

Feferman echoes this distinction between ways of understanding proofs: "it is possible to go through the steps of a given proof and not understand the proof itself. That is a different level of understanding, which when successful, leads one to say, "Oh, I see! In other words, this "really understanding the proof" is a special kind of insight into how and why the proof works." (Feferman 2012, p. 372) He goes on to emphasize the importance of the holistic insight to understanding relations between proofs and – also with Poincaré – for creating new mathematics.

Poincaré and Feferman both draw attention to the importance of the *plan* of the proof; the unity that is found in its "architecture": how the parts are related and connected to one another. Emphasizing unity, and relationships between parts, resonates with Haylock's emphasis on making connections as central to mathematical understanding. In addition, the importance of seeing the whole, or the blueprint, of an argument also calls to mind Michener's view that mathematical understanding enables one to avoid getting lost in details. My claim here is that understanding as seeing the "big picture", the "overview", and (for Michener) the "forest from the trees" resonate with the structuralist viewpoint. Each of these points to a comprehension as an overview, one that is more "broad brush": a perspective that ignores individual "trees" to focus on the overall *shape properties* of a "forest". Structuralism can support these views about mathematical understanding for the following reason. If mathematics is more about the wholes – the structures or forms – than the objects inhabiting the structures, then the view that mathematical understanding requires one to comprehend such wholes, or structures, makes more sense. Structuralism about the subject matter of mathematics thus provides a basis for structuralism about understanding that subject matter.

Other elements of understanding for Poincaré similarly transcend the local, purely deductive, properties of mathematics, and seem similarly supported by the structuralist perspective. Let's consider his claims about the *historical origins* of a definition, the *point* of a question and the ability to *apply* mathematics. One way to think about these is that they all concern context and relations. Poincaré relates appeals to *sense experience, images, analogies* and *historical origins*. (1900, IV–VI) For example, continuous functions are defined via a "complicated system of inequalities", which he complains is impossible to understand without "recalling" the "primitive image" that lies at its basis – that of a smooth line or a point in motion. (1900, V) He compares this primitive image to the temporary supports that are used to construct a physical arch. As with a formal definition in mathematics, once the arch is completed, the supports are of course removed; but one must "recollect" the supports to understand how the arch was constructed. Emphasis on the *scaffolding that supports the construction of a structure*, whether it is a physical arch or a formal mathematical definition, resonates with the structuralist perspective.

Also important to Poincaré are aesthetic properties like *depth, beauty, elegance* and *harmony*, and the ability to perceive them. "It is the harmony of the diverse parts, their symmetry, their happy balance; in a word, it is all that introduces order, all that gives unity, that permits us to see clearly and to comprehend at once both the ensemble and the details. But this is exactly what yields great ideas..." (1908, pp. 372–3) Deep ideas unify disparate areas of mathematics; they do this by introducing, or revealing, shared order or patterns. Further, ideas that provide such bridges lead, he thinks, to deeper and more important mathematical results. Thus Poincaré connects certain aesthetic properties of mathematics, and the ability to perceive them, with mathematical success. The aesthetic properties in question include beauty, elegance, harmony and symmetry. Beauty, for example, is explained in terms of order; and order is claimed to both increase clarity and reveal similarities between different areas of mathematics. This is, in turn, why the perception of beauty leads to mathematical success for Poincaré. But order and relations are the properties central to the structuralist perspective, for it is through relations that create order that structure is revealed. Hence, this perspective helps make sense of why the ability to perceive mathematical beauty is connected to mathematical understanding.

Hardy also stresses the importance of aesthetic properties in mathematics, and with Poincaré he explains them in terms of relations, order, and symmetry: "The mathematician's patterns, like the painter's or the poet's must be beautiful; the ideas, like the colours or the words must fit together in a harmonious way. Beauty is the first test: there is no permanent place in this world for ugly mathematics." (Hardy 1941, p. 14) Hardy not only emphasizes patterns or structure in connection with mathematical beauty; with Poincaré again, he connects these to success: "A mathematician, like a painter or a poet, is a maker of patterns. If his patterns are more permanent than theirs, it is because they are made with ideas." (Hardy, p. 13).[26]

Regarding *creativity*, an important tool for a "maker of patterns" is that of unifying ideas, such as the group concept. If good ideas *unite* different areas of mathematics and reveal order, it seems plausible that the ability to *find* such good ideas involves perceptions of abstract structure. By hypothesis, different areas of mathematics concern different systems, often with different types of objects; structure, and other relational properties, can unite them. Thus the structuralist perspective again makes sense of both the importance of aesthetic properties and how the perception of such properties is connected to mathematical success and understanding.

To sum up, I have tried to add a little flesh to remarks by several mathematicians about important abilities (some, quite elusive) they connect with mathematical understanding and success. Poincaré, Hardy, and Feferman each mention these

---

[26]Thurston is another mathematician who emphasizes the art and beauty of mathematics, and who explores what we mean by mathematical understanding. See his 1994 and 2010.

abilities as indicative of, and leading to, mathematical understanding. I have further argued that the structuralist perspective helps to make sense of why this would be – why these abilities, some of which seem purely aesthetic, have anything to do with the understanding of the successful working mathematician. Now, in urging that the structuralist perspective helps explain and unify these features, it may seem that I am providing weak, merely suggestive, evidence. However, I don't think that is quite accurate. Structuralism yields an important concordance between the *subject matter* of mathematics and these views about what it means to *understand* that subject matter. They are thus, at the very least, mutually supporting views. I next aim to provide a hint of something stronger, arguing that the structuralist perspective is the *best* framework for explaining how certain types of diagrams and other visual tools facilitate mathematical understanding.

8.6 The Structuralist Perspective and Visual Information

Few can doubt that pictures and diagrams are important mathematical tools. The question here is why: why are they so helpful for mathematical explanation and understanding? Even if they are not proofs, appeal to visual information does at times seem to carry epistemic, normative weight. So again, our question is not a psychological one; rather, how do images substantiate, or contribute to the justification of, mathematical beliefs?[27] I believe that structuralist considerations can help explain why images can provide *evidence* in mathematics, and thus why they are not merely *causing,* or *convincing,* us to form beliefs.[28]

Now, a diagram of a triangle can obviously guide our reasoning about triangles. This seems obvious since we are visualizing an example or template of the very type of object about which we are reasoning. Similarly, the various Pythagorean theorem diagrams also provide a more or less direct visual presentation of the information in question, at least when understanding the squaring operation from a geometric perspective. That is, taking the geometric view, there is no real need for a special structuralist explanation of how a diagram such as the following can assist our understanding of why the Pythagorean theorem is true.[29]

---

[27] I hope to provide more argument for this elsewhere. Here I am just addressing *how* or *why* images carry epistemic weight in mathematics, not arguing *that* they can do so.

[28] Tappenden (2005) also appeals to visualization in connection with mathematical understanding, but, as mentioned above, his focus is on whole *theories*. For him, a theory that lends itself to visualization tends to be more understandable and fruitful. (150) He also notes that visualizability counts in favor of the "naturalness" of a theory or framework. (180) Here I focus on a neighboring but different question of how and why visual images function as *tools* for understanding, which Tappenden brackets (155) to pursue his other goals.

[29] This is not to say I accept such diagrams as full proofs or as better than proofs. My point is merely that visual information can assist understanding; my question is why and how.

It is not that this type of visual evidence *violates* the structuralist perspective; rather structuralism does not seem particularly needed to explain why a certain object type is helpful for drawing conclusions about that same object type. Any issues here are less obvious.[30]

Where I think structuralism particularly helps is in explaining why such visual tools can provide evidence for theorems about non-spatial domains, such as number systems. To focus the discussion, I will consider just one simple example of this sort: that of the finite sum theorem, $1 + 2 + 3 + \cdots + n = n^2/2 + n/2$. This example has been treated extensively in the literature on visual thinking in mathematics, picture proofs, etc.[31] I am not attempting to say anything very original about this particular case; my purpose is rather to *use* it, to illustrate the role of structure as providing a bridge between the image and mathematical understanding of the theorem.

Obviously the "traditional" proof of this arithmetic result will be inductive. For the base case, we have $1 = \frac{1}{2} + \frac{1}{2}$. And assuming $1 + 2 + 3 + \cdots + n = n^2/2 + n/2$, then we know $1 + 2 + 3 + \cdots + n + (n + 1) = n^2/2 + n/2 + (n + 1)$, which equals $(n + 1)^2/2 + (n + 1)/2$. Bolzano might say the traditional proof both deduces the theorem and gives its ground, which lies in the inductive nature of the finite numbers. And this arguably makes the traditional proof "proper" for Bolzano, and perhaps explanatory for Steiner.[32]

But this is only one route to *understanding* that the theorem is, and even must be, true. A common "picture" of this theorem takes a staircase form.

---

[30]There are of course some issues. One is to explain how a single diagram can support a general theorem. Another is to connect the geometric and the symbolic expressions of the theorem – to explain how the spatial object supports the numerical equation, $x^2 + y^2 = z^2$. For this latter in particular I think the structuralist lens is important; I focus on another example to make my point more obvious.

[31]See for example, Brown (2008), Mancosu et al. (2005), and Nelsen (1993).

[32]The "object" in question can be seen as the set of natural numbers and its "characterizing property" as induction. (There is of course literature debating whether inductive proofs can explain. For example, Lange 2009 and responses.)

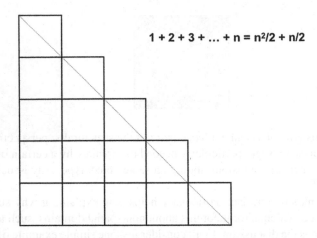

$$1 + 2 + 3 + ... + n = n^2/2 + n/2$$

A significant benefit of the picture is its accessibility. With it, or some physical blocks, a child can come to grasp some of the essence, and even necessity, of the general result. That is, even if she does not grasp the symbolic expression, a child can grasp the fact that each time she adds the next layer or column of blocks, the resulting staircase has exactly one more step than before – provided certain building constraints are satisfied.[33] So she can grasp that when the blocks are organized properly, the shape, or structure, remains the same at each stage: a staircase of different sizes.

The question here is *how*: how does the picture connect with the normal arithmetic symbolism, and can it provide any genuine *justification* for the equation, which can form the basis for *understanding* it.[34] There are several important differences between the picture and the equation, which must first be acknowledged before addressing this question. The first obvious difference is that any picture is of a particular case; here it is n = 5. In contrast, the symbolism states the general theorem; this difference – which is a problem for most diagrams and "picture proofs" – must somehow be overcome for understanding. A second important difference is that the equation is divided into two expressions, which are asserted to be equal. In contrast, there is only one picture, it only has one "side", and it

---

[33] For example, if adding a new "column", one must add it next to the previously "tallest" column and one must make it exactly one block higher. One cannot add that taller column to the middle or the other end of the staircase. If iterating via "layers" one must put a block on top of each existing step, and then add one to the "floor" next to the lowest end or new stack of two. These rules can easily be taught to a child.

[34] Again, such pictures are in my view neither genuine proofs nor better than traditional proofs, except insofar as they are more widely accessible. They do not supplant the equation or its traditional proof. But they are tools that promote understanding, and as such they function like partial justifications. Further, this function explains their normativity (why we are *correct* to believe the theorem on the basis of the picture), and their epistemic contributions (why they lead to *understanding* rather than mere *belief*).

doesn't actually assert anything. We will start here and attempt to progress to an explanation of how the generality problem is (more or less) overcome.

Though the picture is a single object, for it to provide evidence for the theorem, it must first be grasped from two perspectives – as representing *both* the left-hand and the right-hand sides of the equation. And when the one image is understood as representing both sides of the equation, the equality relation will thereby be supported. There are at least four issues that need to be overcome in matching the picture to the equation, which I will differentiate in terms of possible "recognition steps".

A first step for the viewer might be to determine the area of the actual figure. The blocks can simply be counted, but this will not connect the picture to the equation. The "shape" properties – that it *pictures* $5^2/2 + 5/2$ – must be noted for this, and the diagonal dotted line in our version of the picture helps here. The viewer can then see that the geometric area matches the quantity on the right hand side of the equation for the case n = 5.

A second recognition step might connect the diagram to the left-hand side of the equation for the same specific case n = 5. How might this happen? The viewer must be able to match sub-areas of the diagram to the quantities mentioned in the left-hand side of the equation. Those quantities are listed in ascending order and include those mentioned both explicitly (1, 2 and 3 in our statement) and implicitly (4, and 5 to match the diagram). It seems necessary to think of the different parts of the diagram in some patterned order, to match the pattern expressed in the equation. A useful way to do this is to envision the geometric picture being "built" in a sequence. (Physical blocks can also be helpful for this.) Whether or not the diagram is envisioned as constructed sequentially, it should be noted that there is already *pattern-matching* activity here: we are matching the parts of the equation to the parts of the diagram.

Now, in practice it is hard to distinguish the recognition that the picture matches the equation for n = 5 from the realization that the theorem is indeed true for all n. This is because once one grasps the *pattern* of this particular diagram, the general pattern is usually immediately obvious. But there are a couple of logical (if murky phenomenological) steps to be distinguished. To move from the particular case pictured (here n = 5) to the general theorem the viewer must generalize the diagram itself. The generalization of the diagram is thus a kind of inference about it – that the diagram, or its type, supports the right and left hand sides of the general theorem. How might one so infer?

Recall that the plausible second recognition step was to match the portions of the particular diagram to elements of the equation, spelled out on its left hand side, for the case n = 5. A third plausible recognition step is to generalize this second step. Thus the parts of the diagram *and any possible future instantiations* must be correlated with the general pattern of the quantities on the left hand side of the equation. For this example, each succeeding quantity includes all the former cases plus the successor case. So the sequence is: 1, 1 + 2, 1 + 2 + 3, and so on. To see this "and so on" *diagrammatically* one must be able to envision a sequence of such diagrams, noting that in constructing a given diagram, one always adds n squares to the diagram representing the prior case of n−1, whatever n is. This step can be thought of as seeing the generalization of the left hand side in the diagram.

Finally, one must also generalize the first step – the overall shape recognition – in order to connect the diagram to the right hand side of the equation, $n^2/2 + n/2$. That is, no single picture can represent $n^2/2 + n/2$; each individual diagram is of a particular case of n. Thus for the diagram to be grasped as supporting the theorem, the viewer must somehow generalize the right-hand side of the equation as well as the left-hand side. For this, one sees that every iteration of the diagram-sequence preserves the general shape properties it started with: in this case, the staircase shape.

The third and fourth steps, the generalization steps, are essential to understanding why the picture carries normative weight – why it contributes to a *justification* of, rather than a mere *cause of a belief* in, the general claim. As I have said, this normativity is important to the view that such pictures are tools of genuine *understanding* as opposed to mere psychological aids.

I wish to further emphasize two things in connection with the last two recognition steps. First, it is very helpful, perhaps even essential, to envision the diagram as being built in a sequence of construction steps for the generalization-recognition. These construction steps match the sequence of quantities (explicit and implicit) on the left hand side of the equation. Thus, even though the typical picture associated with this theorem is singular, what enables us to understand it as (at least partially) justifying the general result is our ability to envision it as a *sequence* of diagrams – indeed an indefinite iteration of such diagrams.

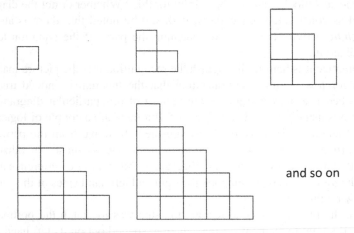

and so on

As stated earlier, whether each instance is viewed as adding a diagonal *layer* of "steps" to the previous iteration, or a vertical *column* of squares to the left of the previous iteration, seeing the diagram as built sequentially seems essential to generalizing it. The indefinite iteration of such diagrams, then, matches the indefinite iteration of quantities on the left hand side of the equation.

Second it is the implicit induction on this sequence of diagrams that justifies the result. One must induce to the general conclusion that for all n, the diagram retains the same shape. And this depends on seeing the constructive relationship between one diagram and the next: that when you construct the $(n + 1)$th case from the nth case (by adding $n + 1$ squares to the nth construction) the shape-type remains the same.

That inductive reasoning on the (perhaps imagined) sequence of diagrams is necessary to back up the general inference means that the lone diagram does not serve as a (full) *proof*.[35] Nor, as I have indicated, is it necessarily better than a proof. However, our question is not whether diagrams *are* proofs, or are *better than* proofs. Our question is more broadly epistemic: whether, *despite not being a proof*, the diagram nevertheless has a normative, justifying role – a role less than that of a proof but more than that of a mere cause, or psychological aid. And further, if the diagram has more epistemic power than a mere psychological aid, from whence does it derive this power?

The answer I have proposed is that such images and diagrams can have normative power, and this power comes from the fact that *they share critical structural properties* with the facts or theorems they represent (as well as with their verbal/symbolic representations). Thus, such diagrams need not be mere *causes*. Though their justifying power falls short of proof, the structuralist perspective provides a good explanation for the justifying power, and therefore the epistemic status, that they do have.

This power, or status, also explains why visual tools are so useful for producing mathematical understanding. That is, because the structuralist perspective can account for the epistemic (v. psychological) power of certain diagrams, it also accounts for their role in promoting understanding. In the case considered here, understanding is produced by seeing that the sequence and pattern of the geometric object *match* the sequence (left hand side) and result (right hand side) of the arithmetic equation. Seeing that two sequences are the same, despite differences of content (visual/spatial vs. symbolic/numerical) involves recognizing their shared structure. Thus the structuralist perspective explains a commonly held view about the importance of visualization, and the power of images, in mathematics, and their effectiveness as a tool for promoting mathematical understanding.[36]

---

[35]Contrary to the views of Nelsen, Brown and others.

[36]I believe other examples can be analyzed similarly to explain why such pictures carry justifying weight. For example, a similar analyses can be given regarding the standard spatial representation of $1 + 3 + 5 \ldots + 2n\text{-}1 = n^2$, or that of the Zeno series (though these further examples are beyond the scope of this paper). As with the case here analyzed, I would argue that the normative status of the diagram depends on shared structural features: that key properties of the picture match key elements of the relevant equations. The shared structure is what explains why they increase our understanding as well as why we take them to reveal the facts (rather than taking them to be mere causes of beliefs that perhaps happen to be true).

8.7 Conclusion

For reasons given above, I do not believe it is possible to articulate a precise *theory* of mathematical understanding. Nevertheless, despite the fact that it is a sprawling, "family resemblance" type of concept, we can improve our understanding of it. I have proposed the structuralist perspective as a framework that assists in this effort. Some abilities connected with understanding clearly carry epistemic weight, such as finding proofs and giving explanations. These are not particularly puzzling. What structuralism contributes is a broad perspective that helps make sense of why other abilities and tools – such as the use of pictures, analogies, seeing relations between different areas of mathematics, perceptions of depth and beauty, etc. – also have epistemic force, and thus why they are relevant for mathematical understanding.[37]

References

Avigad, J. 2008. Understanding proofs. In *The philosophy of mathematical practice*, ed. Mancosu, 317–353. Oxford: Oxford University Press.

———. 2010. Understanding, formal verification, and the philosophy of mathematics. *Journal of Indian Council of Philosophical Research* XXVII (1): 165–197.

Benacerraf, P. 1965. What numbers could not be. *Philosophical Review* 74: 47–73 Reprinted in Benacerraf and Putnam eds., 1983, pp. 272–294.

Benacerraf, P., and H. Putnam. 1983. *Philosophy of mathematics*. Englewood Cliffs: Prentice-Hall.

Bolzano, B. 1810/1996. Contributions to a better-grounded presentation of mathematics. In *From Kant to Hilbert, a Source Book in the Foundations of Mathematics*, ed. William Ewald, vol. 1, 174–224. Oxford: Clarendon Press.

Brown, J.R. 2008. *Philosophy of mathematics: A contemporary introduction to the world of proofs and pictures*. New York: Routledge.

Ewald, W., ed. 1996. *From Kant to Hilbert, a source book in the foundations of mathematics (volumes I and II)*. Oxford: Clarendon Press.

Feferman, S. 2012. And so on . . . : reasoning with infinite diagrams. *Synthese* 186: 371–386.

Frege, G. 1884. *The foundations of arithmetic*. Trans. J.L. Austin. Evanston: Northwestern University Press, 1968.

Friedman, M. 1974. Explanation and scientific understanding. *The Journal of Philosophy* 71: 5–19.

Giaquinto, M. 2007. *Visual thinking in mathematics: An epistemological study*. Oxford: Oxford University Press.

Grosholz, E., and H. Breger, eds. 2000. *The growth of mathematical knowledge*. Kluwer: Boston.

---

[37] I am grateful to an anonymous referee for some constructive feedback on a written draft of this paper, though I fear I have sidestepped rather than answered the hard questions. I am also indebted to participants at two conferences, for the opportunity to try out some of these ideas: the Logic Colloquium 2015 (joint meeting of the *Association for Symbolic Logic* and the *Conference of Logic, Methodology and Philosophy of Science*), at the University of Helsinki, and FilMat 2016 (second international conference of the Italian network for the philosophy of mathematics), at the University of Chieti-Pescara. Finally, thanks go to the editors of this volume for all of their hard work.

Hafner, J., and P. Mancosu. 2005. The varieties of mathematical explanation. In *Visualization, explanation and reasoning styles in mathematics*, ed. P. Mancosu et al., 215–250. Berlin: Springer.

Hardy, G.H. *A mathematician's apology* (London 1941). References to the electronic version. http://www.math.ualberta.ca/mss/.

Haylock, D., and A. Cockburn. 2008. *Understanding mathematics for young children*. Thousand Oaks: Sage Publications.

Hellman, G. 1989. *Mathematics without numbers*. Oxford: Oxford University Press.

Kitcher, P. 1989. Explanatory unification and the causal structure of the world. In *Scientific explanation*, ed. P. Kitcher and W.C. Salmon, 410–505. Minneapolis: University of Minnesota Press.

Kitcher, P., and W. Salmon, eds. 1989. *Scientific explanation*. University of Minnesota Press: Minneapolis.

Lange, M. 2009. Why proofs by mathematical induction are generally not explanatory. *Analysis* 69: 203–211.

Maclaurin, C. 1742. A Treatise of fluxions. Excerpted In: *From Kant to Hilbert, a source book in the foundations of mathematics (volumes I and II)*, ed. W. Ewald (1996), 95–122. Oxford: Clarendon Press.

Mancosu, P. 2000. On mathematical explanation. In *The growth of mathematical knowledge*, ed. E. Grosholz and H. Breger, 103–119. Dordrecht: Kluwer.

———. ed. 2008. *The philosophy of mathematical practice*. New York: Oxford University Press.

———. Explanation in mathematics. In *The Stanford Encyclopedia of Philosophy* (Summer 2015 edn., ed. Edward N. Zalta). https://plato.stanford.edu/archives/sum2015/entries/mathematics-explanation/.

Mancosu, P., K. Jorgensen, and S. Pedersen, eds. 2005. *Visualization, explanation and reasoning styles in mathematics*. Dordrecht: Springer.

Michener, E.R. 1978, August. *Understanding mathematics*. A.I. Memo – 488, LOGO Memo – 50, MIT Artificial Intelligence Lab document.

Nelsen, R.B. 1993. *Proofs without words*. Washington, DC: The Mathematical Association of America.

Netz, R. 2005. The aesthetics of mathematics: a study. In *Visualization, explanation and reasoning styles in mathematics*, ed. P. Mancosu, K.F. Jørgensen, and S.A. Pedersen, 251–293. Dordrecht: Springer.

Poincaré H. 1900. Intuition and logic in mathematics. In: *From Kant to Hilbert, a source book in the foundations of mathematics (volumes I and II)*, ed. W. Ewald (1996), 1012–1020. Oxford: Clarendon Press.

———. 1908. *Science and method*, authorized translation by G. B. Halsted in *The foundations of science*. Washington, DC: The University Press of America, 1982.

Porteous, K. 2008. Understanding mathematics. *Philosophy of Mathematics Education Journal*, 23. http://socialsciences.exeter.ac.uk/education/research/centres/stem/publications/pmej/pome23/index.htm.

Resnik, M. 1997. *Mathematics as a science of patterns*. Oxford: Oxford University Press.

Rota, G. 1997. The phenomenology of mathematical beauty. *Synthese* 111: 171–182.

Šebestik, Jan. Bolzano's logic. In *The Stanford Encyclopedia of Philosophy* (Spring 2016 edn., ed. Edward N. Zalta). https://plato.stanford.edu/archives/spr2016/entries/bolzano-logic/.

Shapiro, S. 1997. *Philosophy of mathematics: Structure and ontology*. New York: Oxford University Press.

———. undated. Mathematical structuralism. *The Internet Encyclopedia of Philosophy*. ISSN 2161-0002. http://www.iep.utm.edu/.

Steiner, M. 1978. Mathematical explanation. *Philosophical Studies* 34: 135–151.

Tappenden, J. 2005. Proof style and understanding in mathematics I: Visualization, unification and axiom choice. In *Visualization, explanation and reasoning styles in mathematics (Synthese library, volume 327)*, ed. P. Mancosu, K.F. Jørgensen, and S.A. Pedersen, 147–214. Dordrecht: Springer.

Thurston, W. 1994. On proof and progress in mathematics. *Bulletin of the American Mathematical Society* 30 (2): 161–177.

———. short essay in article. 2010. Mathematics meets fashion: Thurston's concepts inspire designer. *AMS News Releases*, April 1.

Wittgenstein, L. 1972. *On certainty*. New York: Harper and Row.

Chapter 9
The Explanatory Power of a New Proof: Henkin's Completeness Proof

John Baldwin

Mancosu writes

> But explanations in mathematics do not only come in the form of proofs. In some cases explanations are sought in a major recasting of an entire discipline. (Mancosu 2008, 142)

This paper takes up both halves of that statement. On the one hand we provide a case study of the explanatory value of a particular milestone proof. In the process we examine how it began the recasting of a discipline.

Hafner and Mancosu take a broad view towards the nature of mathematical explanation. They argue that before attempting to establish a model of explanation, one should develop a 'taxonomy of recurrent types of mathematical explanation' (Hafner and Mancosu 2005, 221) and preparatory to such a taxonomy propose to examine in depth various examples of proofs. In Hafner and Mancosu (2005, 2008), they studyf deep arguments in real algebraic geometry and analysis to test the models of explanation of Kitcher and Steiner. In their discussion of Steiner's model they challenge[1] the assertion (Resnik and Kushner 1987) that Henkin's proof of the completeness theorem is explanatory, asking 'what the explanatory features of this proof are supposed to consist of?' As a model theorist the challenge to 'explain' the explanatory value of this fundamental argument is irresistible. In contrasting

---

[1] They write, '[In] Resnik and Kushner (1987, p. 147), it is contended with some – albeit rather vague – reference to mathematical/logical practice that Henkin's proof 'is generally regarded as really showing what goes on in the completeness theorem and the proof-idea has been used again and again in obtaining results about other logical systems?'

J. Baldwin (✉)
Department of Mathematics, Statistics, and Computer Science, University of Illinois at Chicago, Chicago, IL, USA
e-mail: jbaldwin@uic.edu

© Springer International Publishing AG, part of Springer Nature 2018   147
M. Piazza, G. Pulcini (eds.), *Truth, Existence and Explanation*,
Boston Studies in the Philosophy and History of Science 334,
https://doi.org/10.1007/978-3-319-93342-9_9

the proofs of Henkin and Gödel, we seek for the elements of Henkin's proofs that permit its numerous generalizations. In Sect. 9.2 we try to make this analysis more precise through Steiner's notion of characterizing property. And as we will see, when one identifies the characterizing property of the Henkin *proof*, rather than a characteristic property of an object in the *statement* of the theorem, then one can find a variant of Steiner's model which applies in this situation. Key to this argument is identifying the family in which to evaluate the proof. We point out in Sect. 9.3 that this modification of Steiner depends on recognizing that 'explanatory' is not a determinate concept until one fills in the 'who' in 'explanatory for who'. Thus, our goal is to establish that the Henkin proof is explanatory (contra Hafner and Mancosu 2005) and moreover one can adapt Steiner's model to justify this claim.

This paper developed from a one page treatment in Baldwin (2016) and the discussion of Henkin's role in the transformation of model theory in Baldwin (2018). I thank Juliette Kennedy and Michael Lieberman for their comments on various drafts. And I thank Rami Grossberg and Jouko Väänänen for sources on the consistency property and pressing the important contributions to understanding the completeness theorem of Hintikka, Beth, and Smullyan.

9.1 Comparing Gödel and Henkin on Completeness

There are two issues. In a comparative sense, how is Henkin's proof more explanatory than Gödel's? In an absolute sense why would one say that Henkin's proof is explanatory? We begin with the first. For this we analyze the different though equivalent statements of the theorem. We will see that one source of the greater explanatory value of Henkin's argument is simply his statement of the result. Then we examine the actual proofs and come to a similar conclusion. An historical issue affects the nomenclature in this paper. While there are some minor variants on Gödel's argument,[2] Henkin's proof has become a motif in logic. Some of the essential attributes of what I am calling Henkin's proof were not explicit in the publication; the most important were enunciated a bit later by Henkin; others arise in the long derivative literature.

9.1.1   Comparing the Statements

Gödel's version of the completeness theorem for first order logic (Gödel 1929) reads:

---

[2]Robinson (1951) and Kreisel and Krivine (1967).

Theorem 1.1.1 (Gödel formulation) *Every valid formula expressible in the restricted functional calculus*[3] *can be derived from the axioms by a finite sequence of formal inferences.*

But Henkin states his main theorem as

Theorem 1.1.2 (Henkin formulation) *Let S_0 be a particular system determined by some definite choice of primitive symbols.*

If Λ is a set of formulas of S_0 in which no member has any occurrence of a free individual variable, and if Λ is consistent then Λ is simultaneously satisfiable in a domain of individuals having the same cardinal number as the set of primitive symbols of S_0.

We draw three distinctions between the two formulations. Gödel (1929, 75) restates the result to be proved as: '*every valid logical expression is provable*' and continues with the core statement, 'Clearly this can be expressed as *every valid expression is satisfiable or refutable*.' This 'clearly' is as close as Gödel's *proof* comes to a definition of *valid*. More precisely, the effective meaning of 'ϕ is valid' in Gödel's paper is '$\neg\phi$ is not satisfiable' and this double negation is essential. We discuss below the connections between his notion of valid and Tarski's. Henkin makes Gödel's core assertion the stated theorem; the transfer to Gödel's original formulation is a corollary. Thus Henkin's proof gains explanatory value as the argument *directly* supports the actual statement of the theorem.

The last paragraph of Gödel (1929) extends the argument to *applied logic*. Henkin's 'definite choice of primitive symbols' amounts to fixing an applied axiom system. But still the full list of possible relation variables is present in Gödel's context while the modern scheme introduces additional relations only when needed and thus for Henkin, not at all; only constant symbols are added. Thus, he makes the focus what becomes the basic stuff of model theory, first order logic in fixed vocabulary.

Finally, as Gödel observes, his argument is restricted to countable vocabularies; Henkin proves the results for uncountable languages.

As Franks (2014) emphasizes, Gödel make a huge innovation in focusing on a duality between proof and truth rather than defining completeness in terms of 'descriptive completeness' (Detlefsen 2014), or Post completeness,[4] or 'as all that can be proved[5] (by any means)'. Henkin refines this insight of Gödel.

---

[3]The term 'functional variables' used by both Gödel and Henkin are what are now thought of as relational variables (or relation symbols) and the functions that interpret them are relations on a set. They are not functions from the universe of a model to itself.

[4]A set of sentences Σ is Post complete if for every sentence ϕ either $\Sigma \cup \{\phi\}$ or $\Sigma \cup \{\neg\phi\}$ is inconsistent.

[5]Franks (2010) quotes Gentzen's goal as follows, 'Our formal definition of provability, and, more generally, our choice of the forms of inference will seem appropriate only if it is certain that a sentence q is 'provable' from the sentences $p_1, \ldots p_v$ if and only if it represents informally a consequence of the p's. (Gentzen 1932, p. 33). This is an entirely syntactic conception of completeness and so distinct from Gödel.

There is no objection below to the correctness of Godel's proof. There is an objection (not original here) to the casual reading that Gödel proved the theorem in the form (*) below. We draw two lessons from the difference in formulation.

Lesson 1: What hath Tarski wrought? The modern form[6] of the extended completeness theorem reads: for every vocabulary τ and every first order sentence $\phi \in \mathcal{L}(\tau)$

$$(*) \qquad \Sigma \vdash \phi \text{ if and only if } \Sigma \models \phi.$$

is possible only on the basis of Tarski (1935, 1956). It requires a formal definition of logical consequence, $\models$. A conundrum that is often raised but seldom seriously asked is, 'Why did Tarski have to define truth if Gödel proved the completeness theorem earlier?' This query seems to rest on the thought that what Gödel meant by a quantified sentence being true in a structure is just the informal notion generated by the intuition that an existential quantification is true if there is a witness. Indeed on page 69 of Gödel (1929) he says that a formula of the (pure) predicate calculus (that is, there are no fixed interpretations of any relation symbols) is satisfiable in a domain if one can choose relations, constants (and values of propositional variables) on that domain so that the substitution is true in the domain. Nothing is said about how the bound variables in the expression are to be unwound to explicate 'true in the domain'. Further he defines ϕ is valid in a domain I as $\neg \phi$ is not satisfiable in I. That is, for ϕ with relation variables $F_1, \dots F_k$, there is no choice of relations $f_1, \dots f_k$ on I such that $\neg \phi(f_1, \dots f_k)$ is true. He indicates that valid means valid in all domains.

However, in the *actual proof* he only (a) defines truth for atomic formulas in a structure and (b) precisely constructs a satisfying domain for a π_2-sentence that cannot be refuted.

Gödel (1929) writes, 'Here, completeness is to mean that every valid formula expressible in the restricted functional calculus ... can be derived from the axioms by means of a finite sequence of formal inferences.' As we've seen 'ϕ is valid' is taken as '$\neg \phi$ is not satisfiable'. He rephrases the theorem as follows in the published version (Gödel 1930),

> ...the question at once arises whether the initially postulated system of axioms and principles of inference is complete, that is, whether it actually suffices for the derivation of *every* logicomathematical proposition, or whether, perhaps it is conceivable that there are true propositions (which may be even provable by means of other principles) that cannot be derived in the system under consideration. (Gödel 1930)

---

[6]Apparently the first statement of the 'extended completeness theorem' in this form is in Robinson (1951) (Compare Dawson 1993, p. 24). When $\models$ was introduced is unclear. A clear statement recommending its use appears in the preface to Addison et al. (1965) but earlier published uses are hard to find.

In contrast, Henkin gives a semi-formal definition of 'satisfiable in a domain for an interpretation of the relation symbols on I and assignment[7] of variables to elements of I' (by an induction on quantifier complexity). He extends to 'valid in a domain' (satisfied by every assignment), and then defines valid as valid in all domains. He remarks in a footnote that this notion could be made more precise along the lines of Tarski (1935). That is, working in naive set theory, he uses Tarski's inductive definition of truth.

Thus Gödel's formulation of the completeness theorem varies from the post-Tarski version in two ways.

1. Gödel's use in the proof of the notion 'satisfiability in a structure' depends on the ambient deductive system. Specifically, the deductive system must support the existence of a π_2-prenex normal form for each non-refutable sentence. We will observe below various logics satisfying completeness theorems that fail this condition.
2. Gödel does not assert the modern form (*) of the extended completeness theorem (he hasn't defined semantic consequence). Rather he says (Theorem IX), 'Every denumerably infinite set of formulas of the restricted predicate calculus either is satisfiable (that is, all formulas of the system are simultaneously satisfied) or some finite subset is refutable.'

He proves it by using the compactness theorem for *countable* sets of formulas (implicitly in Gödel 1929 and explicitly[8] in Gödel 1930).

Lesson 2: The modern concept of vocabulary. Gödel's emphasis is on the provability of every *valid* formula in the restricted predicate calculus. This calculus has infinitely many relation symbols of each finite arity. In contrast, modern model theory specifies a vocabulary using primitive symbols directly connected to the specific subject. Many mathematicians had adopted this practice (e.g., Pasch, Hilbert, Noether, van der Warden) as geometry became more carefully formalized and notions of groups, rings, and fields developed in the first third of the twentieth century. But the incorporation of this requirement into logic took place later.[9]

Henkin's proof of the completeness theorem was a crucial step towards the modern conception of vocabulary.[10] In general he specifies that a system contains 'for each number $n = 1, 2, \ldots$ a set of functional symbols (relation symbols[11]) of degree n which may be separated into variables and constants'. Henkin makes the modern convention of a fixed vocabulary (e.g. symbols $+, \times, 0, 1$ for rings)

---

[7] $(\forall x) A(x)$ is satisfiable just if $A(a)$ is true for each $a \in I$.

[8] This historical distinction has been emphasized by Franks in (2014).

[9] The abstract notion of 'structure' (for a given vocabulary) was first formalized in 1935 by Birkhoff (1935); both Tarski (1946) and Robinson (1952) refer to that paper. Robinson specifies a vocabulary for the particular topic. Tarski certainly has arrived at the modern formulation by Tarski (1954) and Tarski and Vaught (1956).

[10] I use the word vocabulary; similarity type or one use of 'language' are synonyms.

[11] The article is part of his five year project of revising and publishing his thesis on both second order logic and the theory of types; the function variables appear to accommodate these extensions.

completely explicit in Henkin (1953), which is the published version of the 'algebra' portion of his 1947 thesis, where he used the old notation. In contrast to modern practice, both authors treat the extensional treatment of equality as an 'add-on'; since they are dealing with a purely relational language they avoid the complication of including in the axioms the requirement that equality is a congruence for each function symbol in the language. In Henkin (1953), he applies the completeness theorem to theories with function symbols and equality with no explanation of the shift from a relational language. Modern versions of Henkin's proof require that the equivalence relation of equality is a *congruence* (preserved by operations of the vocabulary). In fact, this extension turns out to be central, for example, in the discussion of the omitting types theorem below.

A key distinction[12] between Gödel and Henkin is that Henkin's proof adds only constants to the vocabulary while to settle the question for a particular sentence Gödel draws on finitely many relation symbols that do not appear in the given sentence.

Before stating the theorem, Henkin restricts the context on page 161 of Henkin (1949) with 'Let S_0 be a particular system determined by some definite choice of primitive symbols.' This apparently minor remark is central to changing the viewpoint from logic as an analysis of reasoning to model theory as a mathematical tool. Henkin emphasizes this aspect of a second difference between his description of the setting from Gödel's.

> In the first place an important property of formal systems which is associated with completeness can now be generalized to systems containing a non-denumerable infinity of primitive symbols. While this is not of especial interest when formal systems are considered as logics–i.e., as means for analyzing the structure of languages– it leads to interesting applications in the field of abstract algebra. (Henkin 1949)

Henkin (Corollary 2) uses the uncountable vocabulary to deduce the full force of the Löwenheim-Skolem-Tarski theorem: a consistent first order theory has models in every infinite cardinality.[13]

9.1.2   Contrasting the Proofs

Hafner and Mancosu (2005) list a number of mottos for explanation that they found in the mathematical literature. Simplifying a bit, they center around the notion of 'deep reason'. In this section we try to identify the 'deep reason' behind Henkin's proof; in the next we fill this out with further explication of its generalizability. Here are outlines of the two proofs.

---

[12]Henkin's procedure preserves for example ω-stability of the initial theory and that must fail for some theories using Gödel's proof.

[13]Note that the proof of completeness for countable vocabularies uses only König's infinity lemma (Smullyan 1966), while the existence of arbitrarily large models via Henkin's argument uses the Boolean prime ideal theorem and the full Löwenheim-Skolem-Tarski theorem requires the axiom of choice.

Gödel:

G1.   Citing the 1928 edition of Hilbert and Ackermann (1938), Gödel notes that an arbitrary formula may be assumed to be in prenex normal form, π_n. That is, n alternating blocks of universal and existential quantifiers followed by a quantifier-free matrix.

G2.   By adding additional relation symbols, Gödel Skolemizes[14] the entire logic and makes every formula equivalent to a π_2-formula.[15]

G3.   Then he shows that every π_2-formula is either refutable or satisfiable in a countably infinite structure.[16]

Note that steps (1) and (2) are entirely syntactic and are in fact theorem schemas – patterns for proofs within the system.

Henkin:

I describe Henkin's proof in more detail using some of the expository enhancements (e.g. the term 'Henkin theory') that have been made in the more than half century the argument has been the standard. T is a *Henkin theory* if[17] for every formula $\phi(x)$, there is a *witness constant* c_ϕ such that

$$T \vdash (\exists x)\phi(x) \rightarrow \phi(c_\phi).$$

H1.   Every syntactically consistent theory can be extended to a Post-complete Henkin theory.

H2.   Every Post-complete syntactically consistent theory with the witness property has a canonical *term model*.

Henkin's vital insight is the separation of the problem into two parts, (1) extend the given theory to one that is complete and satisfies certain additional syntactic properties such that (2) there is a functor from the theories in H1 to structures that realize the original theory. Both of these steps are metamathematical. We report below Henkin's account of his roundabout way of arriving at the necessity of this extension of a theory to a larger one.

Henkin's proof explains the actual argument: finding a model of an irrefutable sentence. The two steps each contribute to that goal.

H1. The extension of an arbitrary consistent T to one satisfying the witnessing property depends precisely on the axioms and rules of inference of the logic.

---

[14]See   http://mathoverflow.net/questions/45487/compactness-theorem-for-first-order-logic   for Blass's outline of a proof using Skolem functions and reduction to propositional logic (nominally for compactness).

[15]Contrary to contemporary terminology where Skolem implies Skolem *function*, each formula $\phi(\mathbf{x}, \mathbf{y})$ is replaced by a relation symbol $F_{\phi(\mathbf{x},\mathbf{y})}(\mathbf{x}, \mathbf{y})$ such that $(\exists \mathbf{x}, \mathbf{y}) F_{\phi(\mathbf{x},\mathbf{y})}(\mathbf{x}, \mathbf{y}) \wedge \forall \mathbf{x}[(\exists \mathbf{x}, \mathbf{y})\phi(\mathbf{x}, \mathbf{y}) \rightarrow (\exists \mathbf{y}) F_{\phi(\mathbf{x},\mathbf{y})}(\mathbf{x}, \mathbf{y}) \wedge \phi(\mathbf{x}, \mathbf{y})]$. This difference is why he can reduce only to π_2 and not universal formulas.

[16]In fact, he constructs the model as a subset of the natural numbers using the arithmetic of the natural numbers for the construction.

[17]Chang and Keisler (1973) say 'has the witness property'.

The Post-completeness is obtained by an inductive construction: add ϕ_α or $\neg\phi_\alpha$ at stage α to ensure that each sentence is decided.

H2. To construct the model, consider the set of witnesses, M and show that after modding out by the equivalence relation cEd if and only $T \vdash c = d$, the structure $M' = M/E$ satisfies T. More precisely, show by induction on formulas that for any formula $\phi(\mathbf{c})$,

$$T \vdash \phi(\mathbf{c}) \text{ if and only if } M' \models \phi(\mathbf{c}).$$

For this he uses both the Post-completeness and the witness property from H1.

Karp (1959) generalized the completeness result to the infinitary logic[18] $L_{\omega_1,\omega}$. This argument has been the source of many generalizations of the result with names that emphasize the goal of step H2 while modifying the argument in H1 to obtain an appropriate 'Henkin theory', such as the *consistency property*[19] (Smullyan 1963; Makkai 1969) and with the theorem renamed as *the model existence theorem* in Keisler (1971).

My original impetus (Baldwin 2016) for discussing Henkin's proof was to give a serious example of how mathematical induction is used in abstract mathematics. There are too many investigations to list of 'explanation' that center around the extremely elementary uses of mathematical induction. But in more advanced mathematics the main use of induction is as proof tool to study objects defined by generalized inductive definition. This includes not only such algebraic constructions as the closure of a set to a subgroup or in a logic, the set of formulas in logic or theorems of a theory, but constructions as in the Henkin proof: truth in a structure, completing a theory and fulfilling the witness property. These constructions are followed by the complementary proof by induction that the canonical structure is in fact a model. A similar pattern of inductive definition is central in Gödel's proof. But, more is on the syntactic side; it is presented as metatheorems on prenex normal form and equivalences between formulas. However, still another inductive definition is a piece of his proof that a π_2-sentence is satisfiable.

There are several fundamental distinctions between the foundational outlooks of Gödel and Henkin.

> Gödel works in a background theory of naive set theory and studies a single system of logic with predicates of arbitrary order; this is essential to the proof. He has a definition of truth for *atomic* formulas, which is extended to π_2 sentences and then by deductive rules to determine truth in a structure for arbitrary sentences.

---

[18]The logic allows countable conjunctions but quantifies over only finite sequences of variable. Proofs can have infinitely many hypotheses.

[19]The name 'consistency property' is apparently introduced in Smullyan (1963). Smullyan remarks that his method takes from Henkin that only constants need to be added to the original vocabulary; Makkai, who is proving preservation and interpolation theorems, carefully lays out the connection of his proof with Smullyan (1963).

Henkin (by 1951) works in a background theory of naive set theory and studies the first order logic of each vocabulary. The proof for each vocabulary adds *only constant* symbols. He has a *uniform definition of truth* in a structure for each vocabulary that has no dependence on the deductive rules of the logic.

This second view underlies modern model theory. While technically, one could incorporate Gödel's argument into the modern framework (by adding the additional predicates ad hoc) this is not only cumbersome but raises the question of what these new predicates have to do with the original topic – something more to explain.

9.2 Generalization and Steiner

In Steiner's seminal (1978), he first rejects the view that 'a proof is more explanatory than another because more general' by specifying more carefully what is involved in an explanatory generalization. Steiner proposed the notion of *characterizing property* to clarify 'explanatory'.

> We have, then, this result: an explanatory proof depends on a characterizing property of something mentioned in the theorem: if we 'deform' the proof, substituting the characterizing property of a related entity, we get a related theorem. A characterizing property picks out one from a family ['family' is undefined in the essay]; an object might be characterized variously if it belongs to distinct families. (Steiner 1978, 147)

Resnik and Kushner assess Henkin's proof as follows"

> The proof is generally regarded as really showing what goes on in the completeness theorem and the proof-idea has been used again and again in obtaining results about other logical systems. Yet again, it is not easy to identify the characterizing property on which it depends. (Resnik and Kushner 1987)

We elaborate below what we think is the source of the 'generally regarded'. To do this, we modify Steiner's notion. He notes on page 143 that the characteristic property refers to a property of the theorem, not the proof, and thus is an absolute rather than relative evaluation of the explanatory value of the proof. This requirement seems not to fit the statement of the completeness theorem where both Godel's statement and the modern statement (*) obscure the crucial point: the construction of a countermodel. Thus, we require that the characterizing property should not be required to be something 'mentioned in the theorem[20]' but of something mentioned in the proof or the theorem. Thus we modify Steiner's model to require the existence of a characterizing property of a proof that appears in a family of arguments to qualify a proof as 'explanatory'. The characterizing property should distinguish this argument from others.

There is an immediate division into two families of proofs of the completeness theorem. In the Gödel-Henkin style the main lemma is a proof that if $\neg\phi$ is

---

[20]One might argue that the deformable objects are the proof systems and the notion of model. But this still seems to miss the crux of Henkin's argument.

not refutable then ϕ has a model. In contrast the main lemma of the Herbrand style asserts that if $\neg\phi$ is not satisfiable then ϕ is provable. Smullyan (1963) introduces the notion of a consistency property, from which he can construct proofs in each style of the completeness theorem.[21] A consistency property Γ is for him a collection of finite sets of sentences such that Γ is closed under a set of operations (e.g. adding a Henkin witness). He shows any consistency property produces a model and distinguishes his argument from Henkin (Footnote 13) because there is no requirement that every sentence is decided. The Herbrand style was developed independently by Beth and Hintikka as the method of semantic tableaux or model sets. Smullyan (1966) introduces his method of *analytic tableaux*, which generalizes and acknowledges (Hintikka 1955) and (Beth 1959). In this line a (natural deduction) proof system is developed such that a proof of ϕ is proved to terminate if ϕ is valid because if the proof does not terminate a model of $\neg\phi$ is constructed. We discuss the Herbrand style no further although further work on this topic would be valuable; our goal here is to distinguish within the Gödel-Henkin style.

Gödel's proof certainly did not arise as a generalization. But, it was clear from the first that first order logic (the restricted functional calculus) admitted of many variants in the formulation of a deductive system: one feature of the theorem is as a precise statement of the equivalence of these variants. The very proof of Gödel was built on Bernays' proof of completeness of propositional calculus. So the idea that distinct logics could be complete was there. Unlike Gödel, Henkin, as he completed the work, already made generalizations to other logics; he proves completeness for three logics: first order, the interpretation of 2nd order with the Henkin semantics, and a theory of types of finite order, again with respect to the Henkin semantics. In the insightful (Henkin 1996), published nine years after (Resnik and Kushner 1987), Henkin explains the relationship between the three results. He reports that he worked for more than a year on a related issue: trying to show (roughly speaking[22]) that in Church's theory of types there was no uniform way to assign a choice function for non-empty sets of reals. After he had almost given up, he realized the key idea of inductively extending both the axioms as well as the collection of functions named. His insight that *both* the axioms and the names must be extended led to the proof of the three completeness theorems. While in the unsolved[23] target problem, functions were to be named, when writing up the completeness results (Henkin 1996, 155) he discovered that in both the first order and the finite type case, he could add only constants – a crucial point for later model theory. Still later, he realized, as is now standard, that the constants can be added first and only their properties defined in the induction.

We observed above that Henkin's vital insight is the separation of the problem into two parts: (1) extend the given theory to one satisfying conditions that (2) permit

---

[21] They are applications B for Gödel-Henkin and C for Herbrand on page 829 of Smullyan (1963).

[22] A more detailed description is given on page 148 of Henkin (1996).

[23] It developed, by work of Gödel and Feferman, that the conjecture Henkin had been attacking was independent.

the construction of a canonical model. Both the sufficient conditions on the base theory and the conditions on the complete theory that supports the construction are quite general and adaptable. In this generality we see that it applies not only to proving completeness for a family of logics but to the other uses of the model construction technique.

Here are the first order conditions. Henkin and later adapters can quickly state rules and axioms for a logic; the key for H1 is to isolate rules that prove that if T is consistent then for any formula $\phi(x)$ and any constant c that does not appear in T, $T \cup \{(\exists x)\phi(x) \rightarrow \phi(c)\}$ is consistent. For H2 the logic must satisfy equality axioms guaranteeing that the equality relation is a congruence in the sense of universal algebra.

In fact, later authors reverted to Skolem's standpoint[24] and give completely semantic versions of the Henkin construction, solely for convenience in Marker (2002) and as a step towards a completeness theorem for infinitary logic in Keisler (1971).

Resnik and Kushner (1987) 'think (though in correspondence Steiner disagrees) that the proof does not make clear that when we apply the proof-idea to second order logic, we must change the sense of model to allow for non-standard models nor that when we apply it to modal logics, we must use many maximally consistent sets, etc.' This seems to misunderstand the effect of Henkin semantics for second order logic. The Henkin semantics interpret second order as a many-sorted first order system and then a similar proof applies. Clearly, there is no completeness theorem for second order logic with the full semantics.

From the view point of a mathematician-model theorist, the witness property appears to be key. But from the standpoint of modal and intuitionistic logic there are a different set of properties of 'propositional logics' which are key. For example, De Jongh[25] identifies the crucial syntactic property to get a canonical model for intuitionistic propositional logics as the 'disjunction property'. As Resnik and Kushner (1987) noted, the concept of model has widened from a first order structure to a Kripke frame; this only emphasizes the depth of Henkin's innovation. Similarly in Cintula and Noguera (2015) the authors stress the role of the *term model* in extending the Henkin proof to many-valued logics and in general to algebraizable logics. Intuitionistic logics do not have a prenex normal form theorem so Gödel's proof could not be adapted to this case.

The characterizing property of the Henkin proof is the systematic extension of the given theory to a complete theory whose canonical model satisfies the original theory. In order to apply Steiner's test, one must step far enough back to recognize the characteristic property. As we will see below as we reformulate this property to

---

[24]Gödel refers to the great similarity between his argument and (Skolem 1967); the distinction is that Skolem ignores the deductive standpoint. See the notes to Gödel (1929).

[25]Slide 27 of his tutorial in the 2008 Logic Days in Lisbon https://staff.fnwi.uva.nl/d.h.j.dejongh/teaching/il/lisbonslides.pdf.

accommodate the wider applicability of the key notion, 'satisfy' can be replaced by a wide variety of stronger conditions (which always include 'satisfy').

An abstract formulation of the completeness theorem was popular in the late 1970s: the set of first order validities is recursively enumerable. This formulation was discussed by the authors but did not make it into Barwise and Feferman (1985). A reason is that it does not persist so nicely to infinitary logics. There are (infinitary) proof systems for the logics $L_{\kappa,\omega}$ and their completeness is proved by adaptations of the Henkin method. This extends to logics that add the Q-quantifier.[26] But these systems do not demonstrate that the validities are recursively enumerable or even Borel.[27] This weakness in the proof system also prevents these infinitary logics from satisfying the full compactness theorem and thus they fail the upward-Löwenheim Skolem theorem. This notion of completeness of course is foreign to Gödel who requires finite proofs and naturally his proof could not be adapted to these contexts.

Henkin already points out that his proof (unlike Gödel's) generalizes easily to uncountable vocabularies. So the first order theory of R-modules can be developed uniformly regardless of the cardinality of the ring R.

The functorial aspect of the Henkin construction is best illustrated by generalizations that require more than mere existence of the canonical model. The *omitting types theorem* is the most basic: add to the requirements in the construction of the Henkin theory that for each $\phi(a, x)$, for each non-principal p there is a p-omitting witness (i.e. the consistency of $\exists x \phi(a, x) \wedge \neg \sigma(x)$ for some $\sigma(x) \in p$). This requirement is easily established as one just has to arrange that each term (in the vocabulary with new constants) omits the type. Somewhat more exotic is the use of Henkin's method in Baldwin and Lachlan (1971) to prove that an $\aleph_1$-categorical but not $\aleph_0$-categorical theory cannot have finitely many countable models. Still more exotic is the modification of the method by Baldwin and Laskowski (2018) to construct atomic models in the continuum by an ω-step Henkin construction. Hodges (1985) provides a plethora of examples in algebra, in exploring the quantifier complexity of definable sets in first order theories, and in various logics.

Earlier than any of these examples, the fundamental idea of the Henkin construction, systematically extend the given theory to a complete theory whose canonical model satisfies a desired property, was regimented by Abraham Robinson's concepts of finite and infinite forcing (Robinson 1970) and expounded for students in Hodges (1985). Very recently, the method is extended to construct counterexamples in functional analysis, e.g. of specific types of C^*-algebras. The authors (model theorists and analysts) of Farah et al. (2016) write, 'We describe a way of constructing C^*-algebras (and metric structures in general) by Robinson forcing (also known as the Henkin construction).'

---

[26] $M \models (Qx)\phi(x)$ means there are uncountably many solutions for ϕ in M.

[27] In fact, the validities of $L_{\omega_1,\omega}$ are Σ_1-definable on the hereditarily countable sets $(\langle H(\omega_1), \in \upharpoonright H(\omega_1)\rangle$ and Σ_2 on $\langle H(\kappa), \in \upharpoonright H(\kappa)\rangle$ for $L_{\omega_1,\omega}$ when κ is uncountable (page 328 of Dickmann 1985).

We have given a myriad of examples where the key idea of Henkin's proof is applied, by deforming (i.e. finding the appropriate) notions of derivation and canonical model but in each case following the Henkin template of extending a given theory to a complete theory admitting a canonical model which satisfies the original theory and perhaps further requirements (e.g. omitting types). In a fundamental sense Henkin's argument is explanatory because he has identified the key features connecting the hypothesis and conclusion, modifying both the syntactic and the semantic component. Moreover, the proof is explanatory by the variant of Steiner's criterion obtained by looking for a characterizing property within the proof.

9.3 Explanation for Who?

The explanatory value of a proof, concept, theory can only be evaluated in terms of the intended audience. We give three accounts of the explanatory value of Henkin's proof, first for undergraduates taking a first course in logic, the second for contemporary research logicians, and the third for research logicians in the early 1950s.

We would motivate the argument for undergraduates by asking, since the goal is to construct a model of a theory T, what properties of an extension of T would allow the equivalence relation (guaranteed by equality axioms) of provable equality on the constants to quotient to a model. The first goal is that the quotient is actually a structure. For this we need the equality axioms for preserving functions. To ensure the structure is a model of T, do an induction on quantifiers, noting the hard case is solved by the witness property. Of course, this will be somewhat more convincing for students with a background in algebra.

The undergraduate argument is reinforced for cognoscenti by observing the various applications that follow the same pattern. See what conditions are needed on a theory T for the logic in question to guarantee that the quotient is a model. For $L_{\omega_1,\omega}$ (logic allowing countable conjunctions) an additional rule of inference is needed; if $\psi \rightarrow \phi$ for each ϕ in a countable set Φ, infer $\psi \rightarrow \bigwedge \Phi$. To build a theory whose 'Henkin' model would satisfy an infinitary sentence, Makkai (1969) extended Smullyan's notion of a consistency property to guarantee the theory was closed under infinite conjunction.

The reader will have noticed that our argument for the identity of the characterizing property depended on this second account and invoked an even wider family of generalized arguments. None of this was known in 1949. Indeed, there is no immediate recognition in the reviews of Henkin's paper of the significance of the change in vocabulary. Only the concrete example of the uncountable Löwenheim-Skolem theorem is even noticed. While the fixed vocabulary approach to first order logic appears in Robinson (1951) and Henkin (1953), it seems to be fully established in Tarski and Vaught (1956) as the fundamental notion of *elementary extension* requires such a stipulation.

In particular, Goodstein's review of Robinson's paper, which provides the first known publication of the completeness theorem[28] in the modern (*) form, does not regard this reformulation of the theorem as a foundational issue.

> ...the Metamathematics of algebra is a book for algebraists not logicians; its claim to a place in the new series of studies in logic and the foundations of mathematics is very slender. Only the first fifth of the book, which is devoted to an extension of Gödel's completeness theorem to non-denumerable systems of statements, has any bearing on the foundations of mathematics, and the remaining four-fifths may be read without reference to this first part which could with advantage have been omitted. (Goodstein 1953)

McKinsey (also noting the uncountable application) and Heyting give straightforward accounts in Mathematical Reviews of the result of Henkin's papers on first order and theory of types respectively with no comments on the significance of the result. Still more striking, Ackermann's review (1950) of Henkin's proof gives a routine summary of the new argument and concludes[29] with, 'The reviewer can not follow the author when he speaks of an extension to an uncountable set of relation symbols, since such a system of notations can not exist.'

As we promised in the introduction, we see that the significance of an explanation depends on the audience; the depth of an argument is often not apparent to the author. Thus, we return to the original remark of Mancosu, explanations that lead to a major recasting of an entire discipline. Henkin's argument was a major component in the turn from model theory as a (mathematical) attempt to understand mathematical reasoning to model theory as a tool in many areas of mathematics. This change in paradigm bloomed with Shelah's classification theory (Baldwin 2018). An essential step was to move from studying a logic which encompassed relations of all orders to the study of theories about particular areas of mathematics in their native vocabularies. Henkin's proof enabled this view and he was one of the pioneers.

However, these ultimately revolutionary features were invisible at the time. This slow reinterpretation of basic notions is a fundamental feature of mathematical development.[30]

References

Ackerman, W. 1950. Reviewed: The completeness of the first-order functional calculus by Leon Henkin. *Journal of Symbolic Logic* 15: 68. German.

Addison, J.W., L. Henkin, and A. Tarski, eds. 1965. *The theory of models*. Amsterdam: North-Holland.

---

[28]See Footnote 6.

[29]This is my very rough translation. The sentence reads, 'Ref. kann dem Verf. aber nicht folgen, wenn er von der Möglichkeit einer mehr als abzählbaren Menge von primitiven Symbolen spricht, da es ein derartiges Bezeichnungssystem doch nicht geben kann.

[30]See, for example, Avigad and Morris (2014), Baldwin (2017), Lakatos (1976), and Werndl (2009).

Avigad, J., and R. Morris. 2014. The concept of 'character' in Dirichlet's theorem on primes in an arithmetic progression. *Archive for History of Exact Sciences* 68: 265–326.

Baldwin, J. 2016. Foundations of mathematics: Reliability and clarity, the explanatory role of mathematical induction. In *Proceedings of the Logic, Language, Information, and Computation, 23rd International Workshop, Wollics 2016*, Puebla Mexico, 16–19 Aug 2016, ed. J. Väänänen, Å. Hirvonen, and R. de Queiroz, 68–82.

Baldwin, J.T. 2017. Axiomatizing changing conceptions of the geometric continuum I: Euclid and Hilbert. *Philosophia Mathematica* 32. https://doi.org/10.1093/philmat/nkx030.

Baldwin, J. 2018. *Model theory and the philosophy of mathematical practice: Formalization without Foundationalism*. Cambridge: Cambridge University Press.

Baldwin, J., and A. Lachlan. 1971. On strongly minimal sets. *Journal of Symbolic Logic* 36: 79–96.

Baldwin, J., and C. Laskowski. 2018. Henkin constructions of models with size continuum. To appear: *Bulletin of Symbolic Logic*.

Barwise, J., and S. Feferman, eds. 1985. *Model-theoretic logics*. New York: Springer.

Beth, E. 1959. *The foundations of mathematics*. Amsterdam: North Holland.

Birkhoff, Garrett. 1935. On the structure of abstract algebras. *Proceedings of the Cambridge Philosophical Society* 31: 433–454.

Chang, C., and H. Keisler. 1973. *Model theory*. Amsterdam: North-Holland. 3rd edition, 1990.

Cintula, P., and C. Noguera. 2015. A Henkin-style proof of completeness for first-order algebraizable logics. *Journal of Symbolic Logic* 80: 341–358.

Dawson, J.W. 1993. The compactness of first-order logic: From Gödel to Lindstrom. *History and Philosophy of Logic* 14: 15–37.

Detlefsen, M. 2014. Completeness and the ends of axiomatization. In *Interpreting Gödel*, ed. J. Kennedy, 59–77. Cambridge: Cambridge University Press.

Dickmann, M. 1985. Larger infinitary logics. In *Model-theoretic logics*, ed. J. Barwise, and S. Feferman, 317–363. New York: Springer.

Farah, I., B. Hart, M. Lupini, L. Robert, A. Tikuisis, A. Vignati, and A. Winter. 2016. Model theory of C*-algebras. Math arXiv:1602.08072v3.

Franks, C. 2010. Cut as consequence. *History and Philosophy of Logic* 31: 349–379.

Franks, C. 2014. Logical completeness, form, and content: An archaeology. In *Interpreting Gödel: Critical essays*, ed. J. Kennedy, 78–106. Cambridge: Cambridge University Press.

Gentzen, G. 1932. Üuber die existenz unabhängiger axiomensysteme zu unendlichen satzsystemen. *Mathematische Annalen* 107: 329–350. In Szabo, *The Collected Papers of Gerhard Gentzen* North Holland, London (1969) as *On the existence of independent axioms systems for infinite sentence systems*, 29–52.

Gödel, K. 1929. Über die Vollständigkeit des Logikkalküls. In *Kurt Gödel: Collected works*, vol. 1, ed. S. Feferman, et al., 60–101. New York: Oxford University Press. 1929 Ph.D. thesis reprinted.

Gödel, K. 1930. The completeness of the axioms of the functional calculus of logic. In *Kurt Gödel, collected works*, vol. I, ed. S. Feferman, et al., 103–123. New York: Oxford University Press. Under the auspices of Association for Symbolic Logic: Providence, Rhode Island;1930c in collected works; first appeared: Monatshefte für Mathematik und Physik.

Goodstein, R.L. 1953. On the metamathematics of algebra by A. Robinson. *The Mathematical Gazette* 37(321): 224–226.

Hafner, J., and P. Mancosu. 2005. The varieties of mathematical explanation. In *Visualization, explanation, and reasoning styles in mathematics*, ed. P. Mancosu, K. Jørgensen, and S. Pedersen, 251–249. Dordrecht: Springer.

Hafner, J., and P. Mancosu. 2008. Beyond unification. In *The philosophy of mathematical practice*, ed. P. Mancosu, 151–178. Oxford: Oxford University Press.

Henkin, L. 1949. The completeness of the first-order functional calculus. *Journal of Symbolic Logic* 14: 159–166.

Henkin, L. 1953. Some interconnections between modern algebra and mathematical logic. *Transactions of the American Mathematical Society* 74: 410–427.

Henkin, L. 1996. The discovery of my completeness proofs. *The Bulletin of Symbolic Logic* 2: 127–158.

Hilbert, D., and W. Ackermann. 1938. *Grundzüge der Theoretischen Logik*, 2nd ed. Berlin: Springer. First edition, 1928.

Hintikka, J. 1955. Form and content in quantification theory. *Acta Philosophica Fennica* 8: 7–55.

Hodges, W. 1985. *Building models by games*. London Mathematical Society, Student Texts. Cambridge: Cambridge University Press.

Karp, C. 1959. *Languages with expressions of infinite length*. Ph.D. thesis, University of Southern California. Advisor, Leon Henkin: http://cdm15799.contentdm.oclc.org/cdm/ref/collection/ p15799coll17/id/140603.

Keisler, H. 1971. *Model theory for infinitary logic*. Amsterdam: North-Holland.

Kreisel, G., and J. Krivine. 1967. *Elements of mathematical logic (Model Theory)*. Amsterdam: North Holland.

Lakatos, D. 1976. *Proofs and refutations*. Cambridge: Cambridge University Press.

Makkai, M. 1969. On the model theory of denumerably long formulas with finite strings of quantifiers. *Journal of Symbolic Logic* 34: 437–459.

Mancosu, P. 2008. Mathematical explanation: Why it matters. In *The philosophy of mathematical practice*, ed. P. Mancosu, 134–150. Oxford: Oxford University Press.

Marker, D. (2002). *Model theory: An introduction*. New York: Springer.

Resnik, M., and D. Kushner. 1987. Explanation, independence, and realism in mathematics. *The British Journal for the Philosophy of Science* 38: 141–158.

Robinson, A. 1951. *Introduction to model theory and to the metamathematics of algebra*. Studies in logic and the foundations of mathematics. Amsterdam: North Holland. 1st edition 1951; 2nd edition 1965.

Robinson, A. 1952. On the application of symbolic logic to algebra. In *Proceedings of the International Congress of Mathematicians*, Cambridge, Aug 30–Sept 6 1950, vol. 1, 686–694. Providence: American Mathematical Society.

Robinson, A. 1970. Forcing in model theory. In *Actes du Congrès international des mathématiciens*, vol. 1, 245–250. Providence: American Mathematical Society.

Skolem, T. 1967. Some remarks on axiomatic set theory. In *From Frege to Godel: A sourcebook in mathematical logic, 1879–1931*, ed. J. Van Heijenoort, 290–301. Harvard University Press. German original published in 1923; address delivered in 1922.

Smullyan, R.M. 1963. A unifying principle in quantification theory. *Proceedings of the National Academy of Science* 49: 828–832.

Smullyan, R.M. 1966. Trees and nest structures. *Journal of Symbolic Logic* 31: 828–832.

Steiner, M. 1978. Mathematical explanation. *Philosophical Studies* 34: 135–151.

Tarski, A. 1935. Der Wahrheitsbegriff in den formalisierten Sprachen. *Studia Philosophica* 1: 261–405. Translated as: The concept of truth in formalized languages in Logic, Tarski (1956).

Tarski, A. 1946. A remark on functionally free algebras. *Annals of Mathematics* 47: 163–165.

Tarski, A. 1954. Contributions to the theory of models, I and II. *Indagationes Mathematicae* 16: 572–582.

Tarski, A. 1956. *Logic, semantics and metamathematics: Papers from 1923–1938*. Oxford: Clarendon Press. Translated by J.H. Woodger.

Tarski, A., and R. Vaught. 1956. Arithmetical extensions of relational systems. *Compositio Mathematica* 13: 81–102.

Werndl, C. 2009. Justifying definitions in mathematics: Going beyond Lakatos. *Philosophia Mathematica* 3: 313–340.

Chapter 10
Can Proofs by Mathematical Induction Be Explanatory?

Josephine Salverda

Abstract In this paper I discuss Marc Lange's argument for the claim that inductive proofs can never be explanatory. I show that several of the assumptions on which Lange's argument relies are problematic, and I argue that there are cases of explanatory inductive proof, providing a number of examples to back up my claim. I finish with a positive proposal on which the examples I put forward can be accounted for by Lange's own account of mathematical explanation.

10.1 Lange's Argument

10.1.1 Introduction

There has been a surge of interest in the recent philosophical literature in the topic of mathematical explanation. One problem faced by philosophers working in this area is that, unlike the case of scientific explanation, it has been difficult to find canonical examples of explanatory proofs against which a proposed account of mathematical explanation can be tested.

Nevertheless, there seems to be a fairly common view in the philosophical literature on the following point: that inductive proofs are not explanatory. Among others, Mark Steiner, Marc Lange, Johannes Hafner and Paolo Mancosu have made this claim (see Hafner and Mancosu 2005; Lange 2009; Steiner 1978).

Indeed, inductive proofs are often seen as a suitable test case. For example, Hafner and Mancosu pose a dilemma for Steiner's account of explanation on the grounds that the account either overgenerates by counting inductive proofs as explanatory, or undergenerates by ruling out a promising example from mathematical practice (Hafner and Mancosu 2005, p. 237). Often the claim that inductive proofs

J. Salverda (✉)
Department of Philosophy, University College London, London, UK
e-mail: josephine.salverda.10@ucl.ac.uk

© Springer International Publishing AG, part of Springer Nature 2018   163
M. Piazza, G. Pulcini (eds.), *Truth, Existence and Explanation*,
Boston Studies in the Philosophy and History of Science 334,
https://doi.org/10.1007/978-3-319-93342-9_10

are not explanatory is made quickly as a side remark or footnote, and not given much evidential support beyond a strong feeling or intuition.

Marc Lange is a notable exception to this trend. In a 2009 paper he gives an argument in support of the claim that inductive proofs can never be explanatory, which is intended to rely only on very general assumptions about explanation. It is interesting to note that in more recent work Lange somewhat modifies his strong claim to the effect that inductive proofs could fall 'somewhere *between* an explanation and a proof utterly lacking in explanatory power. (Explanatory power is a matter of degree)' (see Lange 2014, p. 511, footnote 21, emphasis in the original).

I also am in favour of a more moderate view on which some inductive proofs can indeed count as (somewhat) explanatory. I think this fits more closely with mathematical practice. In this paper, therefore, I will show what goes wrong in Lange's argument for the claim that there are no explanatory inductive proofs; put forward some plausible examples of explanatory inductive proofs; and finally, suggest what might (mistakenly) motivate the seemingly strong and shared philosophical intuition that inductive proofs are not explanatory.

The paper will proceed as follows. In Sect. 10.1, I provide my reading of Lange's argument for the claim that proof by induction is not explanatory, outlining the approach which motivates Lange's argument. In Sect. 10.2, I briefly examine each of the four assumptions on which Lange's argument relies, providing a number of simple examples that cast doubt on the assumptions. I also discuss a problematic 'missing link' in Lange's argument.

In Sect. 10.3, I present some putative examples of explanatory inductive proofs. My conclusion will be that inductive proofs can sometimes be explanatory.

Before we begin: note that although there may be many other kinds of explanation in mathematics, Lange's paper and responses to it revolve around proof, so this paper will be limited to a discussion of explanatory proofs.

10.1.2   What is Proof by Induction?

Many proofs involving mathematical induction are given on the natural numbers, $\mathbb{N}$, and Lange's paper focuses on such proofs. I will touch on induction in other domains in Sect. 10.3.2.

Inductive proofs on $\mathbb{N}$ come in one of a number of different forms, all grounded using the same basic fact about the natural numbers: that they form a well-ordered and non-empty set. Each of the possible variants involves an inductive inference, which is represented by a slightly different logical principle each time. Following Lange, I will list a few of the possible variants of proof by induction, giving the form of inductive inference on which each variant relies.

First, many inductive proofs start from $n = 1$, with the following structure: first, show that the property in question, P, holds for $n = 1$, i.e. $P(1)$. This is called the *base case* of the proof. Then, show that if $P(k)$, then $P(k + 1)$ – known as the *inductive step*. Mathematical induction is an inference from the base case and the inductive step to the claim that the property in question holds for all natural numbers, i.e. that $\forall n\,P(n)$. In this case the inductive inference is represented by the following principle: $P(1) \wedge [\forall k(P(k) \rightarrow P(k + 1))] \vdash \forall n P(n)$. Lange provides the following classic example, where $P(n) \longleftrightarrow 1 + 2 + \ldots + n = \dfrac{n(n + 1)}{2}$:

> Show that for any natural number n, the sum of the first n natural numbers is equal to $n(n + 1)/2$. For $n = 1$, the sum is 1, and $n(n + 1)/2 = 1(2)/2 = 1$. If the summation formula is correct for $n = k$, then the sum of the first $(k + 1)$ natural numbers is $[k(k + 1)/2] + (k + 1) = (k + 1)[(k/2) + 1] = (k + 1)(k + 2)/2$, so the summation formula is correct for $n = k + 1$. (Lange 2009, p. 204)

Second, a proof by induction might start from the base case $n = 5$, say, in which case two inductive steps are needed: one to show that if $P(k)$, then $P(k+1)$, and one to show that if $P(k)$, then $P(k - 1)$ (Lange 2009, p. 207). Following Baker, call a proof which starts from the $P(1)$ base case as above a $P1P$ proof, and a proof which starts from the $P(5)$ base case a $P5P$ proof (Baker 2010, p. 683). A $P5P$ proof involves the following inductive inference: $P(5) \wedge [\forall k(P(k) \rightarrow P(k+1))] \wedge [\forall k > 1\,(P(k) \rightarrow P(k - 1))] \vdash \forall n P(n)$. Note that there is nothing special about the case $n = 5$ here; in principle we can choose any $n \in \mathbb{N}$ for the base case.

Third, note that the variants of mathematical induction discussed so far are known as 'weak' induction, and there is a variant of mathematical induction called 'strong' induction. Strong induction involves only the following inductive step: Show that for every natural number k, if $P(n)$ holds for all natural numbers $n < k$, then $P(k)$ holds. The inductive inference in this case is represented as follows: $\forall k[\forall n < k\,P(n) \rightarrow P(k)] \vdash \forall n P(n)$. Strong and weak induction are logically equivalent in $\mathbb{N}$ in the sense that any proof by weak induction can be transformed into a proof by strong induction, and vice versa. I will say more about strong induction in Sect. 10.2.1.

These are not all of the possible forms of induction on $\mathbb{N}$; for example, another form is known as Cauchy induction after Augustin-Louis Cauchy, and is used to prove results like the arithmetic mean-geometric mean inequality (see for example Dubeau 1991). Nevertheless, the forms cited above are sufficient for the purposes of this paper. We are now in a position to consider Lange's argument for the claim that proof by induction cannot be explanatory.

10.1.3   Lange's Argument

Consider the fixed but arbitrary claim that some property, $P(n)$, holds for all n. Suppose there is some explanatory proof by induction of this general result, and suppose additionally (without loss of generality[1]) that this proof is a $P1P$ proof. Then we can proceed as follows:

1. There is an explanatory $P1P$ proof of the claim that $P(n)$ holds for all n.
2. (Reformulation) For a given $P1P$ proof by induction, there is a PkP proof by induction of the same result for some $k > 1$. [Recall that a $P1P$ proof involves base case $n = 1$, and a PkP proof involves base case $n = k$].
3. (All or nothing) If a $P1P$ proof by induction is explanatory, then the reformulated PkP proof is also explanatory.
4. (Explanatory condition) If a proof is explanatory, then each of its premises partially explains the conclusion.
5. (Asymmetry) For mathematical statements A and B, if A partially explains B, then B does not partially explain A.
6. By 1, 2 and 3, there is an explanatory PkP proof by induction that $P(n)$ holds for all n.
7. By 1 and 4, $P(1)$ partially explains the fact that $P(n)$ holds for all n.
8. Therefore $P(1)$ partially explains $P(k)$.
9. By 4 and 6, $P(k)$ partially explains the fact that $P(n)$ holds for all n.
10. Therefore $P(k)$ partially explains $P(1)$.
11. By 8 and 10, $P(1)$ partially explains $P(k)$ and $P(k)$ partially explains $P(1)$, which contradicts premise 5.
12. Hence, on pain of contradiction, there is no such explanatory proof by induction. See Lange (2009, 207–9).

One immediate problem with the argument is that it's unclear what justifies the inference from step 7 to 8, and from step 9 to 10. Before moving on to consider problems with the argument in Sect. 10.2, I will first say a bit about the motivation behind the argument.

10.1.4   Lange's Approach

It seems in principle possible for it to be the case that some proofs by induction are explanatory while others are not, and that there are further borderline cases, as with many philosophical categories. Lange acknowledges this possibility, writing that 'it

---

[1] Strictly speaking, there is some loss of generality here because (All or nothing) is not formulated as a biconditional. Lange initially states the assumption as a one-way conditional, but goes on to say of the $P1P$ and $P5P$ cases that 'There is nothing to distinguish them, except for where they start' (Lange 2009, p. 209); so if there are any worries about cases where a PkP proof is explanatory while the $P1P$ proof is not, we can simply recast (All or nothing) as a biconditional.

could be that ... some mathematical inductions are explanatory, others are not, and there is no broad truth about what they 'usually' or 'generally' are' (Lange 2009, p. 205, footnote 3).

Nevertheless, Lange writes that his argument 'does not show merely that some proofs by mathematical induction are not explanatory. It shows that none are' (Lange 2009, p. 209). So, when Lange claims that 'proofs by mathematical induction are generally not explanatory' (Lange 2009, p. 205), it seems that we should read 'generally' as meaning 'universally'. This fits with the fact that Lange's argument proceeds by reductio ad absurdum, showing that a contradiction follows from the claim that some given proof by induction is explanatory, given the assumptions made in premises 2–5.

So, Lange chooses to argue for a blanket ban on proof by induction being explanatory, rather than considering cases of proof by induction on an individual basis. Presumably Lange has some motivation for doing so; I think this motivation arises from Lange's attempt to avoid relying on intuitions about specific cases or types of proof. As Lange points out, such intuitions often conflict: 'Philosophers disagree sharply about which proofs of a given theorem explain why that theorem holds and which merely prove that it holds' (Lange 2009, p. 203).

Consider, for example, a proof by induction of the claim that the sum of numbers from 1 to n is $n(n + 1)/2$, as given in Sect. 10.1.2. This claim can also be proved using diagrams, as we will see in Sect. 10.3.1. But, according to Brown, 'a proof by induction is probably more insightful and explanatory than the picture proofs ... I suspect that induction – the passage from n to $n + 1$ – more than any other feature, best characterizes the natural numbers. That's why [the proof by induction] ... is in many ways better – it is more explanatory' (Brown 1997, p. 177, quoted in Lange 2009).

On the other hand, Hanna writes of the same proof by induction, '[T]his is certainly an acceptable proof ... What it does not do, however, is show why the sum of the first n integers is $n(n + 1)/2$ ... Proofs by mathematical induction are non-explanatory in general' (Hanna 1989, p. 10, quoted in Lange 2009).

Disagreements of this kind mean that a philosophical account of mathematical explanation is difficult to test against canonical examples: given the sharp conflict of intuitions, there are no generally accepted examples of explanatory proof against which the proposed account can be tested (Lange 2009, p. 203).

Instead of focusing on examples, therefore, Lange aims to 'end this fruitless exchange of intuitions' altogether, with his 'neat argument' that avoids 'making any controversial presuppositions about what mathematical explanation would be' (Lange 2009, p. 203–5). There seem to be two main components to Lange's approach: 1. Avoid relying on intuitions about specific cases of mathematical proof; and 2. Avoid relying on controversial assumptions about explanation in mathematics.

At first glance, these two conditions place a reasonable restriction on an account of explanation in mathematics. In particular, although our intuitions can be a useful guide, it seems right to say that we should avoid relying solely on intuition in order to build up an account of explanation in mathematics, not least because it is unclear whose intuitions should count.

However, there is a difference between avoiding reliance on our intuitions about examples and avoiding examples altogether. General assumptions can be just as problematic as intuitions about particular cases. Why think that our intuitions about general claims are more likely to be reliable than our intuitions about specific cases? It seems plausible that our intuitions about general cases arise from a (possibly unconscious) generalisation from examples. Such generalisations are in danger of becoming overgeneralisations, if they do not take into account a sufficiently wide range of cases.

In the rest of this paper, I will show that none of Lange's assumptions are uncontroversial, and moreover that attention to examples can help to illuminate where each assumption goes wrong.

Let us now examine Lange's four assumptions in more detail.

10.2 Four Assumptions and a Missing Link

10.2.1 Reformulation

Lange's first and seemingly unproblematic assumption holds that any proof by mathematical induction can be reformulated to start from a different base case. So, for example, any $P1P$ proof can be converted into a $P5P$ proof, though not, as Lange concedes, without some labour (Lange 2009, p. 207). In short, Lange accepts:

> (Reformulation) For a given $P1P$ proof by induction, there is a PkP proof by induction of the same result.

Lange focuses on the case $k = 5$, and shows that a $P5P$ proof can indeed be found for the claim that the sum of numbers from 1 to n is equal to $n(n + 1)/2$.

> For $n = 5$, the sum is $1 + 2 + 3 + 4 + 5 = 15$, and $n(n + 1)/2 = 5(6)/2 = 15$. If the summation formula is correct for $n = k$, then (I showed earlier[2]) it is correct for $n = k + 1$. If the summation formula is correct for $n = k$ (where $k > 1$), then the sum of the first $k - 1$ natural numbers is $[k(k + 1)/2] - k = k[(k + 1)/2 - 1 = k(k - 1)/2$, so the summation formula is correct for $n = k - 1$ (Lange 2009, p. 209).

However, reformulating a proof by induction is not so simple in every case. An example will help us to see that there are cases where it is less clear that (Reformulation) applies. Consider the following proof that proceeds by strong induction.

---

[2]As in Sect. 10.1.2.

Proof A
Claim: For all $n \in \mathbb{N}$, if n is composite, then n is a product of primes, where n is composite
iff $n = b.c$ for $b, c > 1$.
Inductive step: Suppose the claim is true for all $k < n$. Suppose $n = b.c$, where $b, c > 1$.
Then $b, c < n$. If b is not composite, it is prime (since $b > 1$, and only 1 is neither prime
nor composite); if b is composite, it is a product of primes by the inductive hypothesis. If c
is not composite, it is prime (since $c > 1$); if c is composite, it is a product of primes by the
inductive hypothesis. In either case, $n = b.c$ is a product of primes.

Proof A is not a $P1P$ proof, and indeed is not a PkP proof of any kind, since
there is no base case $n = k$ that is handled separately from the rest of the domain.
Hence (Reformulation) does not apply to Proof A, and therefore the assumption
that Proof A is explanatory does not lead to a contradiction. This is not to say that
Proof A must count as explanatory, but rather that the scope of Lange's argument
does not cover all proofs involving induction, as he claims. Instead, it seems that
Lange's conclusion holds only for proof by weak induction, since such proofs must
involve a base case.

This seems an odd result. Recall that strong and weak induction are equivalent
in $\mathbb{N}$, in the sense that any proof by weak induction can be transformed into a proof
by strong induction of the same result, and vice versa. This means there is some
proof, A*, that establishes the same result as Proof A but using weak induction.

To find Proof A*, note that Proof A as given above is a proof of
the claim that $\forall n P(n)$, where $C(k) \longleftrightarrow k$ *is composite*, $\Pi(k) \longleftrightarrow$
k *is a product of primes*, and $P(k) \longleftrightarrow [C(k) \rightarrow \Pi(k)]$. In order
to transform Proof A into a proof by weak induction of the same result, let
$P'(k) \longleftrightarrow P(1) \wedge P(2) \wedge \ldots \wedge P(k - 1)$. We can use weak induction to show
that $\forall n P'(n)$, and since $\forall n P'(n) \longleftrightarrow \forall n P(n)$, we will thereby have shown that
$\forall n P(n)$:

Proof A*
Claim: For all $n \in \mathbb{N}$, if n is composite, then n is a product of primes.
Base case: $n = 1$. $P'(1)$ holds iff $P(0)$ holds, and $P(0)$ holds iff $C(0) \rightarrow \Pi(0)$, which is
vacuously true, since 0 is not composite.
Inductive step: Suppose $P'(k)$. holds, i.e. $P(1) \wedge P(2) \wedge \ldots \wedge P(k - 1)$. Suppose $k = b.c$,
where $b, c > 1$. Then $b, c < k$. If b is not composite, it is prime (since $b > 1$); if b
is composite, it is a product of primes, since $b < k$ and $\forall n < k P(n)$ by the inductive
hypothesis. If c is not composite, it is prime (since $c > 1$); if c is composite, it is a product
of primes, since $c < k$ and $\forall n < k P(n)$ by the inductive hypothesis. In either case, $k = b.c$
is a product of primes, so $P(k)$. Then by the inductive hypothesis $P(1) \wedge P(2) \wedge \ldots \wedge P(k -$
$1) \wedge P(k)$, and therefore $P'(k + 1)$.

Now, we could add two further premises to Lange's argument, as follows:

(Reformulation)* For any proof involving strong induction, there is a reformulated proof by
weak induction of the same result.
(All or nothing)* If a proof that uses strong induction is explanatory, the reformulated proof
by weak induction of the same result is equally explanatory.

Suppose Proof A is explanatory. (Reformulation)* is true, and we have seen there
is a proof, Proof A*, using weak induction and which proves the same result
as Proof A. According to (All or nothing)*, Proof A* is also explanatory. Since

Proof A* uses weak induction, the original assumption, (Reformulation), applies to Proof A*. Lange's argument from Sect. 10.1.3 can then be run to get a contradiction, and we can conclude that Proof A cannot be explanatory. Using this method, the scope of Lange's argument could be expanded to cover both weak and strong induction.

However, it seems to me that (All or nothing)* is quite implausible, since the reformulated Proof A* is often quite contrived. For example, I see no immediately compelling reason to think that Proofs A and A* above must be equally explanatory, and to me Proof A* seems less explanatory than Proof A.

Of course, my intuition about Proofs A and A* is not enough to undermine Lange's argument. But in this section we have seen that the first assumption, (Reformulation), faces a challenge: the premise does not apply to all cases of proof by induction, since proofs involving strong induction need not treat a base case separately. Either Lange's argument does not have the generality he claims, or two further premises covering strong induction (at least one of which seems contestable) must be added to the argument to achieve the general result Lange desires.[3]

10.2.2 All or Nothing

According to (All or nothing), if a $P1P$ proof by induction is explanatory, then the reformulated PkP proof is also explanatory.

This assumption looks implausible in a number of cases. For example, consider a case of induction on the complexity of formulas in propositional logic. Suppose we represent well-formed formulas of propositional logic using parse trees, and define the 'height' of each wff as the length of the longest path of its parse tree (so that the objects of the induction remain in $\mathbb{N}$). We can then prove the following theorem using standard (strong) induction on the height of wffs[4]:

> **Theorem** *Every well-formed formula of propositional logic has an equal number of right and left parentheses.*

I will not give the full proof here for reasons of space, but note that in the standard base case we have $n = 1$ and so we just need to show that the result holds for atoms, which is easy as atoms have no parentheses. If we were to reformulate the proof to start from base case $n \leq 5$,[5] we would need to show that the result holds for all formulas of height up to and including 5. Moreover, there is nothing special about the case $n = 5$: we could reformulate the proof to start from base case $n = 9992414$, say, in which case we would need to show that the result holds for all formulas of

---

[3] In Sect. 10.3 we will see that cases of structural induction in other domains also present a problem for (Reformulation).

[4] I consider structural induction, arguably a more natural way to prove this theorem, in Sect. 10.3.2.

[5] The base case in the (natural) reformulated proof will be $n \leq 5$ rather than $n = 5$ because the original proof uses strong induction.

height (up to and including) 9992414. This would take hours of calculation if the idea is to show the base case by inspection rather than doing a separate inductive step (and if we were to add an extra inductive step, then this extra complexity would presumably give us one way to 'choose between' the $P1P$ and $P9992414P$ proofs).

According to (All or nothing), both the $P5P$ and $P9992414P$ proofs are explanatory if the $P1P$ proof is explanatory. But it looks highly implausible that the base case in the $P992414P$ proof will count as explanatory, if the base case is simply proved by inspection.

Other respondents to Lange's paper have raised similar worries. For example, Cariani also discusses cases of induction on the complexity of formulas and further questions whether the downwards inductive step in a PkP proof can 'support or can be supported by explanatory arguments' (Cariani, p. 6, footnote 6). Baker claims that it is possible to distinguish between the explanatory value of $P1P$ and $P5P$ proofs even in the classic cases considered by Lange (Baker 2010). So it is clear that (All or nothing) is by no means an uncontroversial assumption.

Let us therefore move on to Lange's third assumption.

10.2.3   Explanatory Condition

Lange writes that he will 'presuppose that a mathematical explanation of a given mathematical truth F may consist of a proof (i.e. a deduction) of F from various other mathematical truths $G_1, \ldots, G_n$. ... In such an explanation, the G_i collectively explain why F obtains; each of G_1, G_2 and so forth helps to explain F (e.g. F is explained partly by G_1)' (Lange 2009, p. 206). I have formulated this assumption as follows:

(Explanatory condition) If a proof is explanatory, then each of its premises partially explains the conclusion.

Lange provides no argument for this claim, except to draw an analogy with cases of mathematical explanations of physical phenomena, where, for example, 'Kepler's laws of planetary motion are explained by being deduced from Newton's laws of motion and gravity ... hence, Newton's law of gravity helps to explain why Kepler's laws hold' (Lange 2009, p. 206). More needs to be said about this condition, since the notion of 'helping to explain', or partial explanation, is left unclear. I will say a bit more about partial explanation in the next section.

My main worry here is that Lange's paper starts off by talking about proofs as potential explanatory items, while the moves made in his argument focus on parts of proofs (specific mathematical claims) doing explanatory work. It is not clear that these two concepts of explanation are the same. We might think of a proof featuring as an explanation in an epistemic sense – for example as an argument that convinces an audience – while we might see the explanatory relation between two facts as involving some kind of ontic dependence relation (as in the recent debate on grounding).

A similar kind of worry is pointed out by Baldwin, who asks:

What are the relations between? We have argued that the explanatory object is a proof and such is the title of [Lange's] paper. But he concludes (where P is some property that may or may not hold of a natural number), 'It cannot be that P(1) helps to explain why P(5) holds and that P(5) helps to explain why P(1) holds, on pain of mathematical explanations running in a circle.' This leap exacerbates the reduction of the argument to considering only 'premises and consequence' by conflating the entire explanation with any component of it (Baldwin 2016, pp. 78–9).

So as with the previous assumptions, we see that (Explanatory condition) is not uncontroversial.[6] Let us therefore move on to the final assumption, (Asymmetry).

10.2.4 Asymmetry

Lange writes that he 'presuppose[s] only that mathematical explanations cannot run in a circle ... that when one mathematical truth helps to explain another, the former is partly responsible for the latter in such a way that the latter cannot be partly responsible for the former. Relations of explanatory priority are asymmetric' (Lange 2009, p. 206). Thus Lange endorses:

(Asymmetry) For mathematical statements A and B, if A partially explains B, then B does not partially explain A.

Lange's defence of (Asymmetry) rests on the idea that if (Asymmetry) were false, then 'mathematical explanation would be nothing at all like scientific explanation' (Lange 2009, p. 206). This does not seem a good reason in itself to accept (Asymmetry), especially given Lange's intention to avoid 'making any controversial presuppositions about what mathematical explanations could be' (Lange 2009, p. 203). The assumption that mathematical explanation is like scientific explanation is certainly not uncontroversial. For example, Baker claims that 'the predominant view is that mathematical explanation is qualitatively different both from scientific explanation and from explanation in 'ordinary non-scientific contexts'' (Baker 2012, p. 244). (Asymmetry) thus needs further argument. The issue of whether mathematical explanation is like scientific explanation is important and deserves further attention.

For now, note that further worries can be raised about (Asymmetry). Baker, for example, writes that 'I have some doubt about the inevitability of this condition, especially since Lange's argument is formulated in terms of partial explanation' (Baker 2010, p. 682). I share Baker's worry about (Asymmetry), especially since Lange's notion of partial explanation is left unclear.

If 'A partially explains B' means simply that A contributes to an explanation of B, then (Asymmetry) is unconvincing. Consider the following example: Mary likes

---

[6]Strictly speaking, the quote from Baldwin in (2016) mentions (Asymmetry), but the worry is also relevant for (Explanatory condition).

John, partly because he is kind, and partly because he likes her. John likes Mary, partly because she is witty, and partly because she likes him. It seems plausible that A: 'John likes Mary' partly explains B: 'Mary likes John', and vice versa. Here, then, is an initial reason to doubt (Asymmetry) in general.

Now, the mathematical case must involve non-causal explanation. Nevertheless, it seems plausible that a mathematical counterexample to (Asymmetry) also exists, if partial explanation amounts simply to contribution to an explanation. For example, let A be the intermediate value theorem:

(A) Let $f : [a, b] \rightarrow \mathbb{R}$ be a continuous function, and let u be a real number such that $f(a) < u < f(b)$. Then for some $c\epsilon[a, b]$, $f(c) = u$.

And let B be the intermediate zero theorem:

(B) Let $f : [a, b] \rightarrow \mathbb{R}$ be a continuous function, and suppose $f(a) < 0 < f(b)$ or $f(a) > 0 > f(b)$. Then for some $c\epsilon[a, b]$, $f(c) = 0$.

B is a special case of A. Therefore, it seems that A can contribute to an explanation of B, for we can explain why B holds as follows: B is true because B is a special case of A, where $u = 0$, and A is true. On the other hand, it also seems that B can contribute to an explanation of A, as follows. We can 'see' why B holds when $f(a) < 0 < f(b)$ by looking at a diagram:

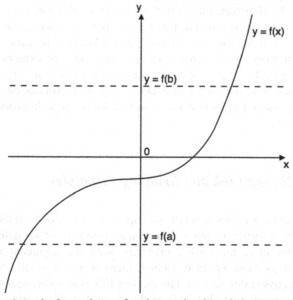

A graph of a continuous function passing through the x-axis

It is clear that a continuous curve on $\mathbb{R}$ passing from below the x-axis to above the x-axis must at some point, c, pass through the x-axis, where $y = f(c) = 0$. But this fact can be used to explain why A holds. For a similar diagram can be given where instead of the x-axis we consider the line $y = u$. The diagram above helps us to understand why a continuous curve passing from below the line $y = u$ to above the line $y = u$ must at some point, c, pass through the line $y = u$. At this point, c, $y = f(c) = u$ as required.[7]

This sketch of an argument does not amount to a proof of A from B, but it does seem that the special case, B, represented in a diagram, helps to explain why A holds. Therefore, it is possible for A to contribute to an explanation of B and for B to contribute to an explanation of A. So Lange's notion of partial explanation cannot involve mere contribution to an explanation; some stronger conception of partial explanation must be intended.

Lange writes that 'when one mathematical truth helps to explain another, the former is partly responsible for the latter in such a way that the latter cannot then be partly responsible for the former' (Lange 2009, p. 206). Unfortunately, this does not clarify the matter, since the idea of A being responsible for B is unclear in a mathematical context, where an appeal to causation is ruled out.

One way to understand the idea of A being responsible for B in the mathematical context might be to appeal to accounts of grounding, understood as some sort of metaphysical dependence relation. However, in recent work Lange argues that grounding and explaining in mathematics are fundamentally distinct (Lange), in which case asymmetry of grounding need not imply asymmetry of explanation.

Another thought might be that A is partly responsible for B if A forms part of a deduction of B. However, this cannot be Lange's intention since (Asymmetry) would then be false: It is possible for A to be part of a deduction of B and for B to be part of a deduction of A. Indeed, this relation will occur whenever two mathematical propositions are shown to be equivalent. For example, in classical propositional logic, $A = \neg p \to q$ can be used to derive $B = p \vee q$, and vice versa.

In the absence of a more detailed account of partial explanation, (Asymmetry) is not uncontroversial. Let us now move on to consider a problematic inference in Lange's argument.

10.2.5   A Missing Link and Bridging Principles

So far, I have put forward some simple examples that cast doubt on the assumptions on which Lange's argument relies. Lange may provide further defence for these assumptions, but there are further problems with the argument that proof by induction is not explanatory. In particular, there is a problematic 'missing link', which I will discuss in this section. The missing link is an inferential gap from steps 7 to 8 in Lange's argument, that is, from the claim that $P(1)$ partially explains the

---

[7]For worries about the reliability of diagrammatic reasoning in this example, see e.g. work by Giaquinto (2011).

fact that $P(n)$ holds for all n, to the claim that $P(1)$ partially explains $P(k)$.[8] Some further principle is needed to justify this inference.

One such principle could be: 'If A partially explains B, and B entails C, then A partially explains C'. However, this principle is incorrect in general, as we see from the following statements.

A. Harry thought that Sally would arrive at 5, and Harry wanted to arrive when Sally did.
B. Both Harry and Sally arrived at 5.
C. Sally arrived at 5.

Assuming that beliefs and desires can play a role in explanation, it seems that A partially explains B, in the sense that A explains one of B's conjuncts. Furthermore, B entails C; and yet A does not partially explain C. So we need a better alternative principle to close the gap in Lange's argument.

Another possible bridging principle would be that 'Any fact which partially explains a universally quantified truth also partially explains every other instance of that truth'. Cariani for example formulates this as a principle of Minimal Closure: 'If $p_1, \ldots, p_n$ explain $\ulcorner \forall x \Phi x \urcorner$, then $p_1, \ldots, p_n$ explain $\ulcorner \Phi t \urcorner$ for any (referring) singular term t in the language (provided, of course, that $\ulcorner \Phi t \urcorner$ is not one of the p_i's' (Cariani, p. 3).

However, in a later paper Lange explicitly denies relying on this principle, writing 'I do not base [the missing link] on the premise that if a fact helps to explain a given universal generalization, then it must help to explain every instance of that generalization' (Lange 2010, p. 326, footnote 15).

Unfortunately Lange does not himself provide an explicit principle to support the inference from $P(1)$ partially explaining $\forall n P(n)$ to $P(1)$ partially explaining $P(k)$. Instead, Lange's support for the inference comes from an analogy:

> Compare a scientific example: Coulomb's law (giving the electrostatic force between two stationary point charges) explains why the magnitude E of the electric field of a long straight wire (of negligible thickness) with uniform linear charge density λ is $2\lambda/r$ at a distance r from the wire. In explaining why for any λ and r, $E = 2\lambda/r$, Coulomb's law explains in particular why $E = 4\, dyn/statcoulomb$ if $\lambda = 10\, statcoulombs/cm$ and $r = 5\, cm$. By the same token, if $P(1)$ explains why for any n, $P(n)$, then $P(1)$ explains in particular why $P(5)$. (Lange 2009, p. 210)

As before, we might question Lange's assumption that an analogy to scientific explanation is appropriate here. But even if mathematical explanation is relevantly like scientific explanation, the problem with the analogy here is that it does not involve partial explanation: instead, Lange suggests that Coulomb's law explains a general relation, $E = 2\lambda/r$, and hence explains any particular instance of that relation. For the analogy to provide a convincing parallel, we need a case where some instance of a given law helps to explain a general relation, and hence helps to explain any other instance of that relation. Lange needs to provide an explanation of the fact that $E = 4d/s = 2(10s/cm)/5\,cm$, say, by another instance of the equation

---

[8]The same gap is present in the step from 9 to 10, i.e. from the claim that $P(k)$ partially explains the fact that $P(n)$ holds for all n, to the claim that $P(k)$ partially explains P(1).

$E = 2\lambda/r$. Without this further case, the situation in Lange's scientific analogy does not seem to match up to the situation in the inductive proof.

How, then, might the missing link be filled? In a recent paper, Hoeltje et al. aim to find a principle on which Lange might implicitly be relying in order to bridge the inferential gap (Hoeltje et al. 2013). Hoeltje et al. put forward the following putative principles of explanation (among others):

Case-By-Rule A universally quantified truth explains its instances.
Rule-By-Case A universally quantified truth is explained by its instances. (Hoeltje et al. 2013, pp. 512–13)

According to Hoeltje et al., Lange must appeal to Case-By-Rule in order to bridge the gap in his argument. But, they argue, Case-By-Rule is unconvincing, because Rule-By-Case has greater appeal and conflicts with Case-By-Rule, if we accept Lange's (Asymmetry) assumption (Hoeltje et al. 2013, pp. 514–15).[9]

In order to find out whether either Case-By-Rule or Rule-By-Case is plausible, we first need to clarify the intended scope of these principles. Hoeltje et al. argue that Case-By-Rule and Rule-By-Case conflict, i.e. it cannot be the case that both 'A universally quantified truth explains its instances' and 'A universally quantified truth is explained by its instances'. Now, (Asymmetry) tells us that a given instance of a universally quantified truth cannot both explain and be explained by that truth. But (Asymmetry) leaves open the possibility that a universally quantified truth explains some of its instances and is explained by some other of its instances, with some instances not involved in an explanatory relation at all. This seems to be a situation in which (limited scope versions of) Case-By-Rule and Rule-By-Case both hold. Since Hoeltje et al. claim the principles conflict, it seems that the intended reading is something like 'A universally quantified truth explains *each of* its instances'.

Furthermore, it seems that Hoeltje et al. intend Case-By-Rule and Rule-By-Case to hold as universal principles of explanation, since they argue that it is plausible that there is a general explanatory principle for the universal quantifier (Hoeltje et al. 2013, p. 515). That is, the intended reading of these principles is something like '*Every* universally quantified truth explains each of its instances'. This reading is backed up by the conclusion drawn by Hoeltje et al. about Case-By-Rule: 'Not only is the latter principle false because it has *some* false instance, *Asymmetry* and *Rule-By-Case* imply that it *only* has false instances' (Hoeltje et al. 2013, p. 515, emphasis in the original). It seems one false instance would have been enough to make Case-By-Rule false, and so we can take it that Case-By-Rule is intended to apply universally.

---

[9]As the authors point out, it is unlikely Lange actually intends to appeal to Case-By-Rule, because together with (Asymmetry) the principle would prove the stronger claim that no instance of a universally quantified truth, T, can explain T. This stronger claim allows for a direct route to Lange's desired conclusion, relying only on (Explanatory condition) and circumventing any worries about the relative explanatoriness of $P1P$ and PkP proofs. We may assume that Lange would have taken this route if he believed it to be a viable one.

For these reasons, I will focus on the versions of Case-By-Rule and Rule-By-Case given below:

Case-By-Rule* Every universally quantified truth explains each of its instances.
Rule-By-Case* Every universally quantified truth is partially[10] explained by each of its instances.

Why think that either of these principles holds? It seems to me that the two principles are in fact highly implausible, both as general principles of explanation and as principles of mathematical explanation.

First, take Case-By-Rule* and consider the question: 'Why yesterday did the sun not set where it rose?'. The fact that the sun did not set where it rose yesterday is an instance of the universally quantified truth that the sun never sets where it rises. But this truth surely seems lacking as an explanatory answer to the question. So, Case-By-Rule* seems incorrect as a universal principle of explanation.

In the mathematical case in particular, a further counterexample can be found. Consider the following universally quantified truth: 'All prime numbers greater than 2 are odd'. The statement quantifies over the domain of prime numbers which are greater than 2, and states that all members of this domain have the property of being odd. But this universally quantified truth surely does not explain the fact that 3 is odd, say, although 3 is an instance of the universal generalisation. Hence there is at least one universally quantified truth in mathematics which does not explain its instances, and Case-By-Rule* is unconvincing as a general principle of mathematical explanation.

On the other hand, Rule-By-Case* does not look very convincing either. Rule-By-Case* holds that a universally quantified truth is explained by its instances. Consider the question: 'Why is everyone in the seminar room a philosopher?'. The answer 'Julia is in the seminar room and she is a philosopher, John is in the seminar room and he is a philosopher, ...' does not provide an explanatory answer to this question, in contrast to the answer 'Because the seminar is being held at a philosophy conference'. So here is a universally quantified truth which is not explained by all of its instances collectively, and is not partially explained by its instances individually. Hence, Rule-By-Case* seems incorrect as a universal principle of explanation.

In the mathematical case in particular, a further counterexample can be found. Consider the following universally quantified truth: 'For all f such that f is a function defined on the natural numbers, $\mathbb{N}$, and $m, n \in \mathbb{N}$, $f(m) \neq f(n) \rightarrow m \neq n$'. Consider an instance of this universal generalisation, namely the constant function f defined as follows: $f(n) = 1$ for all $n \in \mathbb{N}$. The relevant property here holds only vacuously, since the antecedent of the conditional is never fulfilled.

---

[10]Rule-By-Case must involve partial explanation, because it is highly implausible to think that each instance of a universally quantified truth fully explains the truth. Rather, the idea seems to be that a universally quantified truth is explained by its instances collectively, and hence partially explained by each instance. We are still lacking an account of partial explanation in mathematics, but I will not say more about that here.

Although this instance may confirm the universally quantified truth (to the extent that there is confirmation in mathematics), surely the instance does not explain the universally quantified truth. Hence there are universally quantified truths in mathematics that are not explained by each of their instances, and Rule-By-Case* is unconvincing as a general principle of mathematical explanation.

In summary, there are simple counterexamples to both Case-By-Rule* and Rule-By-Case*, if they are taken as universal principles of explanation. I think it is plausible, therefore, that there is no universal explanatory principle linking a universally quantified truth with its instances; rather, I suggest that in some cases the instances help to explain the truth, and in some cases the truth helps to explain the instances.[11] In the same vein, I think it is likely that there are some explanatory and some non-explanatory inductive proofs. (Other respondents to Lange's paper agree; see e.g. Cariani).

In the next section, I will back up this claim by presenting some putative examples of explanatory inductive proofs.

10.3 Explanatory Inductive Proofs

I will present the examples in Sects. 10.3.1 and 10.3.2, saving discussion for Sects. 10.3.3 and 10.3.4.

10.3.1 Two Pictorial Proofs

10.3.1.1 Case 1: The Sum of Numbers From 1 to n

The statement to be proved is the following claim: for all n in $\mathbb{N}$, $1 + 2 + \ldots + n = \frac{n(n+1)}{2}$. This can be proved by induction in (at least) two different ways. I provide two proofs below, and suggest that Proof C is more explanatory than Proof B.

Proof B
Base case
For $n = 1 : 1 = 1 \times \frac{(1+1)}{2}$.
Inductive step
If for $n = k$, $1 + 2 + \ldots + k = \frac{k(k+1)}{2}$, then for $n = k + 1$,

---

[11]It would be interesting to link this to recent debates on grounding. If a universally quantified truth is grounded in but not explained by its instances, can grounding be an explanatory relation?

$$(1 + 2 + \ldots + k) + k + 1 = (\frac{k(k+1)}{2}) + k + 1$$

$$= \frac{k^2 + 3k + 2}{2}$$

$$= \frac{(k+1)(k+2)}{2}$$

Proof C

First, note that the nth triangular number is equal to the sum from 1 to n by definition, ie. $T_n = \sum_{k=1}^{n} k$. The diagram below shows the first four triangular numbers.

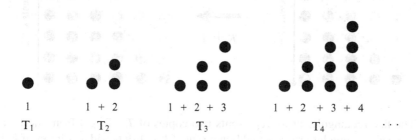

1	1 + 2	1 + 2 + 3	1 + 2 + 3 + 4
T_1	T_2	T_3	T_4

Now, in order to prove our desired result that $1 + 2 + \ldots + n = \frac{n(n+1)}{2}$, we simply need to show that the value of the nth triangular number is equal to $\frac{n(n+1)}{2}$.

<u>Base case</u>

For small values of n, we can see that the value of the triangular number T_n can be found by halving the rectangular array of dots with sides of length n dots and $n + 1$ dots respectively. This is because two copies of T_n fit together to make a rectangular array of dots, as shown in the diagram below:

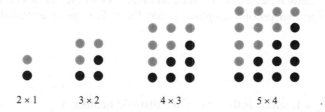

2×1	3×2	4×3	5×4

The first triangular number, T_1, is equal to half of the number of dots in the rectangular array with sides of length 1 dot and 2 dots: $1 = \frac{1 \times (1+1)}{2}$. And since the sum of numbers from 1 to 1 is equal to T_1, we have proved our base case.

Inductive step
Suppose that the value of the triangular number T_k is equal to half of the number of dots in the rectangular array of dots with sides of length k and $k + 1$. We now need to show that T_{k+1} is equal to half of the number of tos in the rectangular array of dots with sides of length $k + 1$ and $k + 2$. But this is easy to see from the following diagram:

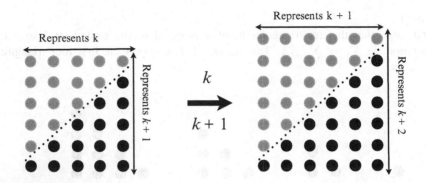

The first rectangular array represents two copies of T_k. To get from the first to the second rectangular array, we add on a row of $k + 1$ dots and a column of $k + 1$ dots. This corresponds to adding on a new row of $k + 1$ dots to each of the triangular components in the first array, so the second rectangular array represents two copies of T_{k+1}. So if $2T_k = k(k + 1)$, then $2T_{k+1} = (k + 1)(k + 2)$.

Therefore, the value of the nth triangular number is equal to $\frac{n(n+1)}{2}$ for all n. Since we saw that the sum of the first n numbers is equal to the nth triangular number, it follows that $1 + 2 + \ldots + n = \frac{n(n+1)}{2}$.

10.3.1.2   Case 2: Sums of Odd Numbers

Claim: For any natural number n, n^2 is equal to the sum of the first n odd numbers. Again, I provide two proofs, suggesting that Proof E is more explanatory than Proof D.

Proof D
Base case
For $n = 1 : n^2 = 1$, which is the sum of the first odd number, 1.
Inductive step
Note that the kth odd number is $2k - 1$, and the $(k + 1)$th odd number is $2k + 1$. Now, suppose that for $n = k$, $k^2 = 1 + 3 + 5 + \ldots + (2k - 1)$.

Then $(k + 1)^2 = k^2 + 2k + 1 = 1 + 3 + 5 + \ldots + (2k - 1) + (2k + 1)$, which is the sum of the first $k + 1$ odd numbers, as required.

Proof E
Base case
For $n = 1 : n^2 = 1$, which is the sum of the first odd number, 1. We can see that the sum of the first n odd numbers forms a square for small values of n by inspection of the following diagrams:

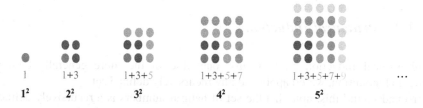

1 1+3 1+3+5 1+3+5+7 1+3+5+7+9 $\cdots$

1^2 2^2 3^2 4^2 5^2

Here n^2 is represented by a square composed of dots, with sides of length n dots. Each time the new 'layer' of dots represents the latest odd number to be added on. The pictures show that when we add on the next odd number in this way we can form a new square, for odd numbers up to 9. In each case the new square has sides which are one dot longer than the sides of the previous square. We see that 1^2 is represented by a square composed of just one dot – the first odd number – since $1^2 = 1$.

Inductive step
The picture given above shows that the sum of the first n odd numbers is equal to n^2 for small values of n. We now need to prove the general result by proving the inductive step. Let $S(k) = k^2$. Suppose that the desired result holds for $n = k$. So $S(k)$ is equal to the sum of the first k odd numbers. We need to show that $S(k + 1)$ is equal to the sum of the first $k + 1$ odd numbers. We can represent $S(k)$ as a square with sides of length k, since this square has area k^2.

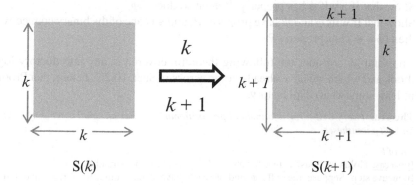

$S(k)$ $S(k+1)$

In the same way we added on dots the base case, we can now add on two rectangles to the square $S(k)$: one with sides of length 1 and $k + 1$, and one with sides of length 1 and k. In this way we can form a square with sides of length $k + 1$. Since this square has area $(k + 1)^2$, it represents $S(k + 1)$. Note that the rectangles added on have an area of $2k + 1$ in total. So $S(k + 1) = S(k) + 2k + 1$. But since the $(k + 1)$th odd number is $2k + 1$, and $S(k)$ is the sum of the first k odd numbers, we see that $S(k + 1)$ is the sum of the first $k + 1$ odd numbers, as required.

10.3.2   Structural Induction

Mathematical induction on $\mathbb{N}$ is just one kind of the more generally defined structural induction, which applies to any recursively defined set.

To understand this, note that the set of natural numbers is a recursively defined set, namely the smallest set containing 1 and closed under succession:

1. 1 is an element of $\mathbb{N}$.
2. If n is an element of $\mathbb{N}$, then so is $n + 1$.
3. Nothing else is an element of $\mathbb{N}$.

Another recursively defined set is the set of well-formed formulas in propositional logic:

1. Every atom is a wff.
2. If ϕ is a wff, so is $\neg\phi$.
3. If ϕ and ψ are wffs, and $*$ is a binary connective (e.g. $\wedge$, $\vee$, $\rightarrow$), then $(\phi * \psi)$ is a wff.
4. Nothing else is a wff.

We can apply structural induction to prove facts about the set of well-formed formulas. In particular, to show that every element of the set has property P, we must:

- Show that each atom has property P.
- Show that if a wff ϕ has property P, then so does $\neg\phi$.
- Show that if wffs ϕ and ψ have property P, and $*$ is one of the binary connectives, then $(\phi * \psi)$ has property P.

As an example, consider the following theorem, covered in any introductory logic textbook and which we have already come across in Sect. 10.2.2. I claim that Proof F is (at least somewhat) explanatory.

Theorem *Every well-formed formula of propositional logic has an equal number of right and left parentheses.*

Proof F
Base case Holds vacuously, since the atoms do not contain any parentheses.
Inductive step Suppose the wffs ϕ and ψ each have equal numbers of right and left parentheses (the inductive hypothesis). That is, suppose 'number of right parentheses in ϕ' = 'number of left parentheses in ϕ'= k, for some k in $\mathbb{N}$, and that 'number of right parentheses in ψ' = 'number of left parentheses in ψ'= m, for some m in $\mathbb{N}$.

The wff $\neg\phi$ has the same number of parentheses as ϕ, so it also has an equal number of right and left parentheses. For each binary connective *, the wff $(\phi * \psi)$ has $k + m + 1$ right parentheses and $k + m + 1$ left parentheses, hence each wff has an equal number of right and left parentheses.

10.3.3   Intuitions

I find Proof C explanatory, and in particular more explanatory than Proof B; I find Proof E explanatory, and in particular more explanatory than Proof D; and I find Proof F quite explanatory. How might we account for this intuition? Various respondents to Lange's paper have proposed that explanatory inductive proofs should be explanatory 'all the way through', as we might put it. For example, Hoeltje et al. suggest that explanatory inductive proofs might exist, 'if they involve explanatory subproofs of both inductive basis and step' (Hoeltje et al. 2013, p. 12), and Cariani proposes a 'Transmission Requirement: a proof by mathematical induction is explanatory only if the arguments for all of its components are themselves explanatory', adding that 'Sometimes one or more of these arguments will be completely trivial; that's enough to pass this requirement' (Cariani, p. 8).

Do my preferred proofs meet this condition? It looks like the pictorial proofs (C and E) do give us some insight into why the result holds for small values of n. For example, Proof C displays the connection between the sum $1 + 2 + \ldots + n$, the triangular number T_n and the term $\frac{n(n+1)}{2}$ in an easily accessible way, so that we can see directly why the result holds in the base case. The diagrams in Proof C also help to display the construction from one case to the next in an accessible way: we visualise a generic instance and the steps are operations on visualised instances, helping us to see directly why the result holds for $k + 1$ if it holds for k.

By contrast, Proof B links the relevant properties through symbol manipulation, using a chain of deductive inferences. Somewhere in the process of symbol manipulation, as we follow Proof B step by step, it is easy to lose track of what the terms mean. The steps are applications of rules of symbol manipulation, rather than operations on visualised instances. This might be one reason to think that Proof B is less explanatory than Proof C.

Note that this suggestion is not uncontroversial. For example, Baldwin denies that keeping track of the meaning of terms is necessary for a proof to have explanatory value:

> The glory of algebra is that one does not need (and possibly cannot) keep track of the meaning of each term in a derivation; nevertheless, the variables have the same interpretation at the end of the derivation as the beginning. The thought seems to be that losing track of the explicit reference of each term means the argument is non-explanatory but mere calculation. We have just seen the fallacy of this assertion . . . (Baldwin 2016, pp. 75–6).

This is no problem for me: if Baldwin is right and Proof B could or should count as explanatory after all, so much the better for my claim that some inductive proofs are explanatory.

Now, what should we make of the case involving structural induction, Proof F? Here we have a trivial proof of the base case, and the inductive step takes us through each possible way of constructing wffs, showing that an equal number of right and left parentheses are added in each construction. What more could we demand as an answer to the question 'Why is the theorem true?' – To my mind, Proof F looks like an explanatory proof.

The reader might (I hope) agree with my claims about the three cases presented, and my intuitions do have some support in the literature. For example, Cariani also presents (a version of) Proof E as a putative example of an explanatory inductive proof (Cariani, p. 10). However, as I noted earlier, Lange emphasises that he wishes to end what he calls the 'fruitless exchange of intuitions' surrounding proof by induction. So more needs to be said in support of my suggestion that some inductive proofs are explanatory.

One might be tempted to press the dialectical point here and simply claim that intuitions about the three cases are more secure than Lange's four assumptions, which we saw in Sect. 10.2 were controversial and sometimes problematic. Indeed, the cases themselves seem to support this point: for example, Proof F provides further reason to think that Lange's argument does not cover all cases of inductive proof, because (Reformulation) with its focus on $P1P/PkP$ cases does not straightforwardly apply to structural induction on the set of well-formed formulas.

However, I think the point is more subtle. The claim is not that we should rely exclusively on intuitions about examples, but that we should use examples to inform the questions we consider and the constraints we place on a successful account of mathematical explanation. In particular, it is important to pay attention to a sufficiently wide range of cases, so that we don't inadvertently jump to overgeneralisations about all inductive proofs from a small number of cases taken from the same domain. A lack of attention to cases from a wide range of mathematical domains probably helps to account for the common philosophical intuition that inductive proofs are not explanatory.

In later papers, Lange is in fact careful to pay attention to a wider range of examples, and he later also somewhat modifies his strong claim about proof by induction. Hence it would be interesting (and could help to support my intuitions) to look at whether the examples I proposed in Sects. 10.3.1 and 10.3.2 fit with Lange's more recent work. I will suggest one positive way of interpreting my examples from the viewpoint of Lange's account of mathematical explanation in the next (and last) section of the paper.

10.3.4   Salient Features

As I noted in the Introduction, Lange has recently somewhat relaxed his view on induction to the effect that inductive proofs could fall 'somewhere *between* an explanation and a proof utterly lacking in explanatory power. (Explanatory power is a matter of degree)' (see Lange 2014, p. 511, footnote 21, emphasis in the original).

This shift seems to have been prompted by his own recent account of mathematical explanation, outlined in the same paper.

Lange suggests that 'The distinction between proofs that explain why some theorem holds and proofs that merely establish that it holds exists only when some feature of the result being proved is salient', where the explanatory proof is one that exploits the salient feature (Lange 2014, p. 507). The idea is that we look for an explanation when some feature of a mathematical problem or theorem seems especially striking to us. Our request for an explanation is satisfied by a proof that exploits this very feature to prove the result. For example, we are sometimes struck by the feature of unity in a result: we wonder why the result holds for *all* members of the domain, and not just some of them. An explanatory proof is one that exploits unity and treats every case alike, in contrast with a proof by cases that treats cases separately and allows for the possibility that the separate cases might hold only as a matter of coincidence (Lange 2014, p. 510).

We might think that inductive proofs are a prime case of unifying proofs, since they show in one swoop that a property holds for every object in the domain. According to Lange, however, 'The inductive proof *nearly* treats every [case] alike. It gives special treatment only to the base case; all of the others receive the same treatment', and hence the inductive proof counts only as an almost-explanation (see (Lange 2014, p. 511, footnote 21), emphasis in the original). I think this qualification is questionable, because as we saw in Sect. 10.2.1, cases of strong induction do not in fact require a base case to be handled separately from the rest of the domain. However, I don't wish to pursue this point here for two reasons: (1) Many cases of (weak) induction do involve a separate base case; and more importantly, (2) I have some doubts that the kinds of why-questions we have about inductively-proved theorems focus on unity. (Take the theorem about the sum of the first n odd numbers: I think a more likely why-question is 'Why is the value of the sum equal to a perfect square, rather than some other value?').

Instead, therefore, I want to explore another way of making sense of (some) explanatory inductive proofs on Lange's account. My tentative suggestion, building on Lange's proposal, is that we could think of induction as a proof strategy that tends to produce explanatory proofs in cases where the *recursive nature of the domain* is particularly salient.

As we saw in Sect. 10.3.2, mathematical induction on $\mathbb{N}$ is just one kind of the more generally defined structural induction, which applies to any recursively defined set. In the case of the natural numbers, the recursive function that defines the set is the successor operation. We are very familiar with this relation and with initial elements in the ordered set from learning to count as children; however, we are rarely explicitly aware at an early age that the set of natural numbers has an important recursive structure. Indeed, the classic symbolic proofs like Proofs B and D are often introduced at a stage of education before students have explicitly entertained any thoughts about the structure of the set of natural numbers.

It is interesting to examine the mathematics education literature on this point. For example, in a study on secondary school students' understanding of mathematical induction, a cohort of 213 students were asked questions about mathematical induction in connection with examples involving natural numbers and real numbers, respectively. The authors note that 'most students stated that MI [mathematical induction] holds only in the set of natural numbers, but they could not recognize the way this claim is justified with respect to the properties of this set. Only three students identified the lack of succession in the set of real numbers, a necessary property for applying the steps of MI' (Palla et al. 2012, p. 1041), which might help to explain the difficulties many students face with understanding inductive proofs: the students have not yet appreciated the crucial feature that distinguishes the set of natural numbers from other number structures.

The authors go on to suggest that 'geometrical patterns can be an effective context for challenging secondary school students ... eventually, to attempt to produce a proof, mainly by appropriately corresponding the recursiveness of MI to the structure of the pattern' (Palla et al. 2012, p. 1043). This would be interesting to investigate in connection with my claim that the pictorial proofs are likely to be explanatory: my suggestion is that the diagrams draw our attention to the recursive nature of the domain by directly revealing a concrete way of constructing one case from the previous one. In situations like the pictorial Proofs C and E, since the recursive nature of the domain is (more likely to be) a salient feature for us, we (are more likely to) find the proof explanatory on Lange's account.

A deeper examination of the mathematics education literature must wait for another paper, but I think the suggestion could also account for the case involving structural induction, Proof F. In cases of more general structural induction like propositional logic, we are usually initially less familiar with the recursively defined set. Our acquaintance with the set is mostly, or solely, via its definition, rather than acquaintance with its elements; or rather, perhaps, our acquaintance with the set involves being explicitly aware of the recursive operation that defines it. Plausibly, then, it is likely that our attention is drawn to the recursive nature of the domain, and hence that the inductive proof is likely to seem explanatory, on Lange's account.

We might have some worries about this suggestion. For example, does it account for the way in which Proof E helps to answers the why-question I posed earlier: 'Why is the value of the sum equal to a perfect square, rather than some other value?' Unfortunately a deeper examination of (possible extensions of) Lange's account of mathematical explanation must also wait for another paper. Here I have simply presented one way in which his account could make room for explanatory inductive proofs.

10.4   Conclusion

In this paper, I have used simple examples from mathematics to suggest that there is reason to doubt Lange's argument for the claim that inductive proofs are never explanatory. In Sect. 10.1, I outlined Lange's argument and the background approach that motivates it. In Sect. 10.2, I examined the four assumptions on which Lange's argument relies. Attention to specific examples helped us to see, for example, that (Reformulation) is problematic and that (Asymmetry) is false under various interpretations. I also discussed a missing link in Lange's argument pressed by Hoeltje et al, suggesting that it is unlikely the link can be bridged by a general principle, and that it is plausible that there is no general rule linking induction and explanatory value. That is, it seems likely that some inductive proofs are explanatory and some are not. In Sect. 10.3, I put forward three putative examples of explanatory inductive proofs, discussing these in connection with Lange's recent account of mathematical explanation.

The general methodological points I have tried to make are that our assumptions about mathematical explanation should stand up to scrutiny of particular examples, and that general intuitions can be just as problematic as intuitions about specific cases. It seems plausible that our intuitions about general cases arise from a (possibly unconscious) generalisation from examples. Such generalisations are in danger of becoming overgeneralisations: I suggest that (some) philosophers have been too hasty to tar all inductive proofs with the same brush, perhaps due to an overfamiliarity with standard arithmetical cases.

Instead, we should be careful to distinguish between different inductive proofs and allow that some of them might be explanatory. Perhaps, moreover, the ones that tend to be explanatory fall into an interesting subclass; building on Lange's recent proposal, it might be those where the recursive nature of the domain is particularly salient that tend to be explanatory.

To develop this proposal further, I think we need to investigate a wider range of inductive proofs from different fields of mathematics. A bottom-up approach of this kind may help us to locate exactly the kinds of canonical examples that are currently lacking.

Acknowledgements   I would like to thank audiences in Cambridge, Chieti, London and Umeå for stimulating and helpful discussion. Particular thanks are due to Luke Fenton-Glynn and Marcus Giaquinto for their insightful comments on earlier versions of the paper. I worked on the paper while I was supported by grants from the UK Arts and Humanities Research Council and then the Royal Institute of Philosophy, Jacobsen Trust; my thanks to both institutions.

References

Baker, Alan. 2010. Mathematical induction and explanation. *Analysis* 70(4): 681–689.
Baker, Alan. 2012. Science-driven mathematical explanation. *Mind* 121(482): 243–267.
Baldwin, John. 2016. Foundations of mathematics: Reliability and clarity: The explanatory role of mathematical induction. In *Logic, language, information, and computation*. WoLLIC 2016, Lecture notes in computer science, vol. 9803, ed. J. Väänänen, Å. Hirvonen, and R. de Queiroz, 68–82. Berlin/Heidelberg: Springer.
Brown, James R. 1997. Proofs and pictures. *British Journal for the Philosophy of Science* 48: 161–180.
Cariani, Fabrizio. Mathematical induction and explanatory value in mathematics. http://cariani. org/files/MIEV.cariani.pdf, manuscript.

Dubeau, Francois. 1991. Cauchy and mathematical induction. *International Journal of Mathematical Education in Science and Technology* 22(6): 965–969.

Giaquinto, Marcus. 2011. Crossing curves: A limit to the use of diagrams in proofs. *Philosophia Mathematica* 19: 281–307.

Hafner, Johannes, and Paolo Mancosu. 2005. Varieties of mathematical explanation. In *Visualization, explanation and reasoning styles in mathematics*, ed. P. Mancosu, K. Jørgensen, and S. Pedersen, 215–250. Dordrecht: Springer.

Hanna, Gila. 1989. Proofs that prove and proofs that explain. In *Proceedings of the 13th International Conference for the Psychology of Mathematics Education*, vol. II, ed. G. Vergnaud, J. Rogalski, and M. Artigue, 45–51. Paris: Laboratoire PSYDEE.

Hoeltje, Miguel, Benjamin Schnieder, and Alex Steinberg. 2013. Explanation by induction? *Synthese* 190(3): 509–524.

Lange, Marc. Ground and explanation in mathematics. *Manuscript*.

Lange, Marc. 2009. Why proofs by mathematical induction are generally not explanatory. *Analysis* 69(2): 203–211.

Lange, Marc. 2010. What are mathematical coincidences (and why does it matter)? *Mind* 119(474): 307–340.

Lange, Marc. 2014. Aspects of mathematical explanation: Symmetry, unity, and salience. *Philosophical Review* 4: 485–531.

Palla, Marina, Despina Potari, and Panagiotis Spyrou. 2012. Secondary school students' understanding of mathematical induction: Structural characteristics and the process of proof construction. *International Journal of Science and Mathematics Education* 10(5): 1023–1045.

Steiner, Mark. 1978. Mathematical explanation. *Philosophical Studies* 34: 135–151.

Chapter 11
Parsimony, Ontological Commitment and the Import of Mathematics

Daniele Molinini

Abstract In a recent paper Alan Baker has argued for the thesis that the use of a stronger mathematical apparatus in optimization explanations can reduce our concrete ontological commitment, and this results in an increase of explanatory power. The import of this thesis in the context of the Enhanced Indispensability Argument is significant because it sheds light on how the Inference to the Best Explanation principle, on which the Enhanced Indispensability Argument crucially depends, may work at the level of concrete and mathematical posits in scientific explanations. In this paper I examine Baker's position and I argue that, although the employment of additional mathematical resources in some explanations can enhance explanatory power, it is highly controversial that Baker's example of cicadas can have a strong import in the platonism vs nominalism debate. I conclude with a general discussion of the way in which a stronger mathematical apparatus may sometimes lead to an increase of explanatory power.

11.1 Introduction

In a recent paper Alan Baker has given a new twist to the analysis of Inference to the Best Explanation (henceforth IBE) in the context of the so called Enhanced Indispensability Argument (henceforth EIA). EIA, as its name suggests, is an improved version of the Quine-Putnam indispensability argument. The basic idea behind the enhanced argument is that since mathematical entities participate in our best scientific explanations on an epistemic par with concrete unobservables, and since IBE supports rational belief in unobservable entities, we also ought to be committed to the existence of those mathematical objects that play an indispensable explanatory role in empirical science (Baker 2009). The debate that has recently emerged in connection with EIA is very lively and it has opened the road to a

D. Molinini (✉)
Centre for Philosophy of Sciences, University of Lisbon, Lisbon, Portugal
e-mail: dmolinini@fc.ul.pt

© Springer International Publishing AG, part of Springer Nature 2018   189
M. Piazza, G. Pulcini (eds.), *Truth, Existence and Explanation*,
Boston Studies in the Philosophy and History of Science 334,
https://doi.org/10.1007/978-3-319-93342-9_11

variety of philosophical analysis that spans a large spectrum of the contemporary philosophy of mathematics and general philosophy of science (for a survey, see Molinini et al. 2016). Among these directions of enquiry, one is particularly important because concerns the main core of the argument, namely the IBE principle on which EIA crucially depends.

IBE is a principle that is commonly related to the resources of scientific realism. Nevertheless, since in EIA it has been plugged into the Quine-Putnam indispensability argument together with the claim that mathematics can play an indispensable explanatory role in science, its scope has become broader and some mathematical realists (notably Alan Baker and Mark Colyvan) have employed it to motivate their realist commitment to the existence of mathematical entities. The question, however, remains about the way in which IBE may (or may not) act at the level of mathematical *and* unobservable physical posits in the same scientific explanation. Indeed, if we accept that IBE seems to offer us the same indirect, inferential epistemic access to concrete unobservables and abstracta (such as numbers) in the very same explanation, it is not clear how the mathematical realist may motivate its use and employ the same epistemic or inferential resources as scientific realists (Busch and Morrison 2016). On the other hand, it is not clear why the mathematical realist should *not* use IBE to get existential claims about mathematical objects. After all, the claim that scientific realists only use a causal model of explanation when they rely upon IBE is not uncontroversial and therefore the possibility that IBE may apply to concrete *and* abstract unobservables is left open (Colyvan 2001).

Few authors have addressed this issue, namely the question of how IBE may act at the level of mathematics and unobservable physical posits in scientific explanations (e.g. Pincock 2012 and Hunt 2016). Among these, Alan Baker has recently analyzed the topic from a novel perspective, thus giving a new twist to the analysis of the import of mathematics in scientific explanations (Baker 2016). More precisely, Baker has linked considerations of ontological parsimony and explanatory power with aspects that concern the use of a stronger mathematical apparatus in a particular class of explanations in science, namely optimization explanations. He has pointed out how ontological parsimony is an indicator of explanatory power, and a stronger mathematical apparatus sometimes reduces our concrete commitments. In tracing this link between ontological parsimony, explanatoriness and the use of mathematics, Baker not only has advanced a novel (though partial) account of how mathematical entities directly contribute to scientific explanations. With his contribution he has also provided a story about how mathematical and physical posits interact in a particular class of explanations, and this has a direct consequence on how IBE may work at the level of concreta and abstract mathematical objects in some scientific explanations.

In this paper I take Baker's analysis as starting point to consider a broader assessment of the connection between ontological parsimony, explanatoriness and the use of mathematics in science. After a short summary of Baker's main claim in the following section, in Sect. 11.3 I will discuss his position and show that it is hard to use his analysis as a lever to support the platonist stance in EIA. And this for two reasons: it is not clear that Baker's more parsimonious explanation fares better in terms of explanatory value; it is not even clear that Baker's explanation is the best explanation we have of the cicada phenomenon. Next, in Sect. 11.4, I will depart from Baker's analysis and put forward some more general considerations that show how the use of a stronger mathematical apparatus in science may lead to an explanation which has more explanatory power, and this without apparent loss (or gain) in concrete ontology. To illustrate my point, I will briefly discuss one example of optimization explanation taken from physics in which the use of more mathematical resources leads to a better explanation. In this case, I claim, the stronger mathematical framework permits to disclose some aspects of the explanandum that were not know before but that are linked to a broader web of knowledge that is relative to the particular scientific phenomenon being considered. In this process of mathematization, mathematics is explanatory because it shows how such connections hold, thus securing our web of knowledge about the scientific fact itself. This is how the 'extra mathematics' can sometimes increase the explanatory power of a scientific explanation.

Before continuing, I would like to stress that I won't, as Baker does in his paper, explicitly locate my analysis in the context of the debate between platonists and nominalists. Though what follows can be considered as an indirect contribution to that debate, I shall not address it in a direct way and my analysis will be more focused on the import of mathematics and ontological parsimony in explanations, together with the consequences that such an import may have on IBE and EIA.[1]

11.2 Parsimony, Mathematics, Explanatoriness

In Baker (2016) the author reconsiders his famous cicada-explanation (firstly proposed in Baker 2005) and offers a refinement of his analysis. A detailed treatment of the cicada-explanation is given in Baker (2005) and therefore I shall repeat here only the essential details concerning this biological example. It turns out that the emergence periods of the periodical cicadas are exactly 13 or 17 years, which are prime numbers. The rest of the time these insects stay dormant underground. In order to explain why it is evolutionarily advantageous for the cicadas to have such a prime-

---

[1]Baker points out that his analysis pushes the platonist case because the nominalist who accepts the more parsimonious explanation must also embrace the stronger mathematics, and it seems too difficult for him to offer a nominalized version of the mathematics used in such an explanation: "For if I am right then there is a general push in such cases towards more sophisticated mathematics, driven by considerations of concrete ontological, ideological, and structural parsimony. This more sophisticated mathematics tends to be more difficult (perhaps impossible) for the nominalist to paraphrase. And this gives more ammunition to the platonist cause, even when it rests on examples of MES [mathematical explanations in science] that look deceptively simple" (Baker 2016, p. 350). Although the possibility of offering a nominalistic reconstruction of Baker's more parsimonious explanation is left open, I do not tackle this issue here.

numbered dormant period we can turn to a number-theoretical result, which guarantees that prime number periods maximize the infrequency of overlap with periodical predators that are assumed to have shorter life periods than the cicadas. The number-theoretical result is contained in the following Lemma (call this Lemma$_1$): If p is any natural number, then p is coprime with every $q < p$ (and hence maximizes the lowest common multiple with every $q < p$) iff p is prime. The resulting explanation is endorsed by some evolutionary biologists (Yoshimura 1997; Cox and Carlton 1998). For simplicity I shall refer to this explanation, using Lemma$_1$, as Expl$_1$.

Baker considers Expl$_1$ as an example of optimization explanation, namely "an explanation of some physical phenomenon, P, that proceeds by situating P as the solution to an optimization problem. An optimization problem is one which seeks to optimize some target feature, T, given certain constraints, C_1, ..., C_m, and certain options, O_1, ..., O_n." (Baker 2016, p. 335). In the case of cicadas, the phenomenon to be explained is the primeness of life cycles and the target feature is the avoidance of periodical predators. The constraints are given by ecological factors (e.g. nutrient availability), while the options are the life-cycle lengths that are biologically possible given these particular constraints.

There are some interesting observations that Baker adds in his 2016 paper and that are not included in his 2005 work. Expl$_1$ *presupposes* that there exist periodical predators of cicadas and that these predators have shorter life periods than the cicada (from 2 years up to 12 years, which is the hypothesized lower bound of cicada periods). This is a presupposition because, as Baker points out, there is no empirical evidence for the existence of such predators. Moreover, using this presupposition in our explanation has a strong ontological import. In fact, if we consider Expl$_1$ as our best explanation of the primeness of the cicada life-cycles, then we should be committed (via IBE) to the existence of predators with shorter life periods than the cicada (the predator's periods may be of every length from 2 years up to 12 years). Applying IBE in this situation, however, seems to raise a problem. The problem with the IBE-generated inference is that biologists seem to think that not all the predator species corresponding to all lower life periods do exist. In other words, scientists believe that the concrete ontology which is needed for this explanation is *too large*. And their intuition seems to be confirmed by the absence of empirical evidence concerning the existence of such periodical predators. As Baker observes, this raises a request of ontological parsimony with respect to the number of types of periodical predators (i.e. predators with periods of 2 years, 3 years, etc.) that we should postulate in our explanation: "is there some smaller range of predator periods for which prime cicada periods remain uniquely optimal?" (Baker 2016, p. 337). Restating the question with the help of mathematics, we want to know what is the minimum upper bound for predator periods that guarantees that prime cicada periods are still uniquely optimal within the range of available cicada periods, which is $12 \leq p \leq 18$.

The answer to our question comes from mathematics, and more precisely from the following Lemma (call this Lemma$_2$): Given two intervals on the natural numbers, $2 \leq q_j \leq z$, and $x \leq p_i \leq y$, such that $z < x$, then the following biconditional holds: no composite p_i is coprime with every q_j (and hence maximizes the lowest common multiple with every q_j) iff the smallest prime, $p^* > z$ is such that $p^* \cdot p^* > y$. If $y = 18$ is the upper bound of the cicada range, this lemma assures us that the minimum upper bound on predator periods that makes a prime cicada period uniquely optimal is the largest prime that is less than or equal to $\sqrt{y}$, which is 3. Therefore, if 3 is the upper bound for periodical predator periods, we can replace Expl$_1$ with an explanation, call it Expl$_2$, in which we only need to postulate the existence of predators with 2- and 3-year life periods. The significance of this is that using the 'stronger' mathematics of Lemma$_2$ we have weakened our concrete commitment (by reducing the number of types of concrete entities, namely postulated types of periodical predators) and accomplished the biologists' request for ontological parsimony. But why should we regard Lemma$_2$ as *stronger* than Lemma$_1$? As Baker notes, Lemma$_2$ is stronger in the sense that it is more general. In fact, Lemma$_1$ can be derived from Lemma$_2$ as a special case (Baker 2016, pp. 338–339).

Hence we have that the greater mathematical resources used in Lemma$_2$ have reduced our concrete ontological commitment, leading to an explanation that is more parsimonious and more explanatory[2]:

> One of my main goals in this paper is to argue for a thesis [...] The thesis is that mathematics can also help to reduce our concrete ontological commitments. In particular, there is a class of mathematical explanations in science (MES) for which strengthening of the mathematical results involved can lead to a reduction in the number and variety of concrete posits that are needed for the explanation to go through (Baker 2016, p. 334)

> the Lemma 2-based MES improves on the original cicada MES both in terms of parsimony and in terms of explanatory power. It seems clear that it is a better explanation [*Ibid.*, p. 339]

Baker's argument, henceforth BA*, can be therefore summarized as follows:

(P_1^*) The hypothesis that requires the assumption of predators with fewer life cycles periods is more explanatory;

(P_2^*) Lemma$_2$ reduces the predator's life cycles to a minimum (2 and 3-year periods are the smaller periods for which prime cicada periods remain uniquely optimal), thus reducing our concrete commitment;

(C^*) Lemma$_2$ yields explanatory power.

Hence the general argument (call this argument BA), which BA* is a particular instantiation of, can be spelled out as:

(P_1) There are cases in scientific practice in which the hypothesis that requires the assumption of fewer types of concrete posits is more explanatory (ontological parsimony is an indicator of explanatory power);

(P_2) A stronger mathematical apparatus, i.e. a more general mathematical result, helps reducing our concrete commitments in some cases of scientific explanation (such as the cicada case);

---

[2]Baker also considers that Expl$_2$ is more explanatory because it permits us to answer various counterfactual questions. More on this point in the next section.

(*C*)   A stronger mathematical apparatus yields explanatory power in some cases of scientific explanations.

According to BA*, Expl_2 (using Lemma_2) should be preferred over Expl_1 because it fares better in terms of explanatory value. In the context of IBE and EIA, this means that if we consider Expl_2 as our *best* explanation of the cicada-emergence, as Baker does, we should be committed only to those unobservables that are needed for this explanation, namely predators having life cycles of 2 and 3 years and those mathematical objects (prime numbers) used in Lemma_2.

The first goal of next section is to discuss P_1^* and show that this premise is hard to be maintained in the way Baker does. I argue that P_1^* needs a more robust defense, and providing such a defense is especially important if Baker wants to use his argument in the context of EIA. Secondly, I provide a discussion of Expl_1 and I show how it is not clear that this explanation is considered by biologists the best explanation of the cicada-emergence phenomenon. In this case too, my considerations cast doubt on the possibility of applying IBE to Baker's explanation(s) and infer existential claims through EIA.

11.3   Cicadas, IBE and EIA

The type of ontological parsimony Baker discusses in his cicada example is qualitative parsimony, namely parsimony with respect to the number of types of entities (number of periodical predators having life cycle of a specific duration in the cicada case). Qualitative parsimony, which should be kept distinct from quantitative parsimony (Lewis 1973, p. 87), is generally seen as a theoretical virtue and its influence on scientific practice seems undeniable.[3]

When faced with a range of alternative scientific hypotheses, scientists seem to prefer the hypotheses that, other things being equal, minimize the number of types of concrete entities postulated. Nevertheless, note that in P_1 (and the same holds for P_1^*), the clause 'other things being equal' is not mentioned. And this is precisely because the idea behind P_1^* is that qualitative parsimony makes a difference in terms of explanatory value. We have two hypotheses, one which uses more types of concrete posits (periodical predators having life cycles of 2 up to 12 years) and the other which uses less (only those periodical predators that have life cycles of 2 and 3 years). These hypotheses not only differ with respect to the ontological assumption

---

[3]With his example, Baker also claims in favor of quantitative parsimony: "strengthening of the mathematical results involved can lead to a reduction in the *number* [emphasis added] and variety of concrete posits" (Baker 2016, p. 334). Although he offers no argument in defense of the claim that the use of additional mathematical resources leads to quantitative parsimony, the idea seems to be that by reducing the number of predator-types P_i to P_2 and P_3 (using Lemma_2), it also follows a reduction of the total number of individual predators postulated. This is obviously true for the case considered by Baker. Nevertheless, it is worth noting that such an argument is not conclusive because, in general, qualitative parsimony does not entail quantitative parsimony.

made about concreta but also in terms of the (greater) explanatory power that one (the more parsimonious) has over the other (the less parsimonious).[4]

Consider now the parsimony criterion on which P_1^* is based. As pointed out by Stathis Psillos, parsimony is a standard of explanatory merits that be spelled out in the following form: suppose two composite explanatory hypotheses H_1 and H_2 explain all data and H_2 uses fewer assumptions than H_1; suppose also that the set of hypotheses that H_2 employs to explain the data is a proper subset of the hypotheses that H_1 employs; then H_2 is to be preferred as a better explanation (Psillos 2009, p. 184). In Baker's example the hypothesis H_2 behind $Expl_2$, in which we only postulate predators having life cycles of 2- and 3-years, is supposed to have more explanatory value than the hypothesis H_1 behind $Expl_1$, in which we postulate all predators that have lower periods that the cicada life cycle. Both hypotheses explain the data, but H_2 has more explanatory power because it involves the postulation of fewer types of concrete entities. Note, however, that this would amount to conflate ontological parsimony with explanatory power, which is something that may be philosophically inaccurate. Daniel Nolan as well, in his influential paper on quantitative parsimony, resists this idea and sees ontological parsimony and explanatory power as two separate notions:

> Of course, it might be possible to define 'explanatory power' in such a way that only the most quantitatively parsimonious postulates were allowed into 'explanation', so defined, but I take it that this would be a verbal victory achieved only by producing an ad hoc redefinition of 'explanatory power'. Better, rather, to have quantitative parsimony expressed as a different principle to the independently plausible principle about 'explanatory parsimony', rather than tying them together in this way (Nolan 1997, p. 339)

It seems natural to think that ontologically parsimonious hypotheses should be preferred because they provide us with better explanations, and not that ontologically parsimonious hypotheses are more explanatory *because* they postulate fewer types of concrete objects. Of course, we might claim for the latter idea, as Baker does, and maintain that quantitative parsimony leads to an increase of explanatory power in some cases (such as the cicada case). Nevertheless, this again raises the question of how explanatory power is increased. We cannot answer this question by simply saying that explanatory power is increased because we postulate fewer types of entities in our explanation. This reply, in fact, would be unsatisfactory to some and, what is worse, it would exhibit a circularity that we may want to avoid (the more parsimonious hypothesis is more explanatory because it is more parsimonious). Independent reasons are therefore needed to secure that the most parsimonious

---

[4]In a previous work, Baker has defended the idea that quantitative parsimony yields explanatory power: "Quantitative parsimony tends to bring with it greater explanatory power. Less quantitatively parsimonious hypotheses can match this power only by adding auxiliary claims which decrease their syntactic simplicity. Thus the preference for quantitatively parsimonious hypotheses emerges as one facet of a more general preference for hypotheses with more explanatory power" (Baker 2003, p. 258).

hypothesis is more explanatory.[5] This is, in fact, something that Baker seems aware of. In analyzing the greater explanatory value of $Expl_2$ (based on hypothesis H_2) over $Expl_1$ (based on hypothesis H_1), in addition to the less (concrete) ontological commitment involved in $Expl_2$ and to some indicators coming from the practice of biologists, Baker also notes that $Expl_2$ permits answering various counterfactual questions, as for instance "what if the cicada range remain the same but there were no 3-year predators, only 2-year predators?" (Baker 2016, p. 339). This counterfactual-gain is, for sure, a trait that makes $Expl_2$ superior, in terms of explanatory power, to $Expl_1$. There is, however, a doubt that may be raised at this point and that concerns the explanatory boost which, according to Baker, $Expl_2$ offers with respect to $Expl_1$. This doubt has to do with a particular feature that a *better* explanation is supposed to have and that corresponds to fit with background data and knowledge about the phenomenon we want to explain.

The explanatory potential of an explanation critically depends on information (e.g. data) available in the background knowledge that concerns the phenomenon itself. Psillos calls "consilience" this feature of IBE which connects the background knowledge with the potential explanation of the evidence: "Suppose there are two potentially explanatory hypotheses H_1 and H_2 but the relevant background knowledge favors H_1 over H_2. Unless there are specific reasons to challenge the background knowledge, H_1 should be accepted as the best explanation" (Psillos 2009, p. 184). It is therefore interesting to evaluate consilience in the cicada's case, namely to examine how the two hypotheses H_1 and H_2 are favored by the relevant background knowledge, and see how it fares when compared to the parsimony standard adopted by Baker.

The relevant background about cicadas includes ecological data (together with constraints) and data about the life cycles of cicadas. As for the presence of cicadas' *periodical* predators, which constitutes part of our background knowledge about the phenomenon, Baker observes that there is "no direct empirical evidence" (Baker 2016, p. 336). Indeed, while it has been largely documented the existence of cicada's predators which are not periodical such as small mammals, birds and parasites (Behncke 2000), no evidence has been found of periodical predators that produce periodic predation pressure on cicadas. Therefore we have to include this information in our background knowledge about the cicada emergence. As a consequence, when considering also consilience as explanatory-standard in the cicada's case, $Expl_2$ does not fare better than $Expl_1$ in terms of explanatory power. This is because our relevant background knowledge about cicadas (which includes well documented observations), and more particularly about the existence of cicadas' periodical predators, not only does not favor $Expl_2$ in any relevant sense

---

[5]Clearly, we may deny that there exists a link between parsimony and explanatoriness, however maintaining that parsimony is a theoretical virtue. In this case, we could still prefer $Expl_2$ over $Expl_1$ (because we consider the first as based on a more theoretically virtuous hypothesis). Although this standpoint is plausible, by adopting it we lose a crucial ingredient of Baker's point, namely the connection between parsimony and explanatory power.

but it also seems to suggest that such periodical predators do not exist. This is a key point. In the case of cicadas, in fact, differently to what happens for the explanation of other scientific phenomena in which we postulate unobservables such as black holes or quarks, we do have a direct empirical access to the phenomenon itself. Moreover, we have an abundance of observations and biological data concerning the 13 and 17 years cicadas and their environment. But what all these observations and data seem to indicate is that there are currently no periodical predators of cicadas, nor have been in their evolutionary past. It is therefore hard to believe that periodical predators have (or had) a strong influence on the cicada's life cycles.[6] For instance, Glenn Webb observes that "there is little field data to argue [for] the existence of perfect 2 and 3 year cycling cicada predators" (Webb 2001, p. 389), while Ryusuke Kon hypothesizes the existence of a periodical predator that produces periodic predation pressure on periodical cicadas but soon after he points out that "It is unlikely that such a predator exists and it is unclear that such a predator has existed" (Kon 2012, p. 856).

How then $Expl_2$ can be said to be a *better* explanation of the phenomenon? Of course, Baker may reply to my observations above simply by adducing considerations of ontological parsimony (i.e. $Expl_2$ is more parsimonious and therefore more explanatory than $Expl_1$). And we might concede, in accord with what Baker claims, that the parsimony standard operates crucially in choosing one explanation over another when the substantive information contained in our background knowledge about the phenomenon cannot discriminate between competing potential explanations of the evidence. But, again, it is not clear in what sense ontological parsimony per se should be an indicator of explanatory power. And this is especially the case when data and observations seem to pull against the existence of the concrete entities (periodical predators) that we should be more parsimonious about. Using an expression that Van Fraassen has coined in his criticism of IBE, it seems that we should regard $Expl_2$ as "the best of a bad lot" (Van Fraassen 1989, p. 143). Moreover, if we consider ontological parsimony as the only metric of explanatory power, we could also stretch Baker's standpoint to a limit and assume that our best explanation of the cicada's emergence should incorporate the information that there are *no* periodical predators of cicadas. After all, this assumption would fit well with our background knowledge and biological observations, which suggest that there are no such periodical predators, and it may turn out to be perfectly plausible from a biological point of view. In this alternative no-periodical-predators explanation, we would have a maximum of ontological parsimony with respect to the postulation of periodical predators.[7]

---

[6]Obviously, evidence for the existence of present, or past predators (for which periodicity cannot be observed from fossil evidence), may be found. On the other hand, lack of direct evidence of periodical predators is certainly a weakness of Baker's explanations, as acknowledged by Baker himself in his last paper on the subject (Baker 2017b, p. 3).

[7]This explanation may be based on features that do not involve the existence of periodical predators, as for instance those discussed later in this section. After all, from a biological point of view the hypothesis that cicadas have no (or had not, in their evolutionary past) predators is perfectly plausible. In biology, a species with no predator is called an 'aphex predator'.

Although the no-periodical-predators explanation I put forward at the end of the previous paragraph may seem ad hoc to some, the points above raise doubt about the greater explanatory power that, according to Baker, $Expl_2$ has over $Expl_1$. More precisely, I showed how the criterion of ontological parsimony adopted by Baker may be insufficient to establish that the hypothesis that requires the assumption of fewer predator's life cycles has more explanatory power (as affirmed in P_1^*). The immediate consequence, in the context of EIA, is that Baker has not conclusively established that $Expl_2$ is the *best* explanation of the phenomenon of cicadas emergence and that it can be used (via IBE) to infer the existence of (some) periodical predators and numbers.

The present discussion raises a more general issue about the existence of alternative explanations of the cicadas' life cycles. Is Baker's $Expl_1$ based on the purported existence of periodical predators (and so the less ontologically expensive $Expl_2$) the only explanation on the market? Is this the *best* explanation we have according to biologists? Providing an answer to these questions is particularly important in the context of IBE and EIA. And this is because, if explanation $Expl_1$ is the best explanation biologists have of the cicadas' emergence, Baker could simply reply to my worries above pointing to the fact that the IBE criterion acts exactly because $Expl_1$ is our best available explanation of the phenomenon. Now, although Baker's $Expl_1$ is largely regarded by philosophers of mathematics as a genuine mathematical explanation of the prime numbered cicadas' life cycles, little attention has been devoted to the issue of whether this is the *best* explanation biologists have of the phenomenon. In the remaining part of this section I want to focus on this question and show how it is not clear that this explanation is considered by biologists the best explanation of the cicada-emergence phenomenon.

When reviewing part of the (huge) biological literature about the cicada-emergence it is easy to see that biologists have proposed, and are still proposing, different explanations for the prime numbered cicadas' life cycles. And this suggests that they are not fully satisfied with the kind of explanation endorsed by Baker, Cox and Carlton, and Yoshimura. Ryusuke Kon, for example, has proposed a different mathematical model to study the dynamic interaction between a periodical cicada species and its hypothetical predator. In his model he assumes the existence of an hypothetical predator whose influence on cicadas' life cycle is periodic and develops a population model for periodical predator and prey. Moreover, contrary to what the mathematician Webb does in his mathematical model, Kon does not base his model on the two assumptions that 'the predator dynamics is independent of the cicada dynamics' and that 'periodical cicadas initially emerge when periodically oscillating predators are abundant' (Kon 2012, p. 856).[8] Although the mathematical details of Kon's model are complex and cannot be reported here, Kon's conclusion that "prime

---

[8]This is relevant to the present discussion because, as observed by Baker, Webb's model assumes only the existence of periodical predators with periods of 2 years or 3 years, which are the periodical predators assumed in $Expl_2$ (Webb 2001, p. 389). Furthermore, Webb's assumption that the predator dynamics is independent of the cicada dynamics is also present in Baker's explanation.

periodicities are not advantageous even under periodic predation pressure" (*Ibid.*, p. 856) explicitly conflicts with Baker's explanations. In Kon's model prey's perfect periodicity is maintained by some mechanisms, e.g. satiation of predators and inter class competition, that are not due to the presence of the hypothetical predator. And this hypothesis, namely to consider that satiation of predators such as birds and resource competition stabilize perfect periodicity, has been explored by other authors as well (Hoppensteadt and Keller 1976; cf. also May 1979; Behncke 2000). Thus there exist other aspects that have been regarded by biologists as relevant to the cicada's emergence phenomenon. Nevertheless, these aspects do not play any role in Baker's explanations.

Other features of the cicada's emergence have been studied and proved to be relevant to the peculiar life cycles lengths of periodical cicadas. For instance, an empirical study by Koenig and Liebhold shows how aviation pressure is related to the life cycle of cicadas (Koenig and Liebhold 2013). In particular, this study reveals that periodical cicada emergences appear to set populations of potential avian predators on numerical trajectories that result in significantly lower potential predation pressure during the subsequent emergence. One of Koenig and Liebhold's main conclusions is that "the extraordinary life cycle of periodical cicadas may engineer bird populations in a way that keeps them from tracking emergences, not by their extreme length or as a consequence of being prime numbered, but rather by setting bird populations on a trajectory such that the subsequent emergence coincides with reduced predation pressure" (Koenig and Liebhold 2013, p. 148). This aspect, again, is not considered by Baker in his explanations. However it has been shown (on the basis of an empirical study) to be relevant to the explanation of the cicada's life cycles.

Life-cycle lengths of 13- and 17-year cicadas are usually considered unvarying, however experimental and observational data of cicadas emerging off-schedule in numbers too large to be plausibly accounted for by mutation have revealed that the 13 and 17 life-cycle lengths may "shift" from the values 13 and 17. Although the shift of 13-year cicadas emerging in 17 years has never been documented, evidence is strong for 1-year premature and/or delayed emergences and for jumps of 4 years (Marshall 2001). Four-years accelerations occur in 17-years cicadas and historical records also reveal that 17-years cicadas sometimes emerge from 6 years early to 5 years late (Marlatt 1907; Maier 1985).[9] According to Marshall, who has extensively studied the cicadas' shifts, these variations in life lengths are due to developmental plasticity (the ability of an individual to modify its development in response to environmental conditions), rather than mutations in life-cycle genes (Marshall 2001). In a later study, Marshall et al. also propose a life-cycle model to account for such shifts (Marshall et al. 2003). Now, Baker's explanations $Expl_1$ and

---

[9]Shift in the periodicities of magicicadas having either 13 or 17 years life cycles has been also considered theoretically by other biologists. For instance, the evolutionary biologist Peter Grant suggests that "these life cycles evolved earlier than the Pleistocene and involved an abrupt transition from a nine-year to a 13-year life cycle, driven, in part, by interspecific competition" (Grant 2005, p. 169). On the fixation of the cicadas' periodicity see also Ito et al. (2015).

Expl$_2$ would still be capable of accounting for the emergence of a 17-year cicada that shifts to a 13-year cycle (in fact, they would still explain why the cycle has remained a prime number, and precisely 13), and shifts of +4 or −4 years can be accounted for by strengthening Baker's original explanation through an auxiliary hypothesis.[10] Nevertheless, it is still an open issue whether or not *all* such unexpected emergences may be properly integrated in Baker's mathematical explanations. Clearly, it may also be contended that such shifts are not stable and that they are due to genetic mutations. And therefore they are consistent with 13 and 17 being stable endpoints (because, by supposition, prime periods are evolutionary advantageous). Note, however, that in this case it is reasonable to think that an explanation of the prime-numbered emergences should incorporate a genetic premise. This is not the case of Expl$_1$ and Expl$_2$. On the other hand, the novel version of Expl$_1$ proposed in Baker (2017b) includes a genetic premise and it is capable to account for 4 years accelerations/delays. It does not, however, cover the well documented shifts of +1 and −1 years. Furthermore, this new formulation of the cicada explanation does not assume the existence of periodical predators with fewer life cycles (indeed, it is a refinement of Expl$_1$, not of Expl$_2$). And therefore, in the context of the present discussion, it is not clear what Baker considers as the best explanation of the cicada emergence. Although these observations do not count as fatal objections, they clearly point to some limitations of Baker's explanation because they leave open the possibility that our best explanation of the cicada phenomenon be an explanation in which the explanatory load is carried by physiological or genetic mechanisms.[11]

Rather than offering explanations that are better (in terms of explanatory potential) than Baker's, the goal of the last paragraphs of the present section has been to show that there is no consensus among biologists that Baker's explanations are the best. In fact, biologist are still looking for models and explanatory stories different from Baker's to account for the cicada life cycles. Some of these models, as for instance Kon's, explicitly contrast with Baker's explanations, while others in-corporate data that have been regarded by some biologists as relevant to the cicada's emergence but that are not mentioned in Baker's explanations. Unsurprisingly, the cicadas' phenomenon of emergence is rather complex and it encompasses a large amount of data and observations. And it is therefore natural to see how Baker's

---

[10]In his more recent paper on the subject, Baker explicitly addresses the phenomenon of period length shifts and proposes a renewed version of Expl$_1$ (Baker 2017b). This version of Expl$_1$ includes the 'genetic' premise "Periodical cicadas are limited by genetic constraints to periods of the form $4n + 1$" and is capable of accounting for shifts of +4 or −4 years.

[11]Nariai et al., for instance, believe that genetic factors have an important role in the life switch of a 17-year cycle population to a 13-year cycle (Nariai et al. 2011). More precisely, they show how life cycle switching by gene introduction appears to be possible under fitness reductions at low population densities of mating individuals ('Allee effect'). But Nariai et al. are not alone in thinking that genetic mechanisms play a decisive role in the cicadas' emergence phenomenon. Lehmann-Ziebarth et al. propose that "the explanation for prime-numbered periods, rather than just fixed periods, may reside in physiological or genetic mechanisms or constraints" (Lehmann-Ziebarth et al. 2005, p. 3200), and yet the sort of explanation that these biologists have in mind seems to considerably diverge from the kind of explanation proposed by Baker.

explanations may appear too simplistic to some biologists, who are still looking for a more decisive and comprehensive explanation.

11.4 Mathematics and Cannonballs

In the previous section I have pointed out how it has not been conclusively established that Baker's explanations are the best. The immediate consequence of my observations is to weaken the impact that Baker's analysis has in the context of EIA. Remember, in fact, that in order to use an explanation in EIA to infer existential claims via IBE we should first establish that such an explanation is the best explanation we have of the phenomenon under investigation.

In this section I want to sketch a different analysis concerning the relation between explanatory power and the use of additional mathematical resources in optimization explanations. More precisely, with the help of an elementary example taken from physics, I argue that the claim "A stronger mathematical apparatus yields explanatory power in some cases of optimization explanations" can be maintained. I provide an analysis of how this is possible, namely an analysis of how the use of more mathematical resources can provide more explanatory power. Note, however, that my defense of the claim that a stronger mathematical apparatus yields explanatory power is offered on the basis of considerations that are different from Baker's. Moreover, in the case I propose, the greater mathematical resources seem to not affect the concrete ontology that we need for the explanation to be successful.

Consider a cannonball that is launched from a cannon at different launch angles θ_i but with the same launch speed v_0 every time. After a series of launches we observe that we obtain the greatest range for the cannonball if the launch angle is $45°$. This is a great surprise and we want to know why the maximum range is obtained for that particular angle. We therefore start analyzing our scenario with the help of mathematics and our knowledge of classical mechanics. The only force that acts upon the cannonball is gravity, which acts downward and can be considered constant during motion. This means that the effect of gravity is independent of the horizontal motion of the object. We can therefore find the kinematic equations for the cannonball motion as those of a body traveling with constant horizontal velocity and constant vertical acceleration, with the two components independent of each other. The components of the initial velocity can be written as $v_x = v_0 \cos \theta$ and $v_y = v_0 \sin \theta$. Hence we can write our kinematic equations: $x(t) = v_0 t \cos \theta$ and $y(t) = v_0 t \sin \theta - \frac{1}{2} g t^2$. Using the typical assumption that the cannonballs do not encounter air resistance, we can find the horizontal range of the cannonball motion and solve an optimization problem to discover how launch angles are related to the maximum range.

If the cannonball is fired from the ground level, the range is the distance between the launch point and the landing point, where the projectile hits the ground. When the projectile comes back to the ground, the vertical displacement is zero. Thus we have $0 = v_0 t \sin\theta - \frac{1}{2}gt^2$. Solving this equation for t we have the two solutions $t_1 = 0$ and $t_2 = 2v_0 \sin\theta/g$. The first solution gives the time when the projectile is thrown and the second one the time when it hits the ground. Inserting the second solution into $x(t) = v_0 t \cos\theta$ we get that the range R of the cannonballs is $R(v_x, v_y) = 2v_x v_y/g$. Furthermore, using the formula $2\sin\theta\cos\theta = \sin 2\theta$, we have the range R only as a function of θ (v_0 and g are constant):

$$R = \frac{v_0^2}{g} \sin 2\theta \tag{11.1}$$

We can now solve a simple optimization problem and find why the range is maximum at $45°$. It is sufficient to observe that R is maximized when $\sin 2\theta = 1$ (the greatest value that sin can have is 1), and $\theta = \pi/4 = 45°$ is the only angle that satisfies $\sin 2\theta = 1$.[12]

The previous optimization method clearly gives us a connection between two physical aspects of our setting (the launch angle and the range of the cannonball) and allows us to gain knowledge about the cannonballs launch. In particular, it shows why the cannonball range is maximum at $45°$ on the basis of the mathematical fact that the sine of an angle cannot have a value bigger than 1. And therefore it should be considered as a non-causal (mathematical) explanation of the physical phenomenon under investigation.[13] Nevertheless, the mathematical approach followed above is not the only path to explain the maximum-range phenomenon. We can also provide an explanation using the more general mathematical resources provided by the method of Lagrange multipliers.

The method of Lagrange multipliers is a powerful tool for solving optimization problems with constraints and it is largely employed in physics, engineering, economics and mathematics. The way in which this method is applied to find the extrema of a function subject to a constraint is fairly simple and it can be illustrated with a short example. Consider we want to maximize or minimize the function $f(x, y)$ which is subject to the constraint $g(x, y) = k$. First, we create the Lagrange function $L(x, y, \lambda) = f(x, y) - \lambda[g(x, y) - k]$, where λ is called 'Lagrange multiplier'. This function is composed of the function to be optimized combined with the constraint function. Then we find the partial derivative with respect to each variable x, y and the Lagrange multiplier λ, namely $L_x = \partial L/\partial x$, $L_y = \partial L/\partial y$ and $L_\lambda = \partial L/\partial \lambda$. Next, we set each of the partial derivatives equal to zero to get

---

[12]For the unconvinced reader: if $1 = \sin 2\theta$, then $\sin^{-1}(1) = 2\theta$; therefore $\frac{\pi}{2} = 2\theta$, which means that $\theta = \frac{\pi}{4}$.

[13]En passant, let me note that my argument in this section does not necessitate that the optimization explanation considered be distinctively mathematical. On the other hand, it seems quite natural to consider it as such because it essentially depends on a mathematical fact which cannot be expressed in causal terms. Moreover, my impression is that this particular explanation can be captured in terms of a particular account of mathematical explanation in science, as for instance the account advocated by Marc Lange in his (2013). However interesting, I shall not pursue this issue here and I leave it for future work.

$L_x = 0$, $L_y = 0$ and $L_\lambda = 0$ (setting the partial derivatives to zero amounts to find the critical points of $L(x, y, \lambda)$). Using $L_x = 0$ and $L_y = 0$, we can now solve for x and for y in terms of λ. Finally, we substitute the solutions for x and y so that $L_\lambda = 0$ is in terms of λ only. It is now easy to solve for λ and use this value to find the optimal values of x and y. Moreover, at this point we can observe that if M is the max or min value of $f(x, y)$ subject to the constraint $g(x, y) = k$, then the Lagrange multiplier λ is the rate of change in M with respect to k.[14]

In our example of the cannonball motion, the function to be optimized is the range function $R(v_x, v_y) = 2v_x v_y/g$ and the constraint is given by the kinetic energy of the cannonballs $m(v_x^2 + v_y^2)/2 = E$, which is fixed (m is the mass of the cannonball). We can therefore write our Lagrange function as $L(v_x, v_y, \lambda) = 2v_x v_y/g - \lambda[m(v_x^2 + v_y^2)/2 - E]$. Taking partial derivatives with respect to v_x, v_y and λ we get the set of equations:

$$2v_y/g - \lambda m v_x = 0 \tag{11.2}$$

$$2v_x/g - \lambda m v_y = 0 \tag{11.3}$$

$$m(v_x^2 + v_y^2)/2 = E \tag{11.4}$$

The solutions are $v_x^* = v_y^* = \sqrt{E/m}$ and $\lambda = 2/mg$. The maximum range is therefore $R(v_x^*, v_y^*) = (2/mg)E$ and $\lambda = dR(v_x^*, v_y^*)/dE$ is satisfied. If we want to read this result in terms of the launch angle θ, it is sufficient to write the maximum range $R(v_x^*, v_y^*)$ in terms of v_0 and confront it with the general expression for the range, which is $R(v_x, v_y) = 2v_x v_y/g$:

$$R(v_x^*, v_y^*) = \frac{2}{mg}\left(\frac{mv_0^2}{2}\right) = \frac{v_0^2}{g} \tag{11.5}$$

$$R(v_x, v_y) = \frac{v_0^2}{g}\sin 2\theta \tag{11.6}$$

Equations 11.5 and 11.6 are equal only if $\sin 2\theta = 1$, and therefore we clearly see that the maximum range is obtained for that angle which satisfies $\sin 2\theta = 1$. As we have seen, this angle is $\theta = 45°$. What, then, is the difference with our initial explanation, in which we did not use the more complex machinery provided by the Lagrangian multipliers? What exactly makes the second explanation different?

The difference is in terms of explanatory value. Indeed, as observed by the physicist Hasan Karabulut, Lagrange multipliers may have a physical interpretation and in the case of the cannonball motion the physical meaning of $\lambda = 2/mg$ is the rate of increase of maximum range $R(v_x^*, v_y^*)$ with energy E (Karabulut 2006). This means that if the kinetic energy E (our constraint) was increased from E to

---

[14]Note that $\lambda = dM/dk$ and therefore λ approximates the change in M resulting in a one unit increase in k.

$2E$ (a one unit increase), then the maximum range would increase by $2/mg$. We therefore have that the more general mathematics used in this second explanation, namely that using the Lagrange multipliers, not only shows why for an angle of 45° we get the maximum range but it also discloses aspects of the phenomenon that we do not 'see' in our first explanation.[15] In this second explanation, indeed, we see how the range of the cannonball is related to its kinetic energy E and how the maximum range would increase if we increase the energy of the cannonball.[16] Both are optimization explanations, however the second should be considered as a better explanation because it discloses more connections between the elements of the web of knowledge relative to the phenomenon itself.

Someone can point out that my analysis of the cannonball-example does not prompt ontological considerations because we are confronted with a case in which we do not need to postulate the existence of additional concrete entities (contrary to what happens with the cicada case). Although I admit that my analysis does not seem to have such an import, this comment does not affect the point that I want to make in this section, namely that the use of more mathematics can sometimes enhance the explanatory power of an explanation in a way that is different from the way that Baker has in mind in his (Baker 2016).[17] Furthermore, I think that a general observation concerning the concrete ontology used in the cannonball-example might be added here. It is sufficient to note that the function of mathematics in the explanation using the Lagrange multipliers does not seem to extend to that of reducing (or increasing) the concrete ontology used in the alternative explanation. Indeed, both the explanations assume the same concrete objects and therefore they do not differ with respect to the concrete ontology they are based upon, but only in their mathematical content and the resulting explanatory value.[18] The explanation

---

[15]The mathematics of Lagrange multipliers is more general in the sense that it embeds the mathematics used in the first explanation, which indeed can be recovered using the Lagrange method, and moreover it includes the larger mathematical resources of multivariable calculus.

[16]The explanation also permits to answer counterfactual questions, as for instance 'What if the mass of the cannonball were the same but the kinetic energy were doubled?'. From $R(v_x^*, v_y^*) = (2/mg)E$ we have that the maximum range will be doubled too. Or 'What if the maximum range remain the same but the mass of the cannonball were doubled?'. In this case we should expect the energy E to be twice its initial value.

[17]Presumably, Baker would be sympathetic to this point. For instance, in another work he explores the issue of how mathematics can provide *topic* generality when used to account for a scientific phenomenon, and how such form of generality is a source of explanatory value (Baker 2017a). He maintains that the cicada explanation exhibits topic generality. Nevertheless, he does not link his analysis to the issues addressed in Baker (2016), namely ontological parsimony and the import of additional mathematical resources.

[18]It may be observed that in the explanation using the Lagrange multipliers we need to consider the mass and the kinetic energy of the cannonball as fixed, and that this assumption alters our concrete ontology. Consider, however, that the mass and the kinetic energy are properties of the object cannonball. And therefore an assumption about these properties does not influence the number of types of concrete objects or the number of individual objects that we should consider in our explanation (although it affects the relations between some properties of the concrete objects that are involved in our explanation).

using the Lagrange multipliers offers more explanatory power and, differently to what Baker wants to show with his example, we are confronted here with a case of optimization explanation in which the extra mathematics used does not seem to have any repercussion in terms of concrete ontology. Although peripheral to my analysis, I think that this observation may be a first step toward exploring cases different from Baker's and in which ontological considerations might play a distinct role.

11.5 Conclusions

Providing a story of how IBE may act at level of concrete and abstract entities is particularly important in the context of EIA. And Baker's 2016 paper *Parsimony and inference to the best mathematical explanation* should be regarded as a significant contribution in pursuing such an investigation. In this paper I have discussed some aspects of Baker's analysis and I showed how its import in the context of EIA is considerably weakened under the light of some considerations coming from the scientific practice of biologists. Even if in my discussion I have mainly limited myself to an analysis of what scientific practice seems to suggest about the explanatory value of Baker's explanations, this does not mean that different approaches cannot be pursued. For instance, it is commonly assumed that 13-year and 17-year cicadas' periodicity is a derived state and some authors believe that the selection for the prime-numbered cycles should have taken place only after the fixation of periodicity (Ito et al. 2015). It is therefore natural to ask how an argument (EIA) for the existence of abstract entities (numbers) may be based on a fact (the cicadas' prime-numbered emergence) that is a pure contingency, or how this argument may receive any support from an analysis of explanation in terms of metaphysical grounding (this aspect has been partially addressed in Liggins 2016). Another philosophical issue concerns the ontological import that the 'extra mathematics' used in Baker's $Expl_2$ has with respect to abstract entities. If mathematical entities participate in our best scientific explanations on an epistemic par with unobservable entities, as claimed by Baker, then it seems that using $Lemma_2$ (in $Expl_2$) we should be committed to more mathematical entities through EIA, and not only to those (prime) numbers that are used in $Lemma_1$. Nevertheless, Baker does not elaborate on this issue. These aspects are central to the debate on IBE and EIA, however in this paper I have not focused on them and I have followed a different approach.

In the final part of my study I departed from Baker's analysis and considered a different context. I presented a case of optimization explanation in which the use of more mathematical resources leads to an increase of explanatory power, thus substantiating further Baker's idea that a stronger mathematical apparatus may yield explanatory power. On the other hand, I showed how this example seems to suggest that mathematics can enhance explanatory value without having any repercussion on the concrete ontology that is needed for the explanation to succeed. If this intuition is correct, mathematics can be said to increase the explanatory power of an explanation in a way very different from what Baker has argued with his example of cicadas, for there are cases in which the additional mathematics does not affect the explanation's concrete ontology. Hopefully, an analysis of similar cases may uncover new connections between ontological parsimony, explanatory value and the import of more (or less) mathematics in our scientific explanations.

Acknowledgements I wish to thank the organizers of the Second FilMat Conference and the members of the audience for useful discussion of this paper. I benefitted immensely from suggestions from Marco Panza, Matteo Morganti, Mauro Dorato, Marc Lange, Davide Vecchi, Stathis Psillos, Andrea Sereni, Michèle Friend, Pierluigi Graziani, Achille Varzi, Francesca Poggiolesi, Mary Leng, Luca Incurvati, Andrew Arana, Josephine Salverda, Claudio Ternullo and Giorgio Venturi. I would also like to thank an anonymous referee for his helpful comments.

References

Baker, A. 2003. Quantitative parsimony and explanatory power. *The British Journal for the Philosophy of Science* 54(2): 245–259.

Baker, A. 2005. Are there genuine mathematical explanations of physical phenomena? *Mind* 114: 223–238.

Baker, A. 2009. Mathematical explanation in science. *British Journal of Philosophy of Science* 60: 611–633.

Baker, A. 2016. Parsimony and inference to the best mathematical explanation. *Synthese* 193(2): 333–350.

Baker, A. 2017a. Mathematics and explanatory generality. *Philosophia Mathematica* 25(2): 194–209

Baker, A. 2017b. Mathematical spandrels. *Australasian Journal of Philosophy* 95(4): 779–793.

Behncke, H. 2000. Periodical cicadas. *Journal of Mathematical Biology* 40(5): 413–431.

Busch, J., and J. Morrison. 2016. Should scientific realists be platonists? *Synthese* 193(2): 435–449.

Colyvan, M. 2001. *The indispensability of mathematics*. New York: Oxford University Press.

Cox, R.T., and C.E. Carlton. 1998. A commentary on prime numbers and life cycles of periodical cicadas. *The American Naturalist* 152(1): 162–164.

Grant, P.R. 2005. The priming of periodical cicada life cycles. *Trends in Ecology & Evolution* 20(4): 169–174.

Hoppensteadt, F., and J. Keller. 1976. Synchronization of periodical cicada emergences. *Science* 194(4262): 335–337.

Hunt, J. 2016. Indispensability and the problem of compatible explanations. *Synthese* 193(2): 451–467.

Ito, H., S. Kakishima, T. Uehara, S. Morita, T. Koyama, T. Sota, J.R. Cooley, and J. Yoshimura. 2015. Evolution of periodicity in periodical cicadas. *Scientific Reports* 5: 14094 EP.

Karabulut, H. 2006. The physical meaning of lagrange multipliers. *European Journal of Physics* 27(4): 709.

Koenig, W.D., and A.M. Liebhold. 2013. Avian predation pressure as a potential driver of periodical cicada cycle length. *The American Naturalist* 181(1): 145–149.

Kon, R. 2012. Permanence induced by life-cycle resonances: The periodical cicada problem. *Journal of Biological Dynamics* 6(2): 855–890.

Lange, M. 2013. What makes a scientific explanation distinctively mathematical? *The British Journal for the Philosophy of Science* 64(3): 485–511.

Lehmann-Ziebarth, N., P.P. Heideman, R.A. Shapiro, S.L. Stoddart, C.C.L. Hsiao, G.R. Stephenson, P.A. Milewski, and A.R. Ives. 2005. Evolution of periodicity in periodical cicadas. *Ecology* 86(12): 3200–3211.

Lewis, D. 1973. *Counterfactuals*. Oxford: Basil Blackwell.

Liggins, D. 2016. Grounding and the indispensability argument. *Synthese* 193(2): 531–548.

Maier, C.T. 1985. Brood vi of 17-year periodical cicadas, magicicada spp. (hemiptera: Cicadidae): New evidence from connecticut, the hypothetical 4-year deceleration, and the status of the brood. *Journal of the New York Entomological Society* 93(2): 1019–1026.

Marlatt, C.L. 1907. *The periodical cicada*. Washington, DC: U.S. Department of Agriculture, Bureau of Entomology.

Marshall, D.C. 2001. Periodical cicada (homoptera: Cicadidae) life-cycle variations, the historical emergence record, and the geographic stability of brood distributions. *Annals of the Entomological Society of America* 94(3): 386–399.

Marshall, D.C., J.R. Cooley, and C. Simon. (2003). Holocene climate shifts, life-cycle plasticity, and speciation in periodical cicadas: A reply to cox and carlton. *Evolution* 57(2): 433–437.

May, R.M. 1979. Periodical cicadas. *Nature* 277(5695): 347–349.

Molinini, D., F. Pataut, and A. Sereni. 2016. Indispensability and explanation: An overview and introduction. *Synthese* 193(2): 317–332.

Nariai, Y., S. Hayashi, S. Morita, Y. Umemura, K.-I. Tainaka, T. Sota, J.R. Cooley, and J. Yoshimura. 2011. Life cycle replacement by gene introduction under an Allee effect in periodical cicadas. *PLOS ONE* 6(4): 1–7.

Nolan, D. 1997. Quantitative parsimony. *The British Journal for the Philosophy of Science* 48(3): 329–343.

Pincock, C. 2012. *Mathematics and scientific representation*. New York: Oxford University Press.

Psillos, S. 2009. *Knowing the structure of nature*. Basingstoke: Palgrave Macmillan.

Van Fraassen, B.C. 1989. *Laws and symmetry*. Oxford: Clarendon Press.

Webb, G. 2001. The prime number periodical cicada problem. *Discrete and Continuous Dynamical Systems – Series B* 1(3):387–399.

Yoshimura, J. 1997. The evolutionary origins of periodical cicadas during ice ages. *The American Naturalist* 149(1): 112–124.

Chapter 12
Applicability Problems Generalized

Michele Ginammi

Abstract In this paper, I will do preparatory work for a generalized account of applicability, that is, for an account which works for math-to-physics, math-to-math, and physics-to-math application. I am going to present and discuss some examples of these three kinds of application, and I will confront them in order to see whether it is possible to find analogies among them and whether they can be ultimately considered as instantiations of a unique pattern. I will argue that these analogies can be exploited in order to get a better understanding of the applicability of mathematics to physics and of the complex relationship between physics and mathematics.

12.1 Introduction

The effectiveness of mathematics in physics has been topic of debate in the philosophy of science in the last decennia (for some reference on the topic, see Wigner 1960; Steiner 1998; Pincock 2012; Bangu 2012). In their attempt to clarify the applicability of mathematics to physics, philosophers usually only focus on cases of applicability of mathematics to physics and ignore *other* kinds of application of (or *to*) mathematics. However, since the application of mathematics to physics is just a part of the more complex interrelation between physics and mathematics, it might be that such an approach is actually too narrow. Maybe, if we better understand how this kind of application (from mathematics to physics) compares to other kinds of application, we might be able to better understand the applicability of mathematics to physics as well.

For example, it might be that by confronting the applicability of mathematics to physics with the applicability of mathematics *to mathematics itself* we can trace useful similarities, that might be of some help in understanding the former kind of

M. Ginammi (✉)
Department of Philosophy (KGW Fakultät), University of Salzburg, Salzburg, Austria

© Springer International Publishing AG, part of Springer Nature 2018
M. Piazza, G. Pulcini (eds.), *Truth, Existence and Explanation*,
Boston Studies in the Philosophy and History of Science 334,
https://doi.org/10.1007/978-3-319-93342-9_12

applicability as well. Hacking (2014) has recently suggested that the applicability of mathematics *to mathematics itself* might help in clarifying the applicability of mathematics to physics, and to science in general.

Another kind of applicability, which is usually not taken into account when dealing with the problem of math-to-physics application, is the application of *physics to mathematics*. This subject has been broadly neglected by the philosophical debate on the applicability of mathematics. The only relevant exceptions, to my best knowledge, are Urquhart (2008a,b). Actually, in contemporary physics and mathematics there is a *fruitful circulation* of methods and representative strategies, in which not only mathematics can be effectively employed to modelize physics, but also physics can be fruitfully 'applied' to mathematics to generate new strategies of mathematical analysis.[1] This (unreasonable?) *effectiveness of physics in mathematics* is still unheeded by the philosophical community and awaits to be explored.

The presupposition that these kinds of applicability are completely different from (and therefore not relevant for) the understanding of the applicability of mathematics to physics might well be wrong. It might be that all these kinds of application (math-to-physics, math-to-math, and physics-to-math) share common features, and that they can be seen as different cases of a unique pattern of application. If this were the case—if there were analogies between these three kinds of application—then we might exploit these analogies in order to offer a generalized account for mathematical application, and to better understand the complex relationship between physics and mathematics.

In this paper I am going to develop this suggestion. In the next sections I will present some examples of math-to-physics (Sect. 12.2), math-to-math (Sect. 12.3), and physics-to-math application (Sect. 12.4). Then, in Sect. 12.5, I will discuss the analogies that can be traced among them, and I will analyse whether these analogies help in clarifying the applicability problems and the relationship between physics and mathematics.

The examples I will take into consideration in the next sections are all examples of what Steiner (1998) calls "heuristic application". He used this terminology only with reference to the math-to-physics application, where he meant the (surprising) circumstance that very often physicsts employ mathematical concepts and structures *not only* to represent or describe physical systems (or phenomena), *but also to discover new physical laws*—and even *new physical entities*. This heuristic application holds incredible surprises, and it is quite difficult to epistemologically justify why this jump from mathematics to reality should be succesfull. Steiner's (1998) book is a great source of examples of this kind. What is particularly problematic, is the fact that often these new discoveries are the result of some purely

---

[1] Also physicists themselves, such as Jean-Marc Levy-Leblond, argue that this topic deserves some closer attention: «This inverse relationship, from physics to mathematics, deserves a full study of its own» (Levy-Leblond 1992, p. 154). Nevertheless, he does not develop this suggestion any further.

formal analogies between mathematical and physical concepts. As Steiner (1998) puts it, physicists «used the relations between the structures and even the notations of mathematics to frame analogies and guess according to those analogies» (p. 5). And again:

> some of the analogies physicists drew were formal, i.e. syntactical: the equations they guessed simply looked like the equations they already had. In such cases, scientists were studying their own representational systems—i.e., themselves—more than nature. (p. 7)

With the proper edits, we can apply this terminology also to the other two kinds of applicability, to refer to cases of math-to-math and physics-to-math application in which considerations in some specific branch of mathematics (or physics) can be exploited *via analogy* to discover new concepts, structures and theories in a completely different mathematical area.

The reason why I will specifically focus on examples of heuristic applicability is that philosophically this kind of application seems to be the most problematic one, and it can be reasonably taken as a good test for the success of a certain account for the topic.

12.2 Math-to-Physics Application

Let's start from the most debated kind of application, the application of mathematics to physics. Many examples can be given, but I will only focus on two examples that I consider particularly relevant: the prediction of the so-called "omega minus particle" and the prediction of the so-called "positron". These examples have been presented by Steiner in his (1998) as a typical case of heuristic (misterious) applicability, and they have been widely discussed by Sorin Bangu, in his (2012) book.[2]

The omega minus particle belongs to the class of the so-called "spin-$\frac{3}{2}$ baryons" and it was completely unknown before 1962. That year, the two physicists Murray Gell-Mann and Yuval Ne'eman independently made the prediction, based on some considerations on SU(3) formalism, that a new particle should exist, and they could even specify the particular properties this new particle had. A wonderful sketch of how they predicted the existence of this new particle is offered by Ne'eman himself and Yoram Kirsh in their Ne'eman and Kirsh (1996) book:

> In 1961 four baryons of spin $\frac{3}{2}$ were known. These were the four resonances Δ^-, Δ^0, Δ^+, Δ^{++} which had been discovered by Fermi in 1952. It was not clear that they could not be fitted into an octet, and the eightfold way predicted that they were part of a decuplet or of a family of 27 particles. A decuplet would form a triangle in the $S - I3$ [strangeness-isospin] plane, while the 27 particles would be arranged in a large hexagon. (According to the formalism of SU(3), supermultiplets of 1, 8, 10 and 27 particles were allowed.) In the same year (1961) the three resonances $\Sigma(1385)$ were discovered, with strangeness -1 and probable spin $\frac{3}{2}$, which could fit well either into the decuplet or the 27-member family.

---

[2]The first example is also discussed in Bangu (2008).

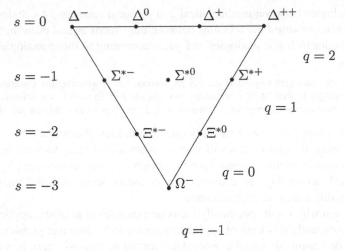

Fig. 12.1 Spin-$\frac{3}{2}$ baryon decuplet. (Credits: http://math.ucr.edu/home/baez/diary/march_2007. html)

At a conference of particle physics held at CERN, Geneva, in 1962, two new resonances were reported, with strangeness -2, and the electric charge -1 and 0 (today known as the $\Xi(1530)$). They fitted well into the third course of both schemes (and could thus be predicted to have spin $\frac{3}{2}$). On the other hand, Gerson and Shoulamit Goldhaber reported a 'failure': in collisions of K^+ or K^0 with protons and neutrons, one did not find resonances. Such resonances would indeed be expected if the family had 27 members. The creators of the eightfold way, who attended the conference, felt that this failure clearly pointed out that the solution lay in the decuplet. They saw the pyramid [in Fig. 12.1] being completed before their very eyes. Only the apex was missing, and with the aid of the model they had conceived, it was possible to describe exactly what the properties of the missing particle should be! Before the conclusion of the conference Gell-Mann went up to the blackboard and spelled out the anticipated characteristics of the missing particle, which he called 'omega minus' (because of its negative charge and because omega is the last letter of the Greek alphabet). He also advised the experimentalists to look for that particle in their accelerators. Yuval Ne'eman had spoken in a similar vein to the Goldhabers the previous evening and had presented them in a written form with an explanation of the theory and the prediction. (Ne'eman and Kirsh 1996, pp. 202–203)

When a two years later, in 1964, experimentalist physicists looked for the Ω^- particle in their accelerators, they found out exactly what Gell-Mann and Ne'eman predicted: the particle was there and it had exactly the predicted characteristics (see Fig. 12.2).[3]

---

[3]The story is actually a bit more complicate than this. They could prove the existence of the Ω^- particle, along with its characteristics—*except* for its spin. Physicists *presumed* that its spin was $\frac{3}{2}$ just because its existence was predicted within the classification symmetry scheme for spin-$\frac{3}{2}$ baryons, but the actual measurement of its spin was unexpectedly difficult. Although this hyperon was discovered more than 40 years ago, a conclusive measurement of its spin has only recently been obtained by Aubert et al. (2006).

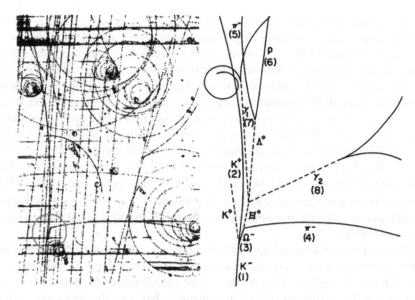

Fig. 12.2 Photograph (left side) and line diagram (right side) of the decay of an Ω^- particle in a bubble chamber. The short track of the Ω^- particle is highlighted by the circle in the low left corner. (Credits: Barnes et al. (1964); Brookhaven National Laboratories)

What seems to be surprising in this example is the fact that Gell-Mann and Ne'eman made their prediction only looking at the mathematical scheme which was supposed to represent and classify the spin-$\frac{3}{2}$ baryons. No new empirical observation or data was used to make the prediction. The new particle manifested itself—so to speak—through mathematical transparency, with no other aid than that. According to Bangu (2008), in order to justify this prediction we should adopt some kind of 'reification principle' (his terminology), something that allows us to say that since X exists in the mathematical structure M, and since all the other elements $Y_1, \ldots, Y_n$ of M have a physical referent, then X must have a physical referent too. The problem with this principle, however, is that it is not a valid principle at all, because it cannot be generalized (sometimes it works, sometimes—many times—it does not), and it seems impossible to justify on purely epistemological terms (at least, not in this form). In Ginammi (2016) I offered an account for the prediction of the omega minus particle that does not rely on this 'reification principle' and also helps in clarifying the role of (mathematical) representation in physics. The key idea is that this prediction has been made possible by some features of the representation which physicists were already working with in the context of spin-$\frac{3}{2}$ baryons classification. In Ginammi (2015, 2016) I defined this representativeness (A's effectiveness in representing B) in terms of structural similarity. More precisely, we can say that A suitably represents B *if* there exists *at least* a *monomorphism* from B to A, where in B we only consider the relevant elements (objects, relations, ...) of the target physical system. Physicists engaged in the spin-$\frac{3}{2}$ baryons classification knew (or, at least, they had good reasons to believe) that the mathematical structure they were adopting (the decuplet scheme) was a good representation of the class of the spin-$\frac{3}{2}$ baryons, i.e. they had good reasons to believe that this class of baryons was *at least monomorphic* to the mathematical structure at issue. But they did not know whether the surplus mathematical structure (the empty place in the decuplet scheme) was also playing a representative role or not. Thus, based on the representative effectiveness of the mathematical structure at issue, they *assumed* that this surplus structure *does* actually play a representative role, they looked for a (natural) interpretation for this surplus, and finally they empirically checked the resulting hypothesis (i.e., that there actually is a tenth spin-$\frac{3}{2}$ baryon out there in the physical world). In other words, they exploited this surplus structure in the representation they already set up for the spin-$\frac{3}{2}$ baryons classification as a *hypothesis generator*—hypothesis that they could then check via empirical control.

This kind of mathematical application in physics—the strategy of inferring new physical laws or entities from only mathematical considerations—is actually at the base of many other physical discoveries. Dirac's prediction of the positron (or "anti-electron", as he initially called it), for example, is based on the very same strategy. At the very base of his discovery there is the attempt to extend the Klein-Gordon equation (a relativistic version of the Schrödinger equation) to the electron. By introducing a higher dimension 4×4 matrices, he found a new equation that both describes the behaviour of the electron and manages to incorporate special relativity:

$$\left[\gamma^\mu \left(i \frac{\partial}{\partial x^\mu} + e A_\mu(x) \right) + m \right] \psi(x) = 0$$

where $A_\mu(x)$ are the electromagnetic potentials (specifying the electric and magnetic fields acting on the electron at every point x of spacetime), and γ^μ with $\mu = 0, 1, 2, 3$ are the so-called "Dirac matrices":

$$\gamma^0 = \begin{bmatrix} 1 & 0 & 0 & 0 \\ 0 & 1 & 0 & 0 \\ 0 & 0 & -1 & 0 \\ 0 & 0 & 0 & -1 \end{bmatrix} \quad \gamma^1 = \begin{bmatrix} 0 & 0 & 0 & -1 \\ 0 & 0 & -1 & 0 \\ 0 & 1 & 0 & 0 \\ 1 & 0 & 0 & 0 \end{bmatrix} \quad \gamma^2 = \begin{bmatrix} 0 & 0 & 0 & i \\ 0 & 0 & -i & 0 \\ 0 & -i & 0 & 0 \\ i & 0 & 0 & 0 \end{bmatrix} \quad \gamma^3 = \begin{bmatrix} 0 & 0 & -1 & 0 \\ 0 & 1 & 0 & 1 \\ 1 & 0 & 0 & 0 \\ 0 & -1 & 0 & 0 \end{bmatrix}.$$

The equation turns out to have four possible solutions, two 'positive-energy' solutions and two 'negative-energy' solutions. However in 1928, when Dirac derived the above equation, no situation was known in which these 'negative-energy' solutions played a role. Instead of discarding these 'strange' solutions, Dirac conjectured that these solutions actually describe a new elementary particle—the positron. No need to say, his conjecture turned out to be right.

12.3   Math-to-Math Application

That mathematics can be (and is) applied to mathematics itself does not seem really surprising. After all—one might say—we are not leaving the domain of abstract entities, concepts and structures that is supposed to be the object of mathematics. However, sometimes mathematicians make *very* surprising discoveries (can we call them "discoveries" without being charged of Platonism?) in some specific area (or structure) of mathematics *just by looking* at some other mathematical area, apparently unrelated to the first one.

The classical example is Descartes' application of algebra to geometry. In his pamphlet *La Géometrié*, included in his *Discours de la méthode* (1637), Descartes offered an innovative program for geometrical problem-solving, based on a particular approach to the relationship between algebra and geometry. As Hacking (2014) comments on this,

> Many difficult problems in geometry can be solved by turning them into arithmetic and algebra, and many problems in the theory of numbers and algebra can be solved by turning them into geometry. This continues [...] from the time of Descartes to the present day. It is as if geometry and the theory of numbers turn out to be about the same stuff. I find this astonishing. (p. 6)

In different words, Descartes discovered and underlined some kind of correspondence between geometry and algebra, some kind of structure-preserving relation that permits us to *translate*—so to speak—propositions in algebra into propositions in geometry, and preserve truth across this translation. As a result, we can rely on this relation and 'export' theorems from algebra to geometry and vice-versa.

As Hacking underlines in the previous quotation, this 'bridge' between geometry and algebra has been constantly enlarged and reinforced along centuries. One of the more recent examples is Ngô Bao Châo's proof of the "Fundamental Lemma" of the Langlands program, based on the application of number theory to geometry. As mathematician Peter Sarnak reportedly said,

> It is very rare that you can take a proof in the geometric setting and convert it to the genuine number theoretic setting. That is what has transpired through Ngô's achievement. Ngô has provided a bridge, and now everyone is using this bridge. What he has done is deep. It is below the surface and it is understanding something truly fundamental. (Devine Thomas 2010, p. 4)

Again, this "bridge" metaphor suggests that what Ngô Bao Châo did in his work was to discover some deep connections between algebra and geometry, along with a way to translate problems and theorems from one language to the other language.

Another surprising exploitation of this algebra-geometry 'bridge' is at the core of another, very famous theorem: Andrew Wiles' proof of Fermat's last theorem. In this case, Wiles found a totally unexpected connection between geometry and number theory:

He 'applied' abstruse features of elliptic functions to a home truth of arithmetic, Fermat's last theorem. It is a source of mathematical joy that structures developed for one purpose, at first following one barely visible track through the woods, should cross a well-known path and suddenly solve its problems. (Hacking 2014, p. 14)

More generally, there is a whole class of theorems in mathematics whose ultimate aim consists in offering a 'bridge' between different mathematical structures: the so-called *representation theorems*. They can be found basically everywhere in mathematics, and their main contribution is that they offer a way to exploit structural similarities among different structures in order to translate theorems about a structure into equivalent theorems about another structure.

As a first approximation, we can compare these examples of math-to-math application to our previous examples of math-to-physics application. In both cases, researchers are trying to exploit some kind of analogy between two different (and – at least in the case of math-to-physics application—even disomogeneous) 'objects' or 'structures'. They are—so to speak—applying some kind of 'analogy' reasoning: if the structure A is analogous (under certain respects) to the structure B, then the object a (or class of objects a_n, or theorem a) in A is analogous to the object b (or class of objects B_n, or theorem b) in B. We can replace the variables in this abstract scheme with the concrete terms of our examples. If the mathematical structure employed by Dirac to describe and represent the electron-like particles' behaviour is analogous to the real electron-like particles' behaviour, then we can infer that the two negative solutions for the equation at object stand for something in the real electron's behaviour. Similarly, in mathematical representation theorems, if a certain mathematical structure is analogous to another mathematical structure, then we can infer that what holds for the first structure will hold for the second structure as well.

12.4 Physics-to-Math Application

A third kind of application that is usually omitted in discussions about the applicability of mathematics is the application *of physics to mathematics*. We are so used to marvel at the extraordinary effectiveness of mathematics in physics, that we hardly notice that very often physics is an incredibly rich source of mathematical knowledge. The word "application" might be a bit out of place here, but for sure the trades between physics and mathematics are much more reciprocal than we normally think.

At the beginning of my research on the present topic, I thought that the examples of effectiveness of physics in mathematics were quite few. I only knew two of them: Newton's and Leibniz's contribution to the infinitesimal calculus and synthetic differential geometry, which originated from physical necessity; and Dirac's δ function. Dirac's δ function is defined as a function on the real line which is zero everywhere except at the origin, where it is infinite, and whose integral between ∞ and $-\infty$ equals 1. This function' turned out to be incredibly effective from a physical viewpoint, but it has a little problem: there is no mathematical function with these properties. In short, if we take it as a function (as it seems to be) it is a *mathematically impossible object.* In order to give this impossible function a place in the mathematical domain, mathematicians tried to interpret it in different ways. The most interesting attempt is Schwartz's, who re-interpreted the δ function as a linear operation on a certain class of functions. Such interpretation proved itself very fruitful in mathematics, since it fostered a new mathematical theory—the theory of distributions. What is quite surprising, in this example, is the fact that Dirac—rougly speaking—'forced' a well known and established mathematical theory (function theory) by adding a new function that the theory considered 'impossible'. The fact that this new impossible function turned out to be incredibly effective in representing and modelizing physical reality, put mathematicians in front of a dilemma: they had either to insist that the δ function was simply an absurdity—then leaving completely misterious its unquestionable physical effectiveness; or to accept that this apparently impossible 'function' was actually not impossible at all—then facing the challenge of offering a coherent mathematical theory in which δ function could be settled home.

As soon as I got deeper into my research, I quickly discovered that cases like these occur rather frequently. Mandelbrot's fractal theory, ergodic theory, dimension theory, Edward Witten's application of topological quantum field theory to the theory of knot invariants[4]—they all originated from physicists' mathematical intuitions that forced mathematicians to develop new theories in order to make this intuitions part of a coherent mathematical structure. In all these cases, the fact that these intuitions turned out to be physically effective put mathematicians in front of a dilemma analogous to the one we saw for Dirac's δ function: either to reject them as mathematically meaningless, thus leaving their effectiveness unexplained, or to accept them and face the challenge to give them a mathematical formulation.

As I previously noticed at the beginning of this section, the label "application of physics to mathematics" might be a bit misleading here. The reader might think that what I am suggesting here is that we can *use* physics or real-world systems as empistemological tools to investigate the abstract realm of mathematics— exactly in the same sense we employ mathematics as an epistemological tool to investigate the natural world. The reader might hold that this kind of 'application' is not different from math-to-physics application: our limitations in the comprehension of the world just force us to make new efforts into the purely abstract domain of mathematics, in order to obtain something useful for a subsequent application to real-world systems. In other words, the only application we have here is the math-to-physics application and there seems to be no need to evoke a physics-to-math application as something different from it. However, in this paper I am only considering cases of what Steiner calls "heuristic application". These cases of heuristic application are characterized by the fact that certain concepts, theories, or structures, are employed not only to represent some other concepts, theories, or structures; but also to

---

[4]As Sir Michael Atiyah said referring to Witten, «[H]e has made a profound impact on contemporary mathematics. In his hands physics is once again providing a rich source of inspiration and insight in mathematics» (Atiyah 2003, p. 525).

discover something new about these other concepts, theories, or structures. In these
heuristic applications, the discovery is typically the result of some kind of analogy
reasoning applied to the concepts, theories, or structures at object. My claim in this
section is that we can find examples of this kind of heuristic application *also from
physics to mathematics*. In other words, we can compare the omega minus case
previously discussed to—just to pick out a paradigmatic example—the role of the
Dirac's delta function in fostering the development (discovery?) of the theory of
distributions. In both cases, we see at work the same kind of 'heuristic mechanism'
that Steiner pointed out: in the omega minus case, the new particle was predicted as
a consequence of a physically non-interpreted object in the mathematical structure;
in the Dirac's delta function case, similarly, the theory of distribution was developed
as a consequence of a *mathematically* non-interpreted object in the physical theory
(delta function). In both cases, what triggered the advancement was some kind of
'parallelism' or 'analogy' reasoning: the non-interpreted objecy has been interpreted
so to mantain the parallelism or the analogy between the two theories.

12.5   Exploiting the Analogies

Before trying to understand whether there are analogies between these three kinds
of applicability (that we might exploit to better understand the boundaries and the
relation between physics and mathematics), let me start with some observations.

First of all, one may notice that these three kinds of application do not share
the same level of problematicity. In particular, the second one (the math-to-math
applicability) does not seem to be particularly problematic, at least for what
concerns the epistemology behind these discoveries. There seems to be nothing
'surprising' or 'problematic' in the fact that we can discover something new about
a certain abstract mathematical structure by showing that structure is similar to
another abstract mathematical structure. If there is some kind of morphism between
the two structures, there is nothing problematic (at least from the epistemological
point of view) in the fact that, depending on the specific kind of morphism, we can
draw conclusions on the latter structure by looking at the former one. The morphism
actually sets up the proper relation to 'export' knowledge from one structure to
another. It might be that the connection is surprising because nobody thought
about this connection before, or because it required the sophisticated intuition
of some mathematical genius in order to see it. But this surprise only concerns
the psychological side of the discovery, not its *epistemological* justification. We
can therefore say that this morphism (be it an isomorphism, a homomorphism,
a monomorphism, or anything else) constitutes the core of the epistemological
justification for the heuristic inference that brings to the discovery, i.e., it is at the
base of the math-to-math heuristic application. It must be noted, however, that this
morphism needs not to be *proved* before the heuristic inference is drawn: such a
morphism can simply stated as a *working hypothesis*, and the heuristic inference
can be drawn just as a consequence of this (still unproved) hypothesis. In other

words, we can just make the hypothesis that the mathematical structure A is—for example—isomorphic to the mathematical structure B, and then we can employ this hypothesis to draw inferences about elements or theorems in B that mirrors elements or theorems in A.

Also the third kind of application does not seem to be epistemologically problematic. Notice that the dilemma (either to reject the absurd mathematical object, or to accept it and look for a coherent theory for it) only arises once the alleged mathematical object has proved itself to be 'physically effective'—i.e., to be physically useful and relevant in representing a certain physical system. If I suggest that a new mathematical (apparently incoherent) object might be fruitfully used within a represention of a certain physical system, but this new mathematical object fails in improving the representation at all, then no mathematician would take that object seriously, and nobody would make the effort to develop a coherent theory to account for it: they would simply laugh at me, while saying "*I told you, you were just wasting time!*". It is the fact that the new mathematical object is 'physically effective' (in the previously specified meaning) that fosters mathematicians to develop a new theory. Why does such a physical effectiveness make a difference? A possible reason is that we assume that nature is *consistent* and thus that physical objects are consistent; so it *must* be possible to give them a mathematical (coherent) formulation. If a certain mathematical object x plays a role in representing a physical system P, there *must* be a mathematical structure S of which P is a model and which X belongs to.

This assumption—that there must be such a mathematical structure that modelizes the physical system—is actually very delicate. Its generalization is what Wilson (2000) calls "mathematical optimism": the idea that «*every real-life physical structure can be expected to possess a suitable direct representative within the world of mathematics*» (p. 297; italics in the original).[5] Opposed to this mathatical optimism there is, according to Wilson, 'mathematical oportunism':

> many of the originators of mathematical physics in the early modern period would never have accepted such a cheery presumption [*mathematical optimism*]; they maintained that it is only when the processes of nature enjoy a special simplicity that mathematics can track its workings adequately. This is the basic thesis I will call mathematical opportunism: it is the job of the applied mathematician to look out for the special circumstances that allow mathematics to say something useful about physical behavior. (p. 297)

For our present aims, we do not need to take a stance on this delicate topic. What is important here is to underline that our previous considerations do not presuppose

---

[5]Mathematical optimism can, according to Wilson, take two different forms: *lazy* and *honest*. The *lazy mathematical optimism* is the idea that «somewhere deep within mathematics' big bag must lie a mathematical assemblage that is structurally isomorphic to that of the physical world before us, even if it turns out that we will never be able to get our hands on that structure concretely» (pp. 296–7). As such, *lazy* mathematical optimism is an a priori thesis. Differently, *honest mathematical optimism* «cannot be plausibly regarded as an a priori truth at all, but, from a physical point of view, it can nonetheless prove quite resilient to refutation» (p. 297). Wilson's thesis is that «'anti-realists' frequently severely underestimate the difficulties in arguing against honest optimism» (p. 297).

any kind of 'mathematical optimism'. We only need to assume that *some* physical systems can be represented in mathematical terms, via some kind of morphism. The key assumption is that when, in representing a physical system of this kind, we need some (apparently) incoherent mathematical tool, we can say that physicists are guessing—so to speak—some pieces of the mathematical structure instantiated by the physical system at issue, and they employ these pieces (in a mathematically non-rigorous way) to represent the physical system they want to represent. Then, it is mathematicians' duty to polish the structure and make it more rigorous. In this process, there seems to be nothing particularly 'unreasonable', at least from the epistemological point of view. Physicists tentatively discover (or create) a new structure that seems to be instantiated in nature, and mathematicians develop it to perfection according to their own standard of rigour. This is not really different from what normally physicists do: they employ mathematical structures to modelize and represent physical systems. The difference here is that the mathematical structure they need is not already available, so they need to create one—so to speak—out of the blue.

Finally, we can try to see whether it is possible to trace some analogies among these three kinds of application. We have seen that the application of mathematics to mathematics can be easily accounted for by means of the notion of "structural similarity". Given that two mathematical structures are structurally similar, we can (depending on the kind of 'similarity' at play) export theorems from one structure to the other, by translating the terms of the first structure into terms of the second one. This 'structural similarity' between mathematical structures has nothing vague or unclear: it can be precisely characterized in terms of morphisms (isomorphism, homomorphism, monomorphism, and so on). Depending on the morphism at play, we can build parallelisms between the two structures. Again, I recall that the morphism do not need to be already proved: we can just make the *hypothesis* that a certain mathematical structure A is—for example—*monomorphic* to another structure B, and then we can employ this assumption as a 'hypothesis generator' (in the sense I used the term in Sect. 12.2; see p. 214) and suggest that some aspects of A should be present in B as well (given the proper interpretation). *Similarly*, we can account for the applicability of physics to mathematics: *by assuming* that a physical system can be represented by a mathematical structure (of which we grasp some pieces, since we can use them to represent the physical system at issue), we can try to reconstruct the whole mathematical structure laying behind it.

Can we do the same also for the examples we saw in Sect. 12.2 (i.e. in case of application from mathematics to physics)? We can just say that at the core of this kind of application there is the assumption that a certain mathematical structure is a good representation of (at least the relevant elements of) a certain physical system, because they are structurally similar. In other words, we can take the notion of representation as eminently structural, and look for an analysis of the notion of representation (at least for what concern the representations involving mathematical structures) in terms of some kind of morphism. As I already mentioned, in Ginammi (2015, 2016) I argued that we can define representativeness (A's effectiveness in representing B) exactly in these terms: we can say that A suitably represents B *if* there exists *at least* a *monomorphism* from B to A, where in B we only consider the relevant elements (objects, relations, ...) of the target physical system. The idea of analysing the notion of scientific representation in terms of structural similarity is not new, and many attempts have already been put forward in the literature (see for example Redhead 1975; French 1999, 2000; Pincock 2012; Bueno and Colyvan 2011), and has been variously criticized (for an overview, see Suárez 2015). A full discussion of these criticisms would exceed the aims and limits of this paper. But notice that the most part of these criticisms point out the difficulties in offering a *purely* structural account for scientific representations, and argue that a general account for scientific models (not only mathematical scientific representations) cannot be given in terms of structural similarities. However, it is important to underline that the account here sketched avoids these main criticisms: first of all, the notion of monomorphism does not exclude that we can have stronger similarities among structures (and hence better representations), simply because isomorphism trivially imply monomorphism, and at the same time do not impose too strong (isomorphism) or too weak (homomorphism) conditions for representativeness; secondly, the direction of the monomorphism (A represents B if B is monomorphic to A) leaves room for a 'structure surplus' (in Redhead's (1975) terms) that can be heuristically exploited; thirdly, the reference to 'relevant elements' leaves room for pragmatical considerations and evaluations about which elements should we consider as 'relevant' for our aims and purposes (in this sense, the account here sketched is not purely structural).

In sum, we can agree on these criticisms, and still try to look for a (partly) structural analysis of the notion of *mathematical* representation. After all, the notion of representation seems to lay at the core of all these three kinds of applicability and, if the notion of representation is the same for all of them, it seems to be quite difficult not to analyze it in terms of *structural* similarity—at least for what concerns the math-to-math application.

If the characterization of representation in terms of monomorphism is appropriate, then we can offer a generalized account for all these three kinds of heuristic application: we *assume* that this monomorphism condition for representation is satisfied for two structures A and B; if we assume that the monomorphism goes from A to B, then there might be some structural surplus in B that is not interpreted over A. At this point we can make the hypothesis that also this surplus structure might be interpretable over A,[6] and then we can work *under this assumption* and check whether it is right or not: we just export knowledge from the one side to the other side, and then we check the rightness of these inferences. In case of math-to-physics application, this check will be mainly empirical, while in the other two cases the check will be mainly a matter of logical and mathematical coherence.

---

[6]This assumption amounts to saying that the structural similarity between A and B is actually wider than the one circumscribed by the original monomorphism, and can coincide—at its limit—with an isomorphism hypothesis.

Let's go back for a moment to the example of the omega minus particle. In this case, Gell-Mann and Ne'eman had good reasons to believe that the mathematical structure they were employing was a good representation of the class of the spin-$\frac{3}{2}$ baryons (in our terms, they had good reasons to believe that this class was *at least* monomorphic to the mathematical structure employed); then they *supposed* that this mathematical structure was actually a *perfectly fitting representation* of the class of the spin-$\frac{3}{2}$ baryons—namely, that the mathematical structure was actually *isomorphic* to it. They exploited this assumption and the surplus mathematical structure made available by this assumption to make a prediction (by simply exporting knowledge from the mathematical structure to the realm of spin-$\frac{3}{2}$ baryons), which turned out to be very precise.

It is interesting, in this regard, to compare the two following quotation, the first (by Bangu) about the omega minus prediction, the second (by Weil) about the role of analogies in mathematics:

> [G]iven the classification scheme for the already known spin-$\frac{3}{2}$ baryons, the unoccupied, apparently superfluous entry in the scheme was taken as a guide to the existence of a new particle. *It was exactly this surplus that suggested the existence of new physical reality* (in the form of new particles, to fill in gaps in multiplets). (Bangu 2008, p. 243; italics mine)

> If one models the theory of functions on the theory of algebraic numbers, one is forced to give a special role, in the proofs, to the point at infinity [...]. In order to reestablish the analogy, it is necessary to introduce, into the theory of algebraic numbers, something that corresponds to the point at infinity in the theory of functions. That is what one achieves, and in a very satisfactory manner, too, in the theory of "valuations". (Weil 2005, p. 339)

In both cases, the strategy seems to be describable in the same way: analogies and similarities among structures. Given the (supposed) existence of a similarity relation among structures, we can export knowledge from one to another.

Notice, by the way, that this account also offers a natural way out in case of failures. As Steiner and Bangu noticed, the problem with this kind of predictions (Dirac's prediction of the positron is analogous under this respect) is that the strategy apparently employed by physicists can hardly be generalized. Sometimes it works, sometimes it does not—but it is hard to say why and when it works, and why and when it does not. The account we just sketched offers an easy way out from this difficulty: this strategy works only when (and because) the *assumption* (that the mathematical structure actually represents the physical system at issue) is true. Physicists employ a certain mathematical structure to represent a certain physical system because they are justified in believing that such a mathematical structure represents the target physical system. They *suppose* that it is so, and then they exploit this supposition in order to gain new knowledge about the target. If, in doing this, they end up with a failure (if, for example, no omega minus particle is discovered) the culprit for this failure *is not the strategy, but the assumption*. The strategy still remains epistemologically and heuristically effective in itself, to the point that—even in case of failure—the strategy has brought us new knowledge: at least, now we know that the mathematical structure we were assuming to represent the target *does not* represent it at all, or does not represent it in the way we thought it did (for example, it might still represent the target, but not in a perfectly fitting way—i.e., not isomorphically).

12.6   Conclusion

In the present paper I argued that our approach to the applicability of mathematics to physics (and, more in general, to the relation between physics and mathematics) might be *too narrow*. I suggested that a wider approach—an approach that takes into consideration also other kinds of mathematical applicability, and other strategies of mathematics-physics application (like the mutual heuristic fruitfulness of physics and mathematics)—might be much more fruitful in offering a general account of how physics and mathematics interact. These interactions define a complex system of trades, in which not only mathematics can be heuristically applied to physics, but also physics can be fruitful for the mathematical development of new theories, structures and concepts. I gave some examples of math-to-physics, math-to-math, and physics-to-math application, and I traced some analogies among them. These analogies suggest that a fruitful approach to the mathematics-physics interactions can be offered in terms of structural similarities and structural representation. These results cannot be considered conclusive, and there is still much work that waits to be done. But I offered some positive reasons to believe that the approach suggested here—to look at these problems from a generalized and unified perspective—is promising, and deserves to be carried on.

Acknowledgements I would like to thank Prof. Gabriele Lolli, for his constant support; Prof. Charlotte Werndl for her suggestions on a previous draft of this paper; Prof. Mario Piazza and Prof. Gabriele Pulcini for their work in making this volume possible; an anonymous referee, for their precious comments. Finally, a special thank to Pauline van Wierst, who read the paper and helped me with constant support and important insights.

References

Atiyah, Michael. 2003. The work of Edward Witten. In *Fields medallists' lectures*, World scientific series in 20th century mathematics, ed. M. Atiyah and D. Iagolnitzer, vol. 9, 2nd ed., 522–526. New Jersey/London/Singapore/Hong Kong: World Scientific.

Aubert, B. et al. 2006. Measurement of the spin of the Ω^- hyperon. *Physical Review Letters* 97: 112001. BABAR collaboration.

Bangu, Sorin. 2008. Reifying mathematics? Prediction and symmetry classification. *Studies in History and Philosophy of Modern Physics* 39: 239–258.

Bangu, Sorin. 2012. *The applicability of mathematics in science: Indispensability and ontology*. Basingstoke (UK): Palgrave Macmillan.

Barnes, V.E. et al. 1964. Observation of a hyperon with strangeness minus three. *Physical Review Letters* 12(8): 204–206.

Bueno, Octavio, and Mark Colyvan. 2011. An inferential conception of the application of mathematics. *Noûs* 45(2): 345–374.

Devine Thomas, K. 2010. The fundamental lemma: From minor irritant to central problem. *The institute letter – Institute for advanced studies (Princeton, NJ)* Summer. Available online at https://www.ias.edu/ideas/2010/fundamental-lemma. Accessed on 23 May 2017.

French, Steven. 1999. Models and mathematics in physics: The role of group theory. In *From physics to philosophy*, ed. J. Butterfield and C. Pagonis. Cambridge: Cambridge University Press.

French, Steven. 2000. The reasonable effectiveness of mathematics: Partial structures and the application of group theory to physics. *Synthese* 125: 103–120.

Ginammi, Michele. 2015. Structure and applicability. In *From logic to practice. Italian studies in philosophy of mathematics*, Boston studies in the philosophy of science, ed. G. Lolli, M. Panza, and G. Venturi, vol. 308, chapter 11. Netherlands: Springer.

Ginammi, Michele. 2016. Avoiding reification. Heuristic effectiveness of mathematics and the prediction of the omega minus particle. *Studies in History and Philosophy of Modern Physics* 53: 20–27.

Hacking, Ian. 2014. *Why is there philosophy of mathematics at all?* Cambridge: Cambridge University Press.

Levy-Leblond, Jean-Marc. 1992. Why does physics need mathematics? In *The scientific enterprise*, Boston studies in the philosophy and history of science, ed. E. Ullmann-Margalit, vol. 146, 145–161. Netherlands: Springer.

Ne'eman, Yuval, and Yoram Kirsh. 1996. *The particle hunters*, 2nd ed. Cambridge: Cambridge University Press.

Pincock, Christopher. 2012. *Mathematics and scientific representation*. Oxford: Oxford University Press.

Redhead, Michael. 1975. Symmetry in intertheory relations. *Synthese* 32: 77–112.

Steiner, Mark. 1998. *The applicability of mathematics as a philosophical problem*. Cambridge (Mass): Harvard University Press.

Suárez, Mauricio. 2015. Representation in science. In *Oxford handbook in philosophy of science*, ed. P. Humphreys. Oxford: Oxford University Press.

Urquhart, Alasdair. 2008a. The boundary between mathematics and physics. In *The philosophy of mathematical practice*, ed. P. Mancosu, 407–416. Oxford: Oxford University Press.

Urquhart, Alasdair. 2008b. Mathematics and physics: Strategies of assimilation. In *The philosophy of mathematical practice*, ed. P. Mancosu, 417–440. Oxford: Oxford University Press.

Weil, André. 2005. Letter to Simone Weil. *Notices of the American Mathematical Society* 52(3):334–341. Translated by Martin H. Krieger.

Wigner, Eugene Paul. 1960. The unreasonable effectiveness of mathematics in the natural sciences. *Communications in Pure and Applied Mathematics* 13(1): 1–14. Reprinted in *Symmetries and Reflections*. 1967. Bloomington: Indiana University Press.

Wilson, Mark. 2000. The unreasonable uncooperativeness of mathematics. *The Monist* 83(2): 296–314.

Chapter 13
Church-Turing Thesis, in Practice

Luca San Mauro

Abstract We aim at providing a philosophical analysis of the notion of "proof by Church's Thesis", which is – in a nutshell – the conceptual device that permits to rely on informal methods when working in Computability Theory. This notion allows, in most cases, to not specify the background model of computation in which a given algorithm – or a construction – is framed. In pursuing such analysis, we carefully reconstruct the development of this notion (from Post to Rogers, to the present days), and we focus on some classical constructions of the field, such as the construction of a simple set. Then, we make use of this focus in order to support the following encompassing claim (which opposes to a somehow commonly received view): the informal side of Computability, consisting of the large class of methods typically employed in the proofs of the field, is not fully reducible to its formal counterpart.

13.1 Introduction

Algorithms, when they do appear in mathematical papers, are often written somehow informally, i.e. without the need of specifying the background model of computation in which they are framed. The goal of this chapter is to understand the philosophical meaning (if any) of this circumstance. In doing so, we provide a philosophical analysis of the notion of 'proof by Church's Thesis', which is classically regarded as the conceptual device that permits to rely on informal methods when working in Computability. We carefully reconstruct the historical development of this latter notion, and we formulate what we shall consider as *the standard view* about it. Then, we challenge this view, and we propose an alternative

The author was partially supported by the Austrian Science Fund FWF through project P 27527.

L. San Mauro (✉)
Institute of Discrete Mathematics and Geometry, Technische Universität Wien, Vienna, Austria
e-mail: luca.san.mauro@tuwien.ac.at

© Springer International Publishing AG, part of Springer Nature 2018   225
M. Piazza, G. Pulcini (eds.), *Truth, Existence and Explanation*,
Boston Studies in the Philosophy and History of Science 334,
https://doi.org/10.1007/978-3-319-93342-9_13

reading. Altough a similar research focus might appear, at first, as a narrow one, we claim that it is strongly connected to a question that shall be of general interest for philosophers of mathematics, i.e.: what is the role of informality in mathematical proofs and constructions?

We begin by recalling few aspects of the Church-Turing thesis (henceforth: CTT). Being CTT the conceptual cornerstone of Computability (and, in fact, of the whole theory of computation), it has been unsurprisingly the central topic of an incredibly vast philosophical literature. It is not our intention to resume this literature or even sketch it.[1] Rather, we simply aim to fix, quite schematically, few ideas of which we make use afterwards. This choice is made possible by the fact that our main focus is on a somewhat neglected topic concerning CTT, a topic that remains almost philosophically untouched (although being well-established among computability-theorists), namely its *practical* use. As a result, while we do not consider the historical roots of CTT, we spend some time in reconstructing the less known history of such practical side.

CTT concerns the 'effective calculability' of the functions on positive integers. In its most general form, it expresses the fact that such notion of calculability – i.e. a pre-theoretic, informal notion – is fully captured by any of the (extensionally equivalent) classical models of computation. Thus, it can be regarded as the amalgamation of several different theses, each one corresponding to a specific model of computation.

CTT
A function is effectively calculable if and only if it is computable in one of the classical models of computation.

As is known, two aspects of the thesis are philosophically hard to reconcile. On the one hand, there is the evidence that CTT is almost universally accepted as true. On the other hand, it is difficult to give a complete account of what makes CTT true. Or, more precisely, it is difficult to say how we *know* that the thesis is true (if so). The problem, as one can immediately see, is epistemological. In any of its versions, CTT links two notions which display different status: a formal one, expressing the computability according to some model of computation, and that certainly describes a specific subclass of the functions on positive integers; and an

---

[1] A classical introduction to CTT can be found in Kleene (1952). See also Church (1936), Turing (1936, 1948), Post (1936), and Gödel (1946). Soare (1987b) contains, in its first part, an accurate reconstruction of the role of Turing's work in the acceptance of the thesis. For recent philosophical work concerning CTT, the reader is referred, for instance, to Olszewski et al. (2006).

informal one, corresponding to the concept of 'being calculable by some effective procedure', which – at least at first sight – appears to be "a somewhat vague intuitive one" (see Kleene 1952).[2]

As we will see, many of these epistemological concerns scarcely overlap the kind of problems we aim to unveil. In fact, we will argue that too much focusing on the epistemology of CTT has corresponded to a certain lack of interest in its practical side. Yet, from all the aforementioned debate, we shall at least note down the following four basic *assumptions* concerning the thesis, on which we will partially rely:

1. In what follows, we do not cast doubts on the validity of CTT. We assume rather that thesis is true, or at least *correct* in some profound sense, being a successful bridge from the informal to the formal side of Computability.
2. We mantain that such correctness is unusual. That is, although the problem of mirroring some pre-formal intuition is arguably at the hearth of most mathematical theories, the case of Computability stands as exceptional. To quote Gödel (1946):

   > With this concept [computability] one has for the first time succeeded in giving an absolute definition of an interesting epistemological notion. (...) In all others cases treated previously, such as demonstrability or definability, one has been able to define them only relative to a given language, and for each language it is clear that the one thus obtained is not the one looked for.

3. We hold that CTT refers to an idealization of *human* calculability. This prevents a somewhat debated way of attacking the thesis, which consists in designing very unconventional constructions relying on stuff that is way out of human reach, such as computations on the edge of a Black Hole.[3] More generally, this assumption expresses the fact that CTT is meaningful inasmuch it refers to (the idealization of) a given human practice. As Wittgenstein (1980) famously puts it: "Turing machines are humans who calculate".
4. Strictly related to this latter assumption is our last one: CTT itself does not embed a problem of reference. Indeed, although the problem of singling out a specific computation by making use of some informal description is a central one (and, in fact, it is deeply intertwined with our main goal of understanding what we call 'the practical side' of CTT), it is important to notice that the standard formulation of the thesis, under any of its possible readings, lies one step further of these problems. That is, whenever we say that any function which

---

[2]The standard interpretation is that CTT is indeed a *thesis*, or, in Post's words, "a working hyphotesis" (Post 1936). That is to say, something that cannot be subject of a mathematical proof. Yet, it has been argued that CTT has not necessarily an hypothetical status, but rather that it can be susceptible of a rigorous mathematical proof, or even that such a proof is already contained in Turing (1936) (for this line of thought see, e.g., Mendelson (1990), Gandy (1988), Sieg (1994), and the discussion contained in Shapiro (2006)). Responses to this latter position can be found in Folina (1998) and Black (2000).

[3]See Welch (2007) for a rich survey on models of transfinite computation. On the other hand, Davis (2006) denies any theoretical significance to "hypercomputationalism" as such.

is intuitively 'calculable' can be formally implemented in a classical model of computation, we always assume that a certain amount of work has been made in order to provide a description of that function which is enough clear to avoid misunderstanding or incorrect reference. We return to this topic after having introduced the announced practical side of CTT.

13.2 The Practical Side of CTT

As already mentioned, most of the philosophical attention revolving around CTT has consisted in clarifying what the thesis is about and how can we possibly ascertain it. Yet, a fundamental topic remains to be considered, namely how the thesis is *used* within the mathematical discourse. In particular, we are interested in its proof-theoretic role.

Definition 1 Call *practical side of CTT* the collection of all the appeals to CTT that are steps of some mathematical proof.

By 'mathematical proofs' we mean concrete examples of mathematical proofs, written in ordinary mathematical language (i.e., a given natural language – in most cases, English – extended with some finite collection of symbols). Therefore, we are referring neither to formal derivations nor to some abstract objects.

It is important to notice that, quite independently from our notion of practice (that we aim to keep as much intuitive as possible), Definition 1 is a rather demanding one. This is because asking for CTT to occur within real proofs is basically the strictest requirement that one can formulate for something to be, in practice, part of the mathematical discourse. So, at least theoretically, it would be perfectly reasonable to expect the practical side, defined as such, to be empty. For instance, CTT could be fundamental in grounding the significance of our theory, or even in setting the agenda of its most important problems, and yet not having any specific proof-theoretic role. To this end, imagine a scenario in which, although acknowledging the validity of CTT, computability theorists would consider the computability (resp. relative computability) of a given function acceptably proved only if the full description of a Turing-machine (oracle-machine) is alleged. What would get lost in a similar scenario? Keep this question in mind.

For the moment, let us just say that the practical side of CTT is far from being empty. In fact, any standard textbook in Computability contains – typically in a very initial segment – a certain amount of occurrences of the following expression: "proof by Church's Thesis".[4] So, the notion is sufficiently widespread to avoid being sensible to possibly idiosyncratic or deviant uses. In fact, the idea of proving

---

[4]For instance, the following is the first proof-theoretic reference to CTT in Rogers (1967):

Theorem 1 *There are exactly $\aleph_0$ partial recursive functions, and there are exactly $\aleph_0$ recursive functions.*

something by Church's thesis is a familiar one among practitioners; then, can it be philosophically grounded? We argue so. Moreover, we aim to show that such practical side of CTT is, in a sense, independent from CTT itself.

In doing so, we proceed with some history of this practical side, focusing on the most relevant landmarks. Actually, we begin with what shall be regarded as the prehistory of the notion of 'proof by Church's thesis'.

13.2.1 Post: "Stripped of Its Formalism"

It is fairly acknowledged that the basic conceptual machinery of Computability (in terms of concepts, definitions, and techniques) derives from Post (1944), which contains, in its opening paragraph, a somewhat curious remark:

> That mathematicians generally are oblivious to the importance of this work of Gdel, Church, Turing, Kleene, Rosser and others (...) is in part due to the forbidding, diverse and alien formalisms in which this work is embodied. (...)
> Yet, without such formalism, this pioneering work would lose most of its cogency. But apart from the question of importance, these formalisms bring to mathematics a new and precise mathematical concept, that of the general recursive function of Hrbrand-Gdel-Kleene, or its proved equivalents in the developments of Church and Turing.
> It is the purpose of this lecture to demonstrate by example that this concept [that of computable function] admits of development into a mathematical theory much as the group concept has been developed into a theory of groups. Moreover, that stripped of its formalism, such a theory admits of an intuitive development which can be followed, if not indeed pursued, by a mathematician, layman though he be in this formal field. (...)
> We must emphasize that (...) we have obtained formal proofs of all the consequently mathematical theorems here developed informally. Yet the real mathematics involved must lie in the informal development. For in every instance the informal "proof" was first obtained; and once gotten, transforming it into the formal proof turned out to be a routine chore.

In a footnote, Post adds:

> Our present formal proofs, while complete, will require drastic systematization and condensation prior to publication.

The whole passage is philosophically striking. What Post is literally saying is that

1. most of the proofs contained in his paper do not meet the standard of formalization fixed by their proper formal field;
2. these proofs, being developed only informally, are – in a very immediate sense – *incomplete*.

Proof All constants functions are recursive, by Church's Thesis. Hence there are at least $\aleph_0$ recursive functions. The Gödel numbering shows that there are at most $\aleph_0$ partial recursive functions.

If taken seriously, these two statements offer an account of proofs which is hardly sound with most ideas concerning the special reliability of mathematical facts. To put it crudely, if Post's proofs are really both incomplete and not formal enough, then why do we trust them? As is clear, to fully answer this latter question, one has to develop some kind of convincing explanation of what makes, in general, mathematical proofs so reliable – a tremendous task. For our part, let us just borrow from Hacking (2014) a convenient distinction between two ideal conceptions of proofs:

> There are proofs that, after some reflection and study, one totally understands, and can get in one's mind 'all at once'. That's Descartes.
> There are proofs in which every step is meticulously laid out, and can be checked, line by line, in a mechanical way. That's Leibniz.

Of course these are just idealizations. Especially in more complex cases, these two conceptions blend together. That is, a global understanding of a complex proof would reasonably encompass both some kind of bird-eye grasping of the proof structure, 'all at once', and an accurate mechanical verification of all delicate details. Nonetheless, they do reflect quite common sensations of which every mathematician has experience when facing many different proofs. Thus, if not completely, such idealizations bring back part of that feeling of "inexorability" that, according to Wittgenstein, characterizes real understanding of mathematical proofs. Then, how well they match with Post's remark? Apparently quite badly.

Most problems of Post (1944) have the following prototypical form: find a computable function f so-and-so.[5] In tackling a similar problem, one has roughly two available approaches. On the one hand, a solution would consists in obtaining a formal implementation of f within some preferred model of computation. On the other hand, it can be regarded as sufficient to provide the description of some procedure that intuitively computes f. Post chooses to adopt the second alternative, and the remark above stands as a sort of methodological disclaimer for that choice.

Now, keep CTT aside for a moment. Where do the proofs of Post (1944) lie in the leibnizian/cartesian spectrum? A leibnizian conception seems to fit way better with the "forbidding, diverse and alien formalisms" of papers such as Church (1936), that precisely represents the standard from which Post is departing. More generally, if the reliability of proofs is due to the possibility of having them presented in the most meticulous fashion – as a Leibnizian view would require – then, of course, to work in the rigid context of a model of computation seems to be the best option.[6]

---

[5]To be fair, Post mainly speaks of computably enumerable sets, there introduced for the first time. But since, by definition, a set is computably enumerable if it is the range of a computable function, then one can trivially translate Post's formulations in instances of our prototype.

[6]It is worth noticing that the Leibnizian ideal is by no means archeological. Quite to the contrary. Hacking reports Voevodsky's opinion that "in a few years, journal will accept only articles accompanied by their machine-verifiable equivalents". More generally – and less radically – research on proof-assistants can be (partially) motivated as a way of improving automatic verification of proofs.

Post's approach might be closer with the Cartesian conception. After all, his announced goal is to show that some fragment of Computability (or, a posteriori, all of the theory) is best suited for an *intuitive* development, instead of being carried out in a mechanical Leibnizian-like way. Nonetheless, the Cartesian idealization asks for completeness, i.e. it requires to see a given proof as a whole – to get it 'all at once' – in order to grasp its truth. In Descartes' words, mathematical proposition are "to be deduced from true and known principles by the continuous and uninterrupted action of a mind that has a clear vision of each step in the process" (Descartes 1628). Thus, no step can be omitted. Is Post's choice of skipping all formal programs, while proving the computability of the objects that he considers, one of such omissions?

Fallis (2003) argues so. In a nutshell, the goal of this latter paper is that of highlighting the presence of "intensional gaps" that mathematicians leave, at times, in their proofs. After having classified several different types of these gaps, Fallis claims that the so-called "universally untraversed gaps" are "examples of justificatory practices that are not captured by the Cartesian story, but that are nevertheless accepted by the mathematical community". As a last example of such type of gaps, Fallis hastily hints to the case we are focusing on, that of describing computable procedures just informally. It is interesting to notice that, while other cases of theoretically significant gaps are sporadic and, by author's admission, somewhat disputable, the case of Computability shows a systematic tendency of leaving intensional gaps. We will extend this consideration in the next sections.

Here, let us stress that Post's opening paragraph mismatches with both ideal conceptions of proofs we have considered. Nevertheless, we already know a possible way out to these problems: CTT. Indeed, the thesis equates the two alternative solutions to our prototypical problem through the following argument:

Church-Turing Bridge (CTB)
If any informal description of an algorithm can be formally implemented in each model of computation (as CTT states) then, in order to prove that something is computable, it is sufficient to describe an informal way to compute it – and then make reference to CTT.

CTB does not appear in this form in Post (1944), and presumably Post's preference for a somewhat informal style in mathematical writing is not mainly motivated by such application of the thesis.[7] Still, some version of this latter argument stands as a necessary theoretical bulwark against the difficulties we have shown above. Indeed, the central notion of Post (1944) is that of "generated set", that in Post's words corresponds "to say that each element of the set is at some time written down, and earmarked as belonging to the set, as a result of predetermined effective processes", hence being an informal concept. The fact that propositions technically proven for generated sets hold also for recursively enumerable sets does certainly require some sort of CTB.

---

[7]For an accurate reconstruction of Post's thought see De Mol (2006)

Let us conclude with an important remark. Post underlines that formal proofs, although being omitted in his presentation, has nonetheless been obtained: "we have obtained formal proofs of all the consequently mathematical theorems here developed informally". We assist here to a sort of conceptual twist. On the one hand, the focus is on the informal development in which "the real mathematics (. . .) must lie". Moreover, from CTB we obtain that informal proofs are acceptable and, in a sense, self-sufficient. But on the other hand, Post warns that for each informal proof presented in the paper a formal counterpart has been derived. The following question naturally arises: how far can one consider the knowledge of an informal proof *independent* from the knowledge of its formal renderings?

As for many other aspects of the foregoing analysis, this latter one will be expanded in Rogers' presentation of the practical side of CTT, to which is devoted the following section.

13.2.2   Rogers: "Proofs Which Rely on Informal Methods"

In the two decades spanning from 1944 to 1967, Computability Theory (called 'Theory of Recursive Functions', at the time) has seen an impressive growth, becoming one of the most active and fruitful area of mathematical logic. Post's informal approach was widely adopted, proving, in Rogers' words, that "the intuitive simplicity and naturalness of the concept of *general recursive function* permitted discourse and proof at a level of informality comparable to that occurring in more traditional mathematics". Such preference towards informality contributed to a significant shift in the mainstream objects of research: from functions computable within some specific model of computation, to sets calculable by some informal procedure, and then to 'degrees', i.e. equivalence classes of sets, intuitively encoding the same complexity (according to a given notion of reducibility). Again, Post (1944) has opened this latter line of research, formulating a famous long-standing problem – solved only in 1957 – of whether there are 'computably enumerable intermediate degrees' for Turing-reducibility, that is, computably enumerable degrees lying strictly between the complexity of computable sets and that of the Halting set.

In this context, Rogers (1967),[8] whose first edition appeared in 1967, accomplishes both a descriptive and a normative goal. On the descriptive side, it offers a quite comprehensive survey of the main results of the field, and its outstanding clarity is among the reasons for which the book became a classic. On the normative side, Rogers enriches the exposition of mathematical results with several profound philosophical insights, aiming to hook the mathematical content with some sort

---

[8]From now on, in describing the practical side of CTT, we will mostly refer to textbooks. This is a natural choice. Since, as already said, there are no philosophical studies concerning the practice of Computability, the most immediate source of observations regarding how such practice has to be intended comes from the kind of expository remarks that abound in books such as Rogers'.

of methodological and pre-formal justification. We are interested in the passage depicting the role of informality in Computability, that is also the one in which the expression "proof by Church's thesis" is firstly introduced:

> A number of powerful techniques have been developed for showing that partial functions with informal algorithms are in fact partial recursive and for going from an informal set of instructions to a formal set of instructions. These techniques have been developed to a point where (a) a mathematician can recognize whether or not an alleged informal algorithm provides a partial recursive function, much as, in other part of mathematics, he can recognize whether or not an alleged informal proof is valid, and where (b) a logician can go from an informal definition for an algorithm to a formal definition, much as, in other parts of mathematics, he can go from an informal to a formal proof. (...)
> Researchers in the area have been using informal methods with increasing confidence. (...) They permit us to avoid cumbersome detail and to isolate crucial mathematical ideas from a background of routine manipulation. (...) *We continue to claim, however, that our results have concrete mathematical status* (...). *Of course any investigator who uses informal methods and makes such a claim must be prepared to supply formal details if challenged.* Proofs which rely on informal methods have, in their favor, all the evidence accumulated in favor of Church's Thesis. Such proofs will be called *proofs by Church's Thesis*.

So, the state of art is clear: one is not committed in supplying formal algorithms, since, for any informal definition, there is a corresponding formal implementation whose existence is guaranteed by CTT (of course this is just a reformulation of CTB). Let us, then, isolate Rogers' definition:

Definition 2   In Computability, a proof is called *proof by Church's thesis* if it relies on informal methods.

Two aspects shall be noticed. First, this definition is so inclusive that it encompasses *almost all* the proofs of Computability. For example, Rogers' proofs are typically formulated without referring to any background model of computation – and so is the case of all the main results of the field. Hence, the practical side of CTT, as defined in Definition 1, is basically as large as possible! Moreover, it is unsurprising that, according to this definition, not only the specification of a given model of computation is omitted, but in most cases the very reference to CTT is left implicit. Indeed, Roger refers to the thesis only while proving quite elementary facts, i.e. those facts for which going from an informal algorithm to a formal one would be really just a matter of taking care of "cumbersome detail". Vice versa, when complex constructions are considered, and the reader is acquainted to this kind of reasoning, CTT does not occur explicitly and it is not entirely clear that informal methods – such as Rogers' version of Friedberg-Muchnik solution to Post's problem – can be *trivially* translated into, say, running Turing machines. In fact, Rogers sketches a sort of 'division of labour', in which going from an informal definition to a formal one is, typically, more a task for a logician than for a mathematician.

This latter observation leads to the second aspect. Recall that Post (1944) states that all the formal proofs, omitted in his presentation, have been nonetheless obtained. Rogers portrays a more delicate scenario. Anyone who employs informal methods "must be prepared to supply formal details if challenged." But, in practice, such a challenge never arises. Suppose one has designed a nontrivial construction for showing that a given object is (relatively) computable. Then, it would be very unconventional – and somehow unacceptable – to write down the construction in the form of a formal program. Rather, with the goal of isolating "crucial mathematical ideas from a background of routine manipulation", what is needed is to make a certain number of sensible choices about which parts of the construction have to be formalized, and to what extent. In other words, one has to set the borders between 'important ideas' and 'negligible details'. Nor the formal side is summoned with respect to the validity of the construction.

Thus, Rogers addresses the following familiar problem. Although we would have rigorous formal languages (embodied by models of computation) at our disposal, the development of Computability is carried out in some informal frame. Gaps, in the sense of Fallis, are everywhere. By introducing the notion of 'proof by CTT', Rogers attaches to the thesis a fundamental theoretical value, that of systematically filling these latter gaps and, in doing so, justifying the adopted convention of working only on the informal side of Computability.

Nonetheless, it is remarkable that this extensive use of CTT, although theoretically significant, does not make any real difference. Computability is, of course, as reliable as any other distinct branch of mathematics, while the Post-Rogers leaning towards informality is after all analogue to that of "more traditional mathematics". We have, here, two partially conflicting views. On the one hand, both Post and Rogers believe that departing from the formal definitions of algorithms is something that needs a solid justification (a justification that, in Rogers' perspective, is fully provided by CTT). But, on the other hand, they both insist that, rather than being a distinctive feature of Computability, informality is the norm of many mathematical fields with no principle like CTT on which one can rely. So, if employing informal tools is generally permitted in mathematics, then we might argue that making use of CTT, in order to make such tools available in Computability, is at best redundant, and at worst conceptually wrong. This is the line of thinking we consider in the next section.

13.2.3 Others: "A Fancy Name to a Routine Piece of Mathematics"

Consider the following passage from Odifreddi (1989):

> There is another avoidable use of the Thesis, in Recursion Theory. Giving an algorithm for a function amounts, by the Thesis, to showing that this function is recursive. Although theoretically not important, and in principle always avoidable (if the Thesis is true), this use is often quite convenient, since it avoids the need for producing a precise recursive definition of a function (which might be cumbersome in details). Strictly speaking, however, this use does not even require a Thesis: it is just an expression of a general preference, widespread in mathematics, for informal (more intelligible) arguments, whenever their formalization appears to be straightforward, and not particularly informative.

Thus, Odifreddi makes use of an observation that is already present in Post and Rogers, that of the fundamentally straightforward nature of any translation from an informal definition of an algorithm to a formal one. In particular, the real cost of providing formal details for all algorithms we are working with would consist in having much less intelligible arguments. The underlying idea is clearly referring to human cognitive limitations. Since we can handle only a fairly limited amount of information at the time, it is convenient to make a distinction between relevant and irrelevant part of a given algorithm – and of course transmit only the former. Furthermore, Odifreddi argues there are no other reasons that motivate – or even, justify – working on the informal side of a given theory. This is the case of most mathematical theories, and Computability is no exception. Hence, according to this perspective, the expression "proof by Church's thesis" would be essentially unnecessary. Preferring informal arguments is a feature common to most part of mathematical endeavour (as we currently know it), and it depends only on some very basic human constraints. If we theoretically put such constraints aside, then no significant difference between formal and informal definition can be revealed.

Epstein and Carnielli (1989) reinforce this latter view, by denying any relevance of the practical side of CTT:

> To invoke Church's thesis when "the proof is left to the reader" is meant amounts to giving a fancy name to a routine piece of mathematics while at the same time denigrating the actual mathematics.

What does it mean to let the notion of "proving by CTT" to be equivalent to saying that "the proof is left to the reader"? In general, it can be regarded as a radical weakening of the notion. First, it expresses the fact that each informal definition of an algorithm has a proper formal referent that can be easily identified by a sufficiently painstaking reader. So, there is no real gain in giving missing formal details. But, on the other hand, such equivalence says also something about the converse relation. That is to say, there is no additional benefit in working with informal definitions, apart from a better understanding of a proof.

Therefore, to sum up, against Rogers (and partially Post), Odifreddi (1989) and Epstein and Carnielli (1989) argue that the preference towards informality in Computability does not require any autonomous justification, being only an instance of the general preference for more intelligible arguments in mathematics.

13.3 The Standard View

In the preceding sections, we have traced some history of the practical side of CTT. We have shown that the idea that mathematical results in Computability can be presented without referring to any background model of computation is rooted in Post (1944). Then, we have highlighted the definition of "proof by Church's thesis" from Rogers (1967), there presented as a necessary justification for the bold role that informality plays in Computability. Finally, we have offered two examples representing a certain tendency of dismissing the importance of such uses of thesis, by claiming instead that the case of Computability is similar to that of more mundane areas of mathematics, and that departing from formal definitions does not really affect the development of the theory. Let us then resume the outcome of this discussion by calling 'standard view (about the practical side of CTT)' the following position:

> **Standard View (SV)**
>
> (a) *CTT allows us to rely on informal methods (by CTB);*
> (b) *Yet, these methods are in the end just a matter of convenience: informal definitions point towards formal ones, and we could theoretically substitute the former with the latter without any significant loss or gain of information;*
> (c) *This operation is analogous to what happens in most parts of mathematics.*

SV has arguably a very large consensus among practitioners. It fits, of course, with Odifreddi (1989) and Epstein and Carnielli (1989). More importantly, it can also explain why the practical side of CTT is so philosophically neglected. Indeed, if supporting SV, one can easily claim that the informal aspects of Computability do collapse onto their formal counterpart. Thus, once justified CTT, there is – philosophically speaking – *nothing more to do*. Let us expand this point.

13.3.1 Proofs vs Derivations, and CTT

First, recall that CTT states that any function that is calculable by some informal procedure will turn out to be computable in any of our (extensionally equivalent) formal frameworks. Then, SV claims, this is all we need in order to bridge the gap between the informal and formal side of Computability. In particular, according to SV, the widespread convention of omitting formal definitions – in any case recoverable, if needed – would be just a matter of convenience, with no particular theoretical significance. Thus, for a philosopher, the main task would be that of justifying CTT (a challenge that many philosophers have indeed accepted), while, on the other hand, considering the way in which functions are informally described, in customary presentations, would be essentially uninteresting.

A noteworthy corollary of this latter perspective is in its connection with a key problem of contemporary philosophy of mathematics, that of relating real mathematical proofs (as they appear, for instance, in ordinary mathematical journals) and the abstract models of proofs studied in formal logic. It is customary to use the word *proofs* for the former, while referring to the latter as *derivations* (see, for instance, Rav 1999). So, here is the problem: what is the nature of this gap between proofs and derivations? Is it a philosophically significant one?

First, one might hastily dismiss our fundamental problem by claiming that any purported gap between proofs and derivations is either trivial or irrelevant. After all, derivations are nothing but *models* of real proofs. Therefore, it is just an immediate consequence of the very idea of modelling that many details are abstracted away, while preserving any whatsoever essential core of the modelled notion. So a gap does certainly exist but, rather than being problematic, the existence of such a gap is basically what modelling is for.

Foundational programs, or rather the kind of philosophical perspectives that stems from them, usually push this latter reasoning forward. Indeed, derivations are not only convenient models of real mathematical proofs, but it is actually fruitful to *identify* the two notions. The main benefit of this identification is of course provided by metamathematics, i.e., to our present concern, the possibility of making use of mathematical tools for answering questions that, to some extent, conceptually belong to the philosophy of mathematics For instance, if we subscribe the identification between formal systems and mathematical theories (which clearly echoes the one between proofs and derivations), then we obtain that limitative results for formal systems, such as Gödel's ones, do apply also to real mathematical theories, even if we choose to understand them in a rather informal way.

The seminal work by Lakatos (1976) has famously pioneered an opposing tradition to these latter identifications. After Lakatos, and in particular in the two last decades, there has been a growing line of research which aims to take account of a large class of problems, emerging from mathematical practice, that are classically neglected by philosophers of mathematics, either because of the influence of formalist positions or for the tendency, within the analytic tradition, of taking very small fragment of elementary mathematics as nothing but case studies for philosophical problems arising elsewhere. In general, this line of research is often referred as Philosophy of Mathematical Practice.[9] Concerning the gap between proofs and derivations, philosophers of mathematical practice obviously endorse the idea that such gap is worth studying. To this end, several examples have been provided of some aspects of proofs, occurring in concrete context, that arguably seem to lie well outside the scope of the formalistic characterization.

All of this certainly echoes the gap between the informal presentation of an algorithm and its formal counterpart obtained by defining the algorithm within some model of computation. Nonetheless, even philosophers that strongly adverse any identification – or, hasty correspondance – between proofs and derivations have conceded, by supporting some version of SV, that CTT permits to match informal definitions with formal ones with almost no distortion. For instance, one of the main

---

[9]The interested reader can consult Mancosu (2008) for an anthology of papers in Philosophy of Mathematical Practice.

thesis of Rav (1999) is that proofs display a certain semantic content that is utterly destroyed when these are translated into derivations. Yet, Rav writes what follows:

> It is has been suggested to name *Hilbert's Thesis* the hypothesis that every conceptual proof can be converted into a formal derivation in a suitable formal system: proofs on one side, derivations on the other, with Hilbert's thesis as a *bridge* between the two. One immediately observes, however, that while Church's Thesis is a two-way bridge, Hilbert's Thesis is just a one-way bridge: from a formalised version of a given proof, there is no way to restore the original proof with all its semantic elements, contextual relations and technical meanings.

Rav does not subscribe to Hilbert's Thesis. Nonetheless, he seems to argue that, when considering Computability, the situation appears to be profoundly different. Being CTT a two-way bridge, it is reasonable to expect that our work is not to be conditioned by the side on which is practically carried out. Notice that, in Rav's view, what is not feasible is to going back from derivations to proofs. On the contrary, CTT would permit "to restore (...) all [the] semantic elements, contextual relations and technical meanings" of a given informal algorithm from any of its formal characterizations. Thus, saying that CTT is a two-way bridge, in the sense of Rav, is quite the same of rephrasing point (*b*) of SV: we can go from informal definitions to formal ones and back with no important distortion.

Therefore, SV embeds a certain conceptual opposition concerning the relation between Computability and other mathematical theories. On the one hand, the original detachment from Roger's definition of the practical use of CTT was partially motivated by the evidence that informal tools are commonly employed in most mathematical theories – and point (*c*) of SV expresses precisely this fact. On the other hand, we have just shown that point (*b*) describes precisely the kind of clear correspondence between informal and formal components that many philosophers argue does not belong to mathematical practice. In order to better grasping this tension – that we eventually aim to solve by claiming that SV is untenable – we have to focus on point (*b*) of SV.

13.3.2   Clarifying Point (b) of SV

Needless to say, of the three points of which SV consists, (*b*) is the most delicate. Two questions are to be answered:

1. What does it mean that informal definitions 'point towards' formal ones?
2. When does a loss (or, a gain) of information count as 'significant'?

As we will see, these two questions are linked. Let us begin with the first one, with the help of some notation. Call $\mathscr{I}$ the class of all the informal definitions for algorithms.[10] Then, let C be a model of computation, and denote by $\mathscr{F}_C$ the

---

[10] $\mathscr{I}$ does clearly correspond to a pre-theoretic object whose formalization would be far from trivial. For instance, there could be a worry concerning a sort of Berry-like paradox, inasmuch we

set of all the programs of C. We think of SV as guaranteeing the existence of a specific map μ from $\mathscr{I}$ to $\mathscr{F}_C$ such that $\mu(x)$ can be intuitively understood as a 'reasonable formalization' of the informal definition x. What can be said about this map? First, it is natural to argue that μ cannot be injective. This is because the grammar of natural languages is much less rigid than that of formal ones. Then, it is easy to imagine that we could make very minor modifications to almost any informal definition x, thus obtaining some x', in such a way that μ would be not sufficiently fine-grained to distinguish $\mu(x)$ and $\mu(x')$. More generally, the injectivity of μ is blocked by the fact that informal algorithms are typically described at a much higher level of abstraction than that of their formal counterparts. For instance, in Blass et al. (2009) are considered two versions of the Euclidean algorithm for finding the greatest common divisor of two positive integers. The basic operation of the first version is division, while the second one is based on repeated subtraction. Then, when considering possible formalizations of these two algorithms, authors point out the following:

> The point we wish to make here (...) is that the use of low-level programs like Turing machines might cut off the debate prematurely by making the two versions of Euclid's algorithm identical. Consider programming these versions of Euclid's algorithm to run on a Turing machine, the input and output being in unary notation. The most natural way to implement division in this context would be repeated subtraction. With this implementation of division, any difference between the two versions disappears.

So, this example shows that the simplicity of the language of Turing machines in many cases can force two different algorithms to be naturally implemented with the same program. In this regard, there is quite a large literature that addresses the problem of finding a formal model better representing the kind of abstraction that is embedded in our informal algorithms. See, for instance, the Abstract State Machines of Gurevich (2000).[11]

Nonetheless, the very fact that there is a gap between the way in which algorithms are informally presented and most of their implementations does not contrast with SV. Arguably, whenever the language of (a fragment of) $\mathscr{I}$ and that of some $\mathscr{F}_C$ do correspond to different level of abstractions – and, again, this is the norm – μ must embeds some sort of distortion. SV does not deny such distortions. Rather, it

---

admit a too relaxed notion on what counts as an informal description for an algorithm. Nonetheless, we can suppose to deal with sufficiently clear descriptions. This is because, although border-cases cannot arguably be expunged, we are more interested, as we will see, in a somewhat global tendency.

[11] All of this is of course related to the philosophical problem of determining if one can possibly formulate a definition for algorithms that would be *correct* in the sense of Shore: "Find, and argue conclusively for, a formal definition of algorithm and the appropriate analog of the Church-Turing thesis. Here we want to capture the intuitive notion that, for example, two particular programs in perhaps different languages express the same algorithm, while other ones that compute the same function represent different algorithms for the function. Thus we want a definition that will up to some precise equivalence relation capture the notion that two algorithms are the same as opposed to just computing the same function"(Buss et al. 2001). See also Dean (2007) for a rich discussion on whether algorithms can be fairly regarded as abstract mathematical objects.

expresses the thesis that, for most x, some gap between x and $\mu(x)$ does certainly exist, but it turns out to be practically not significant. This clearly leads to our second question: what does it mean to be 'not significant' in the present context?

The guiding intuition is that, in accordance with SV, we want CTT to be a two-way bridge. Hence, the idea is that one should be able to restore all the information distorted by μ. We might represent this latter situation as follows. SV stases that, in addition to μ, there is some sort of inverse map μ^{-1} such that the following holds:

$$\mu^{-1}(\mu(x))Ex, \tag{13.1}$$

where E is a binary predicate, defined on $\mathscr{I}$, which corresponds to 'being essentially the same algorithm'. Let us stress that these symbols are for illustrative purposes only. E is a very pre-theoretic notion, and arguably E contains several contextual and normative aspects – in several cases, two algorithms are the same if we *choose* to consider them the same – that would make the goal of developing a definitive formalization for such predicate very problematic, or even not really feasible.[12] Yet, fortunately, our notions can be left to some extent vague. Indeed, SV concerns more the global picture of Computability than particular instances of elements of $\mathscr{I}$. In particular, SV says that a scenario as the one expressed by (13.1) is *usually* correct. That is, according to SV, Computability does not extensively rely on some kind of informal devices that, for some reason, cannot be formally translated.

To shed light on this latter point, let us consider again a question we have asked on page 228. Consider the following hypothetical scenario. Suppose computability theorists did not adopt Post's proposal and Computability had developed with the need of providing formal implementations for each algorithm considered. Now, there are no doubts that a similar theory would appear to be different – or rather, distorted – with respect to the standard Computability we currently deal with. For one thing, proofs would be far less readable. Yet, the main question is the following: such alternative Computability would differ from the usual one in some profound sense, shifting the general meaning of the theory? A coherent supporter of SV, we argue, has to answer: no (otherwise, the possibility of regarding CTT as a two-way bridge would be severely challenged, and informality would have some independent, theoretical value). For our part, in the rest of this chapter, we will defend a positive answer.

---

[12]Indeed, Blass et al. (2009) argue that "one cannot give a precise equivalence relation capturing the intuitive notion of 'the same algorithm.'"

13.3.3   A Case-Study: The Existence of a Simple Set

So, let us focus on a rather easy case in which some informal procedure is actually involved. First, we need to recall what follows.

There are several ways to define a *standard numbering* of partial computable functions, all essentially equivalent. Fixed one of such numberings, it is customary to denote by φ_e its eth element. A numbering ψ is said to be *acceptable* if there are two computable functions f, g such that: (i) $\varphi_{f(x)} = \psi_x$; (ii) $\psi_{g(x)} = \varphi_x$. It is immediate to see that, for any acceptable numbering ψ, there is a computable permutation of ω, π, such that, for all x, $\varphi_x = \psi_{\pi(x)}$. Acceptable numberings are important since they define the scope of most results of Computability: indeed, a theorem due to Rogers shows that a numbering is acceptable iff it satisfies both the Enumeration and s-m-n Theorems (hence preserving all the results based on these latter).[13] To our purpose, it is interesting to notice that numberings can be regarded as very abstracts ways to speak about models of computation, leaving aside all intensional aspects of such models. In particular, any natural coding of a given classical model of computation leads to an acceptable numbering, since one can always effectively translate this latter model into that of Turing machines.[14]

These properties of acceptable numberings have a practical consequence that largely overlaps our main topic. In Soare's words (Soare 1987a):

> Since most natural numberings are acceptable and two acceptable numberings differ merely by a recursive permutation, it will not matter exactly *which* acceptable numbering we chose originally.

As is clear, not specifying the background numbering on which are work is based echoes – and, in fact, subsumes – the kind of practical use of CTT that we have extensively discussed. To better illustrate such analogy, and see why SV fails, consider the following classical notion introduced in Post (1944):

Definition 3 We say that a c.e. set S is *simple* if is it co-infinite and, for each infinite W_e, $A \cap W_e \neq \emptyset$.

Thus, a set of positive integers is simple if it meets all infinite c.e. sets while remaining co-infinite. Simple sets do exist:

Theorem 2 (Post) *There is a simple set.*

Proof Let $S = \text{range}(f)$ where:

$$f(i) \text{ is the first element} \geq 2i \text{ enumerated in } W_i.$$

---

[13]The reader is referred to Rogers (1967) for the proof of this fact, and to Odifreddi (1989) for additional results concerning numberings.

[14]Some equivalence between classical models of computation can be found in Odifreddi (1989).

Since f is partial computable (by CTT), we have that S is c.e. Let W_i be an infinite
c.e. set. By construction, it is immediate to see that the first element $\geq 2i$ enumerated
in W_i – which does certainly exist being W_i infinite – belongs to S. Thus, S intersects
any infinite c.e. set. Then, notice that, if $x < 2i$ and $x \in S$, then there must be k such
that $x = f(k)$. Hence, $|S \cap \{y \mid 0 \leq y < 2i\}| \leq i$. So S is co-infinite, and therefore
simple.

We have built a simple set S. That is, our proof informally describes a way to
computably list all elements of S. So far, so good. But what do we *know* about S
(and thus of the algorithm by which S is listed)?

For instance, consider the following two questions:

1. Does 7 belong to S?
2. Or, is S infinite?

It is worth noticing that these two questions, in Computability, do not share the
same epistemological status. On the one hand, one can immediately check that S has
to be infinite, since there are infinitely many infinite c.e. sets. But, on the other hand,
whether 7 (or any other positive integer) would belong to S is not determined by the
construction, for it essentially depends on *how* we list c.e. sets, and thus it rests on
the choice of the numbering ψ. Therefore, our construction does not extensionally
fix a single set, but only up to a given (acceptable) numbering.

Now, suppose that, in accordance to SV, we want to translate our informal
definition into a formal one, by writing instructions for a Turing machine that lists
S. Of course, this is feasible (and CTT, if true, precisely guarantees that we can do
so), nevertheless in order to complete our implementation we have to specify some
numbering. Otherwise, without knowing in which order the c.e. sets are enumerated,
the computation will be blocked and no S would be generated. The problem is that,
under such a specification, we would consider a much less general version of S,
one limited by the choice of a formalism. Thus, a significant distortion between the
informal construction of S and any of its formal renderings, of the kind not permitted
by SV, lies in wait.

A possible way-out for a supporter of SV would consist in claiming that S, as
defined above, is essentially incomplete. In this perspective, omitting to specify a
background numbering, while constructing S, would be just another gap that we
leave behind for reasons of convenience. In other words, the way in which the proof
is presented would be another expression of the widespread preference for clarity
and generality in mathematics. Yet, from a theoretical point of view, the proof has
to be intended as a sort of a prototype, to be completed by specifying a certain
numbering ψ. That is, our informal proof would correspond to a general method
for describing, for each acceptable numbering, a corresponding simple set. To some
extent a similar interpretation is certainly correct, but it is also too limited. Indeed,
the following fact trivially holds:

Fact 1 *Let ψ be an acceptable numbering and let S be the simple set constructed
as above with respect to ψ. Then, S is a simple set also with respect to any other
acceptable numbering.*

So, we have the two following facts hold:

1. The proof of Theorem 2 provides a method for building, for each acceptable numbering, a corresponding simple set;
2. Yet, any simple S built by specifying an acceptable numbering ψ in Post's construction is also simple with respect to any other acceptable numbering.

Thus, in a sense, the theory of simple sets is *invariant* with respect to the acceptable numbering we choose to work with, making this very choice superfluous. So, against SV, it seems, first, that S does not refer to any of its formal definitions. But furthermore, S has not to be regarded as incomplete: although our informal proof does provide a method for producing, for all numberings, a given simple set, as already said, nevertheless to collapse the meaning of such proof to this method would correspond with claiming that an implicit reference to numberings is somehow needed to make complete sense of our construction of S – and this is precisely what is denied by Fact 1. Rather, the notion of simplicity is better understood as an *absolute* one, i.e. independent from the chosen formalism.

Most properties studied in Computability do share this character of absoluteness. For one thing, the notion of Turing-degree of a set (probably the main notion of the classical theory) is of course independent of the way in which we enumerate partial computable functions. Here, let us notice that, instead of being limited to properties of sets, this idea of working up to an acceptable numbering – and so without referring to any specific formal background – is embedded in most of the methods by which sets are typically constructed. In general, the success of the Post-Rogers paradigm could have led at least theoretically to a messy class of informal descriptions in the definitions of algorithms. In a sense, this circumstance would have been sound with SV: if informal algorithms are only conceptual shortcuts pointing towards formal ones, then there would be no reasons for expecting any independent grammar framing them. But, historically, that was not the case. In fact, these informal constructions rapidly converged towards an acknowledged standard in the form and the logic of their exposition, and their generalizations gave rise to what are called "methods". We do not wish to enter in a discussion concerning the philosophical significance of such methods (although our current analysis might be regarded as preliminary work for this latter investigation). For the moment, let us just make one more example related to SV.

Many constructions in Computability embed what shall be regarded as an effective version of Cantor's diagonalization. Generally, the goal of these constructions is to build one or more object (maybe by accessing to some oracle) in such a way that an effective list of *requirements* is eventually satisfied. Most notably, the Friedberg-Muchnik solution to Post's problem consists in constructing, by steps, two c.e. sets A and B, ensuring that, for any computable functional Φ_e, the two following requirements would eventually hold:

$$\chi_A \neq \Phi_e^B; \quad \chi_B \neq \Phi_e^A,$$

where χ_Y of course denotes the *characteristic function* of Y.

It is important to notice that A and B so defined do not refer to any specific enumeration of the functionals Φ_e. But again, this does not mean that the construction is to be considered as incomplete – as a certain reading of SV would require – but rather that we do refer to a kind of absolute version of A and B, i.e. independent from the choice of a formalism. In the next section, we aim to provide a better characterization of these *absolute objects*.

13.3.4 Indifference choices

In discarding SV, one has to clarify how objects such as the simple set S, or the Turing-independent c.e. sets A and B, are to be thought if not as objects to be completed with missing formal details. In doing so, we borrow the notion of *indifference* from Burgess (2015). Indifference, in Burgess' view, is best understood in contrast with *structuralism* in philosophy of mathematics. Under this heading, there is a variety of positions – and a large body of work[15] – corresponding to different philosophical characterizations of, roughly, the same basic idea: mathematicians are not concerned with the nature of the objects they deal with, but rather with *structures* involving them. So that, if two classes of objects display the same structure, then it does not really matter which one we choose to work with. Structuralism, at least in this naive formulation, does certainly reflect a common tendency of mathematical practice. For instance, Euclid's definitions – such as "A *point* is that which has no part" – are irrelevant to his proofs, and what really counts as the mathematical meaning of the basic objects of Euclidean geometry (points, lines, angles, etc.) are the relations they entertain with each other, as expressed by the axioms. Of course, this interpretation has been strengthened in Hilbert (1899), that represents both "the culmination of a trend toward structuralism within mathematics" (Shapiro 2010) and one of the most influential starting point for the distinctive emphasis on structures that characterizes contemporary mathematics obviously much more than ancient one.

Philosophers have put considerable effort in trying to propose a coherent account embracing the role that structures play in mathematics. However, most proposals develop a kind of ontological framework (motivating the existence or, alternatively, the nonexistence of structure themselves) that is quite far from our present perspective.[16] A more practice-oriented approach builds on the tempting inclination of considering the structural resemblance – up to which class of objects shall be regarded as the same – as being formally captured by isomorphism relation. As Awodey (2014) puts it:

The following statement may be called the Principle of Structuralism:

---

[15]For a classical defense of structuralism in philosophy of mathematics, the reader is referred to Resnik (1997).

[16]Two noteworthy exception being Carter (2008) and McLarty (2008).

Isomorphic objects are identical.

From one perspective, this captures a principle of reasoning embodied in everyday mathematical practice: (...)

• The Cauchy reals are isomorphic to the Dedekind reals, so as far as analysis is concerned, these are the same number field,

(...) Within a mathematical theory, theorem, or proof, it makes no practical difference which of two "isomorphic copies" are used, and so they can be treated as the same mathematical object for all practical purposes. This common practice is even sometimes referred to light-heartedly as "abuse of notation," and mathematicians have developed a sort of systematic sloppiness to help them implement this principle, which is quite useful in practice.

A principle such as PS encounters unsurprisingly many difficulties, because part of the specific value of considering two isomorphic objects as the same comes from the possibility of distinguishing them when needed (and, indeed, the goal of Awodey's paper is to show that PS is incompatible with the standard set-theoretic foundation, promoting rather the so-called Univalent Foundations). To our interests, it is important to notice that PS does not represent an available option in the context of Computability. After all, one of the key feature of this latter theory is that of enriching the study of mathematical structures by means of considering some (possibly invariant) computational aspects, to which classical mathematics is insensitive. Consider, for instance, the case of two noncomputably isomorphic presentations of the same structure, that immediately would reject PS.

One might obtain a better candidate for Computability by replacing isomorphism, in the PS principle, with computable isomorphism. Nonetheless, it is just false that, in the theory, any two computably isomorphic object are identical – or, even, that are regarded as identical for all practical purposes. For instance, *index sets*, e.g. sets containing all indices of a given computable function, do clearly depend on some background numbering. Thus, it is trival, given some index set I, to define a computable permutation of ω, π, such that the set $\{\pi(i)|i \in I\}$ is not an index set. More generally, the problem here is that the distinctive focus on absolute notions in Computability seems to be a *choice*, and hence it would be very hard to explain it if referring only to the formal side of the theory.

Burgess' notion of indifference is precisely an attempt of making sense of the phenomenon highlighted by structuralist philosophers, without being committed to heavy-duty principles such as PS:

It should be emphasized that (...) various structuralist philosophers of mathematics have performed a real service by pointing out to philosophers a real phenomenon, a kind of *indifference* on the part of working mathematicians, namely, an indifference to exactly how one got to the point from which their own investigations begin.

According to Burgess, there is a process – that he considers a process of "rigorization" – by which a certain amount of the work that has been made in order to develop a given formal notion can thereafter be dropped by remaining indifferent with respect to *how* this very notion has been introduced. So, we can define real numbers as Cauchy sequences or as Dedekind cuts, and then, while doing analysis, being perfectly indifferent with respect to which is the underlying formalization of reals. It is not our purpose to review Burgess' long defense of this notion, that, he argues, incorporates several different phenomena emerging in mathematical practice.

Let us just notice two aspects that we find appealing of Burgess' proposal. First, contrary to most structuralist philosophers, he does not refer to some peculiar nature of mathematical objects while explaining the fact that different classes of objects are regarded as the same. Thus, he claims that the correct focus is not on the ontology of mathematics, but rather on the reasons for which mathematicians do something (i.e., identifying such classes). Secondly, Burgess acknowledges that this operation of remaining indifferent towards parts of the mathematical discourse "tells us that any aspect of old work not needed or useful for new work can be disregarded, but it does not tell us *which aspects of old work these are likely to be.*" Therefore, to some extent indifference shall be regarded as a *choice*, to be accepted or rejected as part of a given mathematical theory. This reading, as we will see, has noteworthy consequences for our context.

As is now clear, our thesis is that the Post-Rogers detachment from the need of defining formal algorithms is well-represented as an instance of such indifference. The main benefit of this interpretation is two-fold. First, it takes account of the emphasis on absolute notions in Computability – an emphasis completely disregarded by SV – without referring to some peculiar ontological feature that objects of the theory would display. That is, it is not our simple set S that is absolute (in some ontological sense to be clarified), but rather it is our way to refer to a simple set that remain indifferent with respect to which acceptable numbering we make use.

Secondly, indifference is perfectly compatible with the evidence that absolute notions, although characterizing most notions of the theory, do not represent them all (recall the case of index sets). This is because indifference, in Computability, corresponds to a choice, that of having particular focus on such absolute notions – nevertheless, this focus can be suspended or disregarded, if it is fruitful doing so.

13.4   Final Remarks: Back to CTT

We have shown that SV fails to represent a central phenomenon of Computability, that of conceiving most constructions as absolute, i.e. independent from the background formalism and yet not to be regarded as incomplete. Furthermore, we have argued that this feature of absoluteness can be better grasped by appealing to a certain notion of indifference. More generally, the idea embedded in SV that working on the informal side would not shift the general meaning of the theory appears to be untenable. Let us then conclude by spending a very few words on how this latter perspective can shed some light on the aforementioned problem of relating Computability with other mathematical theories. In doing so, we have to turn back to CTT.

As already said, SV trivializes the 'proofs vs derivations problem', in the context of Computability, by interpreting CTT as a two-way bridge. That is, according to SV, Computability – contrary to almost any other mathematical theory – would permit, via CTT, to deny any significance to our practice, and with almost no philosophical cost. For our part, one of the main goal of this chapter has been precisely that of separating CTT from its practical side, i.e. we have defended the following claim: although CTT, being valid, does certainly fix a unique extensional notion of calculability, on the other hand this does mean that CTT fixes also a unique practice. Rather, the practical side of Computability (as always in mathematics) relies on a collection of choices and omissions, altogether amounting to the kind of indifference that we have sketched above. What is peculiar of Computability is that this process of highlighting and neglecting parts of formal discourse – that defines how a given theory has to be intended – is much more explicit than in other mathematical contexts. Indeed, first Post and then Rogers had to somehow justify an overall conventional level of informality in their expositions because dealing with the very unconventional fact that a fully formal approach to their theory was, at least in principle, available.

Thus, once abandoning the perspective, embedded in SV, that *after CTT* there would be philosophically nothing to say about Computability, we have arrived to the hypothesis that Computability might be even taken as an ideal context in which studying practice-oriented questions, for the relation between theory and practice (and the corresponding one between formal and informal) does appear at a very high level of clarity. Of course a lot of further work has to be done in order to confirm such hypothesis.

Acknowledgements A preliminary version of this paper appeared as a chapter of my PhD thesis. I would like to thank my supervisors, Gabriele Lolli and Andrea Sorbi, for their guidance and support. I have presented this work at several conferences. In particular, I am grateful to the participants of APMP 2014, in Paris, and of FilMat 2016, in Chieti, for their comments. Finally, Richard Epstein's remarks were fundamental in rethinking the organization of the present material.

References

Awodey, S. 2014. Structuralism, invariance, and univalence. *Philosophia Mathematica* 22(1): 1–11.
Black, R. 2000. Proving Church's thesis. *Philosophia Mathematica* 8(3): 244–258.
Blass, A., N. Dershowitz, and Y. Gurevich. 2009. When are two algorithms the same? *Bulletin of Symbolic Logic* 15(02): 145–168.
Burgess, J.P. 2015. *Rigor and structure*. Oxford: Oxford University Press.
Buss, S.R., A.S. Kechris, A. Pillay, and R.A. Shore. 2001. The prospects for mathematical logic in the twenty-first century. *Bulletin of Symbolic Logic* 7(02): 169–196.
Carter, J. 2008. Structuralism as a philosophy of mathematical practice. *Synthese* 163(2): 119–131.
Church, A. 1936. An unsolvable problem of elementary number theory. *American Journal of Mathematics* 58(2): 345–363.
Davis, M. 2006. Why there is no such discipline as hypercomputation. *Applied Mathematics and Computation* 178(1): 4–7.
De Mol, L. 2006. Closing the circle: An analysis of Emil Post's early work. *Bulletin of Symbolic Logic* 12(02): 267–289.

Dean, W.H. 2007. *What algorithms could not be.* PhD thesis, Rutgers University-New Brunswick.

Descartes, R. 1628. Rules for the direction of the mind. In *Selections.* Trans. R.M. Eaton. New York: Charles Scribner's Sons, 1927.

Epstein, R.L., and W. Carnielli. 1989. *Computability: Computable functions, logic, and the foundations of mathematics.* Pacific Grove: Wadsworth & Brooks/Cole.

Fallis, D. 2003. Intentional gaps in mathematical proofs. *Synthese* 134(1): 45–69.

Folina, J. (1998). Church's thesis: Prelude to a proof. *Philosophia Mathematica* 6(3): 302–323.

Gandy, R. 1988. The confluence of ideas in 1936. In *The Universal Turing machine: A half-century survey,* ed. R. Herken, 55–111. Wien/New York: Springer.

Gödel, K. 1946. Remarks before the Princeton bicentennial conference on problems in mathematics. In *Kurt Gödel: Collected works,* ed. S. Feferman, J. Dawson, and S. Kleene, vol. II, pp. 150–153. Oxford: Oxford University Press.

Gurevich, Y. 2000. Sequential abstract-state machines capture sequential algorithms. *ACM Transactions on Computational Logic (TOCL)* 1(1): 77–111.

Hacking, I. 2014. *Why is there philosophy of mathematics at all?* Cambridge: Cambridge University Press.

Hilbert, D. 1899. Grundlagen der geometrie. In *Festschrift zur Feier der Enthüllung des Gauss-Weber-Denkmals in Göttingen,* 1–92. Leipzig: Teubner.

Kleene, S.C. 1952. *Introduction to metamathematics.* Amsterdam: North Holland.

Lakatos, I. 1976. *Proofs and refutations: The logic of mathematical discovery.* Cambridge: Cambridge University Press.

Mancosu, P. 2008. *The philosophy of mathematical practice.* Oxford: Oxford University Press.

McLarty, C. 2008. What structuralism achieves. In Mancosu (2008).

Mendelson, E. 1990. Second thoughts about Church's thesis and mathematical proofs. *The Journal of Philosophy* 87(5): 225–233.

Odifreddi, P. 1989. *Classical recursion theory,* vol. I. Amsterdam: North Holland.

Olszewski, A., J. Wolenski, and R. Janusz. 2006. *Church's thesis after 70 years.* Frankfurt/New Brunswick: Ontos Verlag.

Post, E.L. 1936. Finite combinatory processes–formulation. *The Journal of Symbolic Logic* 1(03): 103–105.

Post, E.L. 1944. Recursively enumerable sets of positive integers and their decision problems. *Bulletin of the American Mathematical Society* 50(5): 284–316.

Rav, Y. 1999. Why do we prove theorems? *Philosophia Mathematica* 7(1): 5–41.

Resnik, M.D. 1997. *Mathematics as a science of patterns.* Oxford: Oxford University Press.

Rogers, H., Jr. 1967. *Theory of recursive functions and effective computability.* New York: McGraw-Hill.

Shapiro, S. 2006. Computability, proof, and open-texture. In Olszewski et al. (2006), 420–455.

Shapiro, S. 2010. *Mathematical structuralism.* Internet Encyclopedia of Philosophy. https://www.iep.utm.edu/m-struct/. 25 June 2018.

Sieg, W. 1994. Mechanical procedures and mathematical experience. In *Mathematics and mind,* ed. A. George, 71–117. Oxford: Oxford University Press.

Soare, R.I. 1987a. *Recursively enumerable sets and degrees.* Perspectives in mathematical logic, omega series. Heidelberg: Springer.

Soare, R.I. 1987b. Interactive computing and relativized computability, In *Computability: Turing, Gödel, Church, and beyond,* ed. B.J. Copeland, C.J. Posy, and O. Shagrir, 203–260. Cambdrige: MIT Press.

Turing, A.M. 1936. On computable numbers, with an application to the entscheidungsproblem. *Proceedings of the London Mathematical Society* 2(1): 230–265.

Turing, A.M. 1948. Intelligent machinery. In *Collected works of A.M. Turing: Mechanical intelligence,* ed. D.C. Ince, 107–127. Amsterdam: North-Holland

Welch, P.D. 2007. Turing unbound: Transfinite computation. In *Computation and logic in the real world,* ed. B. Löwe, B. Cooper, and A. Sorbi, 768–780. Berlin: Springer.

Wittgenstein, L. 1980. *Remarks on the philosophy of psychology.* Oxford: Blackwell.

Chapter 14
Existence vs. Conceivability in Aristotle: Are Straight Lines Infinitely Extendible?

Monica Ugaglia

Abstract Aristotle is committed to finitism in mathematics. But there are certain uses of infinity in mathematics which are indispensable, even in the mathematics of his day. So we had better understand Aristotle's finitism in a way that is compatible with the mathematics of his time (unless we are willing to ascribe complete naivete to him about these mathematics).

In particular, we have to address the issue of infinitely extendible lines, which are used in Greek mathematics – for example, in Euclid's definition of parallel lines. Aristotle denies that such lines exist: like any other object which exceeds the fixed and finite size of the cosmos, infinitely extendible lines are definitely excluded from Aristotle's physics. Moreover, due to Aristotle's immanentism, they are excluded from his mathematics, too.

But how can Aristotle do mathematics without infinitely extendible lines? In the following I will suggest a possible solution, based on an analysis of the procedure of *converse increasing*, which Aristotle introduces and discusses in *Ph.* III 6. This procedure is both infinite – it is infinitely iterable – and does not require the existence of any infinite (or infinitely extendible) magnitudes.

The procedure is interesting, but it must be handled with care. On the one hand, if one fails to acknowledge its subtle mathematical content, one also risks compromising the philosophical interpretation, by ascribing to Aristotle gratuitous naivieties. On the other hand, if one overstates this mathematical content, one risks ascribing to him anachronistic, 'non-Euclidean' intents. At the end of the paper I will discuss two cases in point.

M. Ugaglia (✉)
Department of Philosophy, University of Florence, Florence, Italy

© Springer International Publishing AG, part of Springer Nature 2018   249
M. Piazza, G. Pulcini (eds.), *Truth, Existence and Explanation*,
Boston Studies in the Philosophy and History of Science 334,
https://doi.org/10.1007/978-3-319-93342-9_14

14.1 Introduction

In *Physics* III 5-6, Aristotle denies the existence of anything which is actually infinite, and heavily constrains the existence of anything potentially infinite.[1] As he shows in *De Caelo* I-II – and as I will discuss in Sect. 14.2 below – the denial of the actual infinite has an immediate and evident effect on physics: in Aristotle's universe, which is finite in size, only finitely many objects actually exist, and they are finite in size too.

Moreover, as a consequence of Aristotle's immanentism, the size of the cosmos limits not only physics, but also mathematics: in Aristotle's mathematics no actually infinite sets of elements exist, nor actually infinite lines, nor infinitely extendible finite lines. As I will discuss in Sect. 14.3, Aristotle's mathematical finitism has apparently damaging consequences: how can there be infinitely many numbers, if there are no infinite sets of elements to count? And what about parallel lines, or asymptotic properties, if there are no infinitely extendible curves?

One way of solving the problem is to challenge a strict interpretation of Aristotle's immanentism, which entails the extension of physical constraints, like finiteness, to mathematical objects. After all, while Aristotle says that there is a perfect coincidence between the objects of study of the physicist and those of the mathematician, he still allows for a difference in perspective: while the physicist handles natural objects in their totality, the mathematician only works with them after a preliminary act of 'removal' (*aphairesis*), which puts all physical change aside.

Since Aristotle is rather vague about this preliminary operation, it has been a widespread practice to interpret the operation as something more than a simple subtraction.[2] By removing physical properties the mathematician creates a pure mental object, on which he can mentally act, forgetting its physical origin and its physical constraints. For instance, after having removed all physical features from a natural line, which of course is finite, and having obtained a mathematical segment, he can mentally act on this segment by extending it without limit, as required, for instance, by Euclid's parallel postulate. This is an easy way to save Aristotle's mathematics from a blind finitism, but it is not what Aristotle says, and it makes him contravene his own rules; as he says at the end of his discussion of the infinite, indeed, "it is absurd to rely on thought".[3]

---

[1] In this paper I develop, from the perspective of the philosophy of mathematics, the interpretation of the procedure of converse increasing I offered in Ugaglia (2009, pp. 210–211) and Ugaglia (2012, pp. 35–40). A preliminary version of this work has been presented at *The 89th Joint session of the Aristotelian Society and the Mind Association* in Warwick, in July 2015. On this occasion I profited greatly from my discussions with Adrian Moore, whom I thank.

[2] On these questions see notes 22–28 below.

[3] *Ph.* III 8, 208a14–15. The whole passage is discussed in Sect. 14.3 below.

In Sect. 14.3.2 I will show that there is another solution, simply following Aristotle's own texts, for saving both mathematics – a finite mathematics – and Aristotle's coherence. The idea is to take Aristotle seriously when, in *Physics* III 7, he introduces the procedure of *converse increasing* (ἀντεστραμμένη πρόσθεσις). According to the interpretation I offer, this procedure, far from being the marginal notion to which it is usually reduced,[4] is in fact the trick which saves Aristotle's finitism from naivete. Indeed, even where the mathematicians appear to use procedures which require ignoring physical constraints – infinitely extending lines, for instance – their arguments can be reformulated in terms of converse increasing, and traced back to firm physical properties. Indeed, converse increasing is both infinite – it is infinitely iterable – and does not require the existence of any infinite (or infinitely extendible) magnitudes: it only requires continuity, so that its existence is perfectly compatible with Aristotle's finite cosmos.

In order to distinguish between the unproblematic kind of infinite involved in the procedure of converse increasing, which ultimately reduces to the infinite divisibility of the continuum (see Sect. 14.2.2 below), from the infinite by extension, which Aristotle firmly denies, even in its potential form (see Sect. 14.2.3 below), I will introduce the opposition between *l*-infinite and *e*-infinite, where *l* stands for 'local', and *e* for 'extended'.

14.2 Finite Physics

Aristotle's cosmos is ultimately characterized by two properties: finiteness and continuity. A significant portion of Aristotle's physical works is devoted to discussing these properties and to exploiting their consequences, against both advocates of the actual infinite and atomists.

The consequence I am interested in is the noteworthy fact that, starting from continuity, Aristotle succeeds in building a peculiar notion of the infinite, which he calls the potential infinite, that is perfectly compatible with the finiteness of the cosmos. Indeed, as I will show in this section, the potentiality of Aristotle's infinite ultimately reduces to its being completely independent from objects. For Aristotle the infinite exists potentially in the sense that it is a feature of iterable *processes*, specifically those that satisfy the conditions D1–D3, which I discuss in Sect. 14.2.1. In Sect. 14.2.2 I give an example of such a process – the infinite decreasing of magnitudes – which in Sect. 14.2.3 I contrast with the apparently-similar case of increasing magnitudes, a procedure that for Aristotle cannot be infinite.

---

[4] An exception is the recent (Cooper 2016), where the importance of the procedure is emphasized, but its content is spoiled by a deeply problematic mathematical interpretation. Since – apart from Ugaglia (2009, 2012, 2016), – this is the only work I know which specifically affords the topic, I will consider its major critical points in Sect. 14.3.3.2.

14.2.1 Potential Infinite

Aristotle's intricate analysis of the infinite, which occupies Chaps. 4, 5, 6, 7 and 8 of
Book III of his *Physics*, is ultimately steered by the following question: is it possible
to construct a concept of the infinite which is compatible with a finite cosmos?

The answer is yes, under certain conditions, and studying these conditions reveals
the really interesting aspects of Aristotle's discussion of the infinite. Aristotle's
attempt to reconcile the notion of the infinite with the constraint of a finite cosmos
leads him to conceive a very peculiar form of being, which has implications of
central importance for reconstructing his philosophy of mathematics.

The crucial point is presented at the beginning of Chap. 6: immediately after
having shown what absurdities the hypothesis of an actual infinite leads to – whether
it is conceived as a 'something' that exists by itself, or as an attribute of 'something'
that exists by itself[5]– Aristotle points out the similar difficulties that would follow
from an unconditional negation of the infinite. This suggests the need to adopt an
intermediate position, granting the infinite some form of partial being, but not full
being. To quote Aristotle, the infinite "in a sense is, in another it is not".

Now, the way in which the infinite "is not" is the one just stated: the infinite is
not in actuality. In more concrete terms, the cosmos, which is to say the sum of all
existing things, is finite in magnitude – it is a sphere of fixed, finite diameter – and
in multiplicity – it contains an actually finite number of objects.[6]

But if the infinite it is not in actuality, by a process of elimination it must be
potentially, and now we need to understand – and this is no simple task – what the
phrase "to be potentially" is meant to signify when referred to the infinite. Indeed, as
Aristotle explicitly states, one must not take the notion of potentiality here according
to its standard meaning, for the potentiality of the infinite does not presuppose any
actual form of being:

> We must not take potentiality here in the same way as that in which, if it is possible for this
> to be a statue, it actually will be a statue, and suppose that there is an infinite which will be
> in actuality (*Ph.* III 6, 206a18–21).[7]

Instead, to be potentially is the only form of being of the infinite, a 'processual' form
of being,[8] which Aristotle tries to elucidate by analyzing the way in which a day or a
contest can be said to exist. A day or a contest, and more generally a process, never

---

[5]He argues that the term 'infinite' (τὸ ἄπειρον) cannot be employed as a subject – 'the infinite' per
se does not exist (*Ph.* III 4, 202b36–203a16; cf. III 5, 204a8–34; III 4, 203a6–16 and III 1 200b27),
nor as the attribute of a subject – 'an infinite thing' does not exist as well (III 4, 203a16-b3; cf III
5, 204b4–206a7).

[6]On the finiteness of the cosmos and its parts see *Cael.* I 5-7. On its spherical shape see *Cael* II 4.
On its fixed diameter *Cael.* II 2.

[7]Translations of *Ph.* III are Hussey's, sometimes with major modifications.

[8]On the identification between potentiality and processuality as the proper form of being of what
is infinite see Ugaglia (2016).

exists in its entirety: what exists, over and over again, is a single moment of the day, or a single stage of the contest, or in general a single step of a process.[9]

Using Aristotle's words, one can say that a day or a contest is, not because it is a well defined 'something', but because it is 'over and over again something else'. At this point one realizes, however, that the analogy with a day or contest cannot be maintained all the way through. Although these things have a processual way of being, they admit of an outcome, which is not the case with the processes that manifest the infinite. While the day or the contest may also be considered in their totality, and to do so it is sufficient to place oneself after the deadline represented by the last step of the process – whether it is the sunset or the victory – there is no way to place oneself after an infinite process. To use Aristotle's words again, the infinite is not that 'of which there is nothing beyond' – indeed, that would be a whole – but rather that 'of which there is over and over again something beyond'. This brings us to Aristotle's definition of the infinite:

> **D**: the infinite is in virtue of [**D1**] another and another thing being taken, over and over again (*aei*); and [**D2**] what is taken is finite, over and over again (*aei*); but [**D3**] it is a different thing, over and over again (*aei*). (*Ph.* III 6, 206a27–29)

As is clear, **D** is not a conventional Aristotelian definition, but an operative characterization. It does not describe an object but an action – 'taking' something – and the iteration of the action, signalled by the adverb *aei*. For each thing we take (*first step*), there is another thing to take (*second step*), and another to take beyond that (*third step*), and another beyond, and so on, over and over again (*aei, next steps*).[10] This is the first requirement, which I called **D1**.

As we have seen, the day and the contest are useful examples because they give an immediate and easily-graspable idea of a processual way of being, but they do not satisfy **D1**; indeed, both have an end, a limit which one cannot go beyond.

To find a good example of a properly infinite process, Aristotle turns to something less familiar: the procedure of division of the continuum, which is characterized as 'divisible into the always (*aei*) divisible'.[11] Take a continuous segment, for instance:

---

[9]Since to be potentially in the sense of processuality does not precede or imply any other form of being, this is the only way for the infinite to manifest itself. For this reason one could say that in some sense it is also its way of being 'in act': "And it is in act in this way, as we say that the day and the games are, and in power, in the way in which matter is, and not in itself, as the finite is" *Ph.* III 6, 206b13–16. On the processual way of being of the day see Ugaglia (2012, pp. 162–163). On the connection between being separated and being a 'something' in act see also *Ph.* III 5, 204b7. A different interpretation, on which actuality and potentiality are not simply two different ways of describing the infinite's unique processual form of being, but there is instead an infinite 'in power', before the process starts, and an infinite 'in activity', during the process, is provided in Cooper (2016).

[10]On the necessity of taking the adverb *aei* in its iterative meaning, typical of mathematics, see Ugaglia (2009).

[11]Aristotle defines continuity as a concept of relation: two parts of a whole are said to be continuous when they are in contact and their ends therefore become one (*Ph.* V 3, 227a11–12; see. *Ph.* V 4, 228a29–30; *Cat.* 6, 4b20–5a14). Accordingly, Aristotle calls continuous a whole composed of continuous parts, that is to say parts that have an end in common (*Ph.* VI 1, 231b5–6; *GC* I

it is divisible (*first step*), and every segment one obtains from its division is still divisible (*second step*), and every segment one obtains from this second division is still divisible (*third step*), and so on (*aei, next steps*). The absence of any inner hindrance, which characterizes the continuum, permits us to continue dividing over and over again, going beyond any reached division, as required by **D1**.

Condition **D2** adds a clarification required in the case of those iterative procedures which produce an actual thing at every step.[12] Given Aristotle's denial of the actual infinite, this thing must be finite.

Requirement **D3** discriminates between iterations that are infinite in a proper sense and iterations that are infinite because they are periodic (let me call them *c*-infinite): in the latter case, one comes back to something already taken after a certain number of steps.[13]

To sum up:

- as **D1** clearly shows, Aristotle's characterization of potential infinite applies primarily to processes, and only secondarily to objects: when the processes do not occur in a substrate – think about heaven's motion, or the generation and corruption of human beings – the attribute belongs just to the procedure; when they take place in a substrate – think about the division of magnitudes or their increasing – the substrate too is called infinite.

- An infinite conceived in this way cannot be 'reached': it can be grasped only in the iterative process which defines it, and from which it is not separable. As Aristotle says in *Metaph.* Θ 6, 1048b15–17: "the fact that the division does not stop ensures that one can say about this activity that it is in power but not separate". In other words, the infinite exists insofar as the process exists.[14]

6, 323a3–12; *Metaph.* Δ 13 1020a7–8). But in order to have an end in common, parts must be congeners (*Ph.* IV 11 220a20–21; *Ph.* V 4, 228a31-b2) so that they must lose their individuality, and form a homeomeric whole, free of inner limits (*Ph.* 5 IV, 212b4–6; *Metaph.* Z 13 1039a3–7 see Δ 26 1023b33–34; Z 16, 1040b5–8). This leads to Aristotle's peculiar characterization of the continuum as "infinitely divisible" (*Ph.* III 1, 200b18; *Ph.* I 2, 185b10–11; VI 6, 237a33; VI 8, 239a22) or "divisible into the always (*aei*) divisible" (*Ph.* VI 2, 232b24–25; see IV 12, 220a30; VI 1, 231b15–16; VI 6, 237b21; VIII 5, 257a33–34; *Cael.* 1 I, 268a6–7). Indeed, if there are no individual parts, and no actual limits, then there are no parts in act (*GC* I 2, 316a15–16) and the continuum can be divided everywhere. A deep and clear analysis of the *GC* passage is contained in Miller (1982, pp. 87–100).

[12] Aristotle mentions the procedures of shortening and enlarging a magnitude, where at every step a persisting object is taken, and contrasts them with time and the process of coming-to-be and passing-away of human beings, where the taken 'objects' cease to be.

[13] The main example of *c*-infinite is the uniform circular motion of the heavens (see in particular *Ph.* VIII), whose number, which is to say whose time, is infinite in the proper sense: in fact, it is both *l*-infinite and *e*-infinite (*Ph.* IV 14).

[14] In this sense Aristotle can say that the infinite is similar to matter (*Ph.* III 6, 206b13–16, cf. note 9 above): matter too can be grasped only in the form that defines it, and from which it is not separable.

- What makes the process infinite, namely the possibility of going beyond any given step, is a local property: even if it characterizes the process 'globally', it concerns the single action of the process. Such a local characterization of the infinite – which does not depend on any factual result – is what ultimately makes the infinite capable of existing in the finite Aristotelian cosmos. For this reason I will call l-infinite an infinite which satisfies **D**.
- More generally, no final result can be reached: of course, if the procedure stops – namely, if **D1** does not hold, and there is a last step, say the n-th – a result is obtained, but this result has nothing to do with the infinite, regardless of how great the number n is.[15]
- Every iterative procedure that is *merely conceived* satisfies **D1**, whose possible failure is a matter of physics. In other words, every existing infinite is potential (i.e., it can be read as an attribute of an iterative procedure), but not every potential infinite which is conceivable exists (i.e., not every conceivable iterative procedure gives rise to an existing infinity).

More concretely, Aristotle argues that it is always possible to infinitely reduce a magnitude, but the size of the cosmos imposes a limit on the possibility of infinitely enlarging it. In the next two sections I will display the situation in a simplified graphical way, taking – as Aristotle does – unidimensional magnitudes.

14.2.2   Potential l-infinite: Decreasing

It is possible to infinitely reduce a magnitude because of a real feature of Aristotle's cosmos: its continuity. As we have seen, from Aristotle's point of view, continuity means the absence of inner limits, so that if one starts dividing a continuous magnitude one can go on indefinitely, for the action of dividing does not encounter any obstruction.[16] If after any division one of the two portions is subtracted, the procedure of infinite division can be read as a procedure of infinite decreasing.

Take a segment, say AB, less than (or equal to) the size of the cosmos, divide it in half and subtract one half:

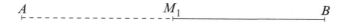

---

[15]In this sense potentiality is nothing but the possibility of repeating each step. For this reason I usually prefer to speak in term of processuality, instead of potentiality, when it is referred to the infinite. See Ugaglia (2016).

[16]On Aristotle's potential infinite and the division of the continuum see the fundamental (Wieland 1970, ch. 17), the debate (Hintikka 1973) – (Lear 1979), and the more recent (White 1992). For a deep analysis of Aristotle's theory of continuity and his argument against atomists, see Miller (1982) and White (1992, pp. 23–28 in particular).

Then divide the remaining half and subtract:

$$A \;\vdash\!\text{-----------------------}\!\!\underset{M_1}{\vdash}\text{-------------}\!\underset{M_2}{\vdash}\text{----------}B$$

Then divide the remaining half and subtract:

$$A \;\vdash\!\text{-------------------------}\!\!\underset{M_1}{\vdash}\text{-----------}\!\underset{M_2}{\vdash}\text{--}\underset{M_3}{\vdash}\;B$$

Then divide the remaining half and subtract ... and so on, over and over again.

The magnitudes produced at any step: $AB, M_1B, \ldots, M_nB, \ldots$ decrease as a geometrical sequence,[17] and as it is easy to see, the procedure of decreasing satisfies all of Aristotle's requirements: **D1** holds in virtue of the continuity of the starting segment – which is to say, for a physical reason – while **D2** and **D3** hold in virtue of the mathematical definition of the procedure. In particular, **D2** is satisfied because, for every n, the segment M_nB, obtained at the n-th step, is a part of the initial segment AB, which is finite. **D3** is satisfied because M_nB is a proper part of $M_{n-1}B$, and hence they are different.

14.2.3 Potential e-infinite: Unconstrained Increasing

While the possibility of infinitely dividing a magnitude is ensured by a real feature of the cosmos, no natural features provide a foundation for procedures of infinite increasing. On the contrary, there is an inescapable physical property of Aristotle's cosmos which makes some such procedures impossible.

Take again a segment AB, less than the size of the cosmos:

$$A \;\vdash\!\text{------------------------}\!\;B$$

---

[17] $1, \frac{1}{2}, \frac{1}{2^2}, \frac{1}{2^3} \ldots$ I have confined myself to considering the dichotomical procedure of division, partly for convenience, and partly for historical reasons, given Zeno's choice, in his paradox, to divide in half. Nothing changes with another kind of division, provided that the ratio is fixed: "For if, in a finite magnitude, one takes a definite amount and goes on taking in the same proportion [...] one will not traverse the finite magnitude" (*Ph.* III 6, 206b7–9). In modern notation, Aristotle accepts every division which follows a geometrical sequence: $1, \frac{1}{a}, \frac{1}{a^2}, \frac{1}{a^3} \ldots$ with $a > 1$ but not necessarily $a = 2$.

but now instead of shortening it, try enlarging it. Following Aristotle's example,[18] add to it the same fixed finite magnitude; for instance, a segment one half the length of AB:

And add one half of AB:

A B N_1 N_2

And add one half of AB:

A B N_1 N_2 N_3

And so on, over and over again:

A B N_1 N_2 N_3 N_n

At first glance, the situation appears to be the same as before: one has an action (adding), and the instruction of repeating it, exactly as in the previous case one had an action (dividing), and the instruction of repeating it.

The problem is that, while in the first case the repetition was warranted, we are now facing an obstruction. However small the starting segment may be, in a finite number of steps the size of the cosmos is reached, and there the process will have to stop[19]:

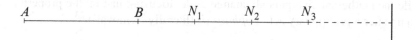

Notice that the reason why the procedure stops is not that there is no place to go beyond the cosmos, but that there is no 'beyond' the cosmos: once the size of the cosmos has been reached, the action of going beyond loses all meaning. Therefore, the step that reaches the size of the universe must necessarily be the last, and the

---

[18]See T2 in Sect. 14.3.2.1 below.
[19]See T5 in Sect. 14.3.2.1.

condition **D1** fails: *the iteration is broken*; reaching a last step means there is an outcome of the process, and this means that the increasing is not infinite.[20]

Let me emphasize the fact that the situation is radically different from the one occurring in the case of decreasing. In the case of decreasing, even if no final result can ever be reached, it is does not involve any physical impossibility: it does not exist only because it would be the 'result' of a process that by definition has no result. But the process itself exists, so that Aristotle can say that whereas there is no l-infinite in act (i.e., no result of the infinite process of shortening), there is an l-infinite in power (namely, the infinite process of shortening). In more concrete terms, there are no infinitely small magnitudes, but any magnitude is infinitely reducible.

In the case of increasing, on the contrary, the hypothetical final result cannot be reached for physical reasons: not only because it would be the 'result' of a process that by definition has no result, as was the case with the actually l-infinite magnitude, but because no magnitudes exist bigger than the fixed size of the universe. As Aristotle repeatedly states, there is no e-infinite: neither in act (i.e., the result of an infinite process of increasing), nor in power (i.e., an infinite process of increasing).[21]

To sum up, a procedure of unconstrained increasing – let me call it an e-increasing – can be conceived, but it does not exist: in Aristotle's cosmos there are no infinitely extended magnitudes, but neither are there infinitely extendible ones.

14.3   Finite Mathematics

In the introduction I claimed that for Aristotle, finitism in mathematics is a consequence of finitism in physics, and I suggested that Aristotle's immanentism explains this consequence. Before focusing on finitism, and his way of getting around the problems it presents, I should say a few words about the relationship between immanentism and finitism, following Aristotle's texts.

As Aristotle stresses on several occasions, there is perfect coincidence in his system between the objects of study of the physicist and those of the mathematician:[22] they are the same objects – physical objects – with the only difference being that the physicist handles them in their totality, focusing on their being subjects of change, while the mathematician puts all change aside, focusing just on the properties that remain after a preliminary act of *aphairesis* (literally 'removal').[23]

---

[20]It is worth observing that, depending on the mathematical structure of the procedure, **D2** and **D3** still hold (until the process goes on). In particular, **D2** is satisfied because the segment AN_n obtained at the n-th step is the sum a finite number of finite amounts, and hence it is finite. **D3** is satisfied because AN_{n-1} is a proper part of AN_n, and hence they are different.

[21]See T6 in Sect. 14.3.2.1.

[22]*Metaph.* M 3, 1077b22–1078a9; *Ph.* II 2, 193b23–194a12; cf. *de An.* I 1, 403a15–16; *Metaph.* N 2 1090a13–15.

[23]*Metaph.* E 1, 1026a14–15 *et seq.* On *aphairesis*, which is a technical mathematical term usually improperly translated as 'abstraction', see Mueller (1990) and the bibliography quoted there.

Aristotle is rather vague about this crucial preliminary operation, but one point is clear: what is removed is not the mathematical object, but the physical properties of the natural object – i.e., what has to do with change. Indeed, no matter what method is adopted, removing something requires knowing it; but accepting that the mathematical objects are known before the physical ones, and independently of them, means endorsing a Platonic point of view, totally alien to Aristotle.

Since a mathematical object is part of a physical one, and since it is obtained from that simply by putting change aside, its properties are limited to those of the natural object which cannot be removed in this way.[24] Finiteness is one such property. Thus, if the physical universe contains a finite number of things, and if numbering things means counting them, then there is no way for the Aristotelian mathematician to define a sound mathematical notion of infinite number. Analogously, if the universe has a finite size, and a mathematical straight line is a physical rectilinear object deprived of changeable features,[25] all mathematical lines are finite, and there is no way to define a sound mathematical notion of an infinite line.

The problem is not that the mathematician is unable to imagine a number greater than the number of existing objects, or a magnitude greater than the dimension of the cosmos; he can indeed do so, just as the physicist can imagine a void, or a human being exceeding the size of a town:

> It is absurd to rely on thought: the excess and deficiency are not in actual things but in thought. Thus one might think of each of us as being many times as large as himself, increasing each of us ad infinitum; but it is not for this reason, because someone thinks so, that anyone exceeds the town, or the particular size we have (*Ph.* III 8, 208a14–19).[26]

Rather, the problem is that just as one cannot do physics with voids, or giants – which are merely imaginary objects, like sphinxes, or hircocervi – one cannot do mathematics with infinitely extendible lines.[27]

---

[24]For instance, Aristotle says that a bronze ball touches a straight line at a single point (*de An.* 1 I, 403a12–15; as for the apparently contradictory passage in *Metaph.* B 2, 997b34–998a6, the aporetic context suggests a reference to the opinions of Aristotle's opponents, and not to his own point of view). This interpretation of Aristotle's philosophy of mathematics is maintained in Lear (1982); among the many supporters of the opposite view, see for example Mueller (1970). More generally on mathematics in Aristotle see Cleary (1995) and Mendell (2008).

[25]"Geometry studies physical lines, but not *qua* physical". (*Ph.* II 2, 194a9–11).

[26]The statement is weakened in the text edited by Ross, where the reference to the town has been excised. I follow the manuscripts and the commentators in reading it (see Ugaglia 2012, p. 175).

[27]Aristotle himself makes use of infinite straight lines when he hypothetically assumes that the universe is infinite, in order to disprove it. See for example *Cael.* I 5, 271b28–272a7, where Aristotle adopts the hypothesis of an infinite sky and infinitely extends a line within such a model (cf. *Top.* 148b30–32). Analogously, he assumes the void and its (absence of) properties in *Ph.* IV 6–9.

In other words, since the mathematician inherits his constraints from the physicist, he must only deal with objects whose conceivability is not in conflict with the physical structure of the cosmos. In our case, he must deal with procedures that satisfy **D1** not only in thought, but also in the real world.[28]

14.3.1 Infinite Number

A universe finite in extension contains only a finite number of objects. Hence one cannot reach a mathematically sound notion of infinite number simply by imagining going on in the action of counting objects.

But what about counting actions, instead of objects? Take again a segment, and start dividing it, but now imagine attaching a label to each division; in plain words, imagine counting the steps of the process.

Divide our segment in half and say: "one". Then divide the result in half and say: "two". Then divide the result in half and say: "three"... and so on, over and over again.[29] Since the process of division is infinite, so is number. But like the process which produces it, number is only potentially (that is, processually) infinite: since each division implies the possibility of a next division, each number implies the possibility of a next number, over and over again, without end. This is how Aristotle explains the existence of infinitely many numbers:

> But in the direction of more it is always possible to conceive of <a larger number>, since halvings (αἱ διχοτομίαι) of a magnitude are infinite. So it is potentially infinite, but not actually infinite; but the thing taken always exceeds any definite number.[30]

This explanation works because it ties the mental procedure of numbering to the physical procedure of dividing, and this is essential for including the numbers among the objects of mathematics. Counting, then, is not just a matter of imagining

---

[28]The false claim that a straight line can be indefinitely extended, thus obtaining a mathematical object in the Aristotelian sense, is expressed in a very clear form already in Simplicius and Philoponus (Diels 1882, 512.19–36); (Vitelli 1887, 482.28–483.14). It seems that a completely different opinion must be ascribed to Alexander, but unfortunately the very few extant passages (Diels 1882, 511.30–512.9) do not allow us to settle the question. See however (Rashed 2011) on Alexander's mathematical ontology, which indeed seems to require such an interpretation.

[29]On the relation between number and division see Lear (1979, pp. 195–198). A very long explanation is offered in Cooper (2016, pp. 175–182), for whom it is not sufficient to potentially infinitely divide a magnitude, and to count the increasing multiplicity of cuts – or of produced segments – in order to obtain the potential infinity of numbers. Indeed, in order to account for the fact that numbers increase *by addition* of a unity, Cooper needs the fact that "any physical magnitude is 'infinite' not only 'by division' but also 'by addition'." (p. 181). On this issue, see my Sect. 14.3.3.2.

[30]*Ph.* III 7, 207b10–13, Lear translation modified. See b1-15 for the whole argument.

oneself endlessly engaging in numeration – which does not warrant **D1** – but depends on continuity, which is a well-defined physical property and ensures **D1**.[31]

Aristotle's notion of infinite number is not separable from the process that defines it. If one stops the procedure of division – namely, if one actually invalidates **D1** – of course one obtains a number, but this number is always a finite one. Once it has been reached, this finite number can be separated from the process,[32] thought of independently and put in relation with something else: the set of segments produced by division, for example, but also any other set of things – horses, colors, actions, or anything else.

Of course, by going on with the process one can produce, and separate from it, larger and larger finite numbers. But the action of separating them, making them actual numbers, is subordinate to the action of reaching the corresponding step in the division, and since the procedure is infinite, there is neither a final step to reach, nor a final number to associate to it. So, like every other infinite in Aristotle's system, number is infinite only in power, and not in act.[33]

---

[31] When I say that the procedure of division, in order to be an acceptable mathematical operation, must be grounded on continuity, I do not mean that the infinite *already is* in the continuum. I mean that it manifests itself (ἐμφαίνεται) in the continuum, as Aristotle maintains at the very beginning of *Ph*. III 1. Indeed, I think that the absence of inner limits, which I have discussed in Sect. 14.2.1, is enough to warrant the possibility of repeating the action of dividing, in which the infinite ultimately consists. For this reason I do not accept the interpretation advanced by Lear (1979), which not only grounds the procedure on the structure of the continuum, but locates the infinite primarily in this structure – the continuum is made of infinitely many potential parts – and only secondarily in the procedure of actualizing them. This approach faces the question of how potential parts can be defined prior to, and independently of, the action of dividing.

[32] One can conceive of the separation of a number from the process of division by analogy with the separation of a statue from a block of marble: of a Hermes from the stone, to use a typical Aristotelian example. As far as the Hermes is in the stone, we do not have a body in the proper sense, but only a body in power, for according to Aristotle's definition, a body is separate when it is actually 'delimited', and not only limitable, by a surface. In the case of a number, it is 'separate' when it is actually counted, and not only countable.

[33] A very similar situation holds in the case of time, which is defined by Aristotle as "the number of motion" (*Ph*. IV 11, 220a24–25). Since all sublunar natural motions are rectilinear, and hence *e*-finite, their number too is finite. Of course, one can imagine going on counting beyond (after or before) the actual extension of a finite motion, exactly as one can imagine going on counting beyond the actual number of a finite set of objects, but the notion of infinite time one obtains in this way is not a scientific object. Exactly as in the case of number, to obtain a sound notion of (potentially) infinite time it is necessary to ground it in a physical (potentially) infinite process. In the case of number the process was the *l*-infinite division of the continuum; in the case of time it is the *c*-infinite motion of the heavens (*Ph*. IV 14, 223b29–31). In this case, no converse increasing is needed. Even less is it necessary to divide the potentially infinite time of the heavens in order to obtain a potentially infinite number of 'bits' of time, which can be eventually (conversely?) added, in order to (re)obtain an infinite time, as in Cooper (2016, pp. 182–186). See the analogous example of the (re)construction of a half-line, which I discuss in Sect. 14.3.3.2.

14.3.2 Infinite Converse Increasing

In Sect. 14.2.3, I discussed Aristotle's denial of the procedure of infinite increasing of magnitudes. To be more precise, what Aristotle denies is a procedure of unconstrained infinite increasing, or e-increasing, that tends, as its hypothetical result, towards an e-infinite magnitude. The actual size of the cosmos prevents this kind of procedure from fulfilling **D1**.

But not all procedures of increasing are of this kind: in fact, there is a very crucial exception that wraps up the whole story. Before I quote and discuss Aristotle's own texts, let me briefly present the process – which Aristotle calls *converse increasing* (ἀντεστραμμένη πρόσθεσις)[34] – in the usual simplified graphical way.

Take a segment AB, less than (or equal to) the size of the cosmos, and divide it into two halves:

$$A \vdash\!\!\!-----------\vdash-----------\dashv B$$

but now keep both halves: the first to your left, and the other to your right.

$$A \vdash\!\!\!-----\!\!M_0 \qquad\qquad M_0\!\!-----\dashv B$$

Now, divide the segment on your right into two halves, subtract one of them, and add it to the segment on your left:

$$M_1 \quad B$$

$$A \vdash\!\!\!----------\!\!M_0 \quad M_1$$

Again, divide the segment on your right into two halves, subtract one of them, and add it to the segment on your left:

$$M_2 \quad B$$

$$A \vdash\!\!\!--------\!\!M_0 \qquad M_1 \quad M_2$$

---

[34]From the verb *antistrephein* (cf. *Ph.* III 6, 206b27; 207a23), which in *Prior Analytics* is used in a technical sense for the procedure of conversion of a premise.

and so on, over and over again.

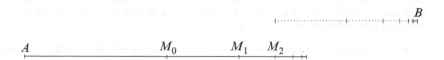

It is clear that we have two l-infinite processes, neither of which involves any e-infinite magnitudes. On the right, a segment M_0B becomes smaller and smaller, and on the left, an initially equal segment AM_0 becomes larger and larger. As much as you subtract on the right, so much are you adding on the left. Since the process on your right is infinite, the one on your left will be infinite too, but since the total amount you can subtract on the right – that is, the whole M_0B segment – is finite, the total amount you can add on the left is finite too. In other words, the procedure of converse increasing is a local infinite procedure 'tending towards' a finite magnitude (namely, the sum $AM_0 + M_0B$, which is your initial segment AB). Let me call it an l-increasing.

Following our usual path, we can easily see that a procedure of l-increasing satisfies all of Aristotle's requirements for being infinite. To begin with, the proofs for **D2** and **D3** are the same as in the case of an e-increasing: **D2** is satisfied because the segment AM_n obtained at the n-th step is the sum of a finite number of finite amounts, and hence is finite. **D3** is satisfied because AM_{n-1} is a proper part of AM_n, and hence they are different.

Unlike in the case of an e-increasing, however, **D1** also holds here, because the continuity of AB implies that the division has no end, and the fact that AB is less than (or equal to) the size of the cosmos implies that the addition also has no end, for it never reaches the size of the cosmos.

14.3.2.1 Textual Analysis

Aristotle does not offer many details about the procedure of converse increasing, and he does not give any diagram, but his description is very clear – especially if, in the following passage, one imagines replacing Aristotle's 'something' with 'the segment AB':

> **T1** The <infinite> by increasing is in a sense the same as that by division. For in that which is finite it comes to be by increasing, conversely: just as something is seen as being divided *ad infinitum*, in the same way it appears to be increased to a definite amount (*Ph.* III 6, 206b3–6).

Having described the procedure, Aristotle contrasts a process of converse increasing with a process of unconstrained increasing. He observes that, whatever magnitude the procedures operate on, an l-increasing can go on, over and over again, while an e-increasing necessarily reaches a stop after a finite number of steps. In other words, **D1** holds for the first process, but not for the second one:

T2: for if, in a finite magnitude, one takes a definite amount and goes on taking in the same ratio, not taking a magnitude which is the same with respect to the whole, one will not traverse the finite magnitude; but if one raises the ratio so that one always takes the same particular magnitude, one will traverse it, because every finite quantity is exhausted by any definite quantity whatever (*Ph.* III 6, 206b7–12).

A more metaphysical remark follows, about the way in which a process – either decreasing or *l*-increasing – can be said to be infinite: potentially and 'by way of reduction' (*locally*, in our terminology).

T3: The infinite, then, is in no other way, but is in this way: potentially and by way of reduction (*Ph.* III 6, 206b12–13).

Then Aristotle restates **D1**, specifying that the possibility of taking new segments, longer and longer, does not mean that their length will exceed every length: the infinite steps of an *l*-increasing stay within the fixed size of the starting segment.

T4: The infinite by increasing, too, is potentially in this way; this infinite, we say, is in a way the same as the infinite by division, for it will be possible to take something beyond, over and over again, though it will not exceed every magnitude in the way in which, in division, it exceeds every definite quantity and will be smaller and smaller, over and over again (*Ph.* III 6, 206b16–20).

An *e*-increasing, on the other hand, would require an actually *e*-infinite body. Not an infinite in itself – a separate substance, as conceived by Plato or by the Pythagoreans – but a physical body, of which the infinite is a necessary attribute, but not the substance. The clarification is needed because the existence of the first kind of actual *e*-infinite would be sufficient to ensure the conceivability but not the physical possibility of the process:

T5: To exceed every <magnitude> by increasing is not possible even potentially unless there is something which is actually infinite, accidentally, as the natural philosophers say that the body outside the cosmos, of which the substance is air or some other such thing, is infinite. But if it is not possible for there to be a perceptible body which is actually infinite in this way, it is manifest that there cannot be one even potentially <infinite> by increasing, except in the way that has been stated, conversely to the division (*Ph.* III 6, 206b20–27).

The situation is explained in a clearer way in the subsequent chapter, where Aristotle makes an explicit reference to the actual size of the cosmos:

T6: In the case of magnitudes the contrary is true: the continuum is divided into infinitely many parts, but there is no infinite in the direction of 'greater'. For a magnitude in actual operation may exist of any size, of which a magnitude may potentially exist. Since therefore no perceptible magnitude is infinite, there may not be an exceeding of every definite magnitude – for then there would be something greater than the world (*Ph.* III 7, 207b15–21).

To sum up: in Aristotle's cosmos the infinite exists only potentially, in the form of infinite iterative procedures. Nevertheless, not every conceivable procedure corresponds to an effective manifestation of the infinite, for the procedures that involve physical impossibilities cannot go on without end. An infinite by addition can only arise from a procedure of *l*-increasing, which keeps within the finite size of the cosmos.

14.3.2.2   The Procedure of *l*-increasing Lines

In Sect. 14.3.1, we saw how Aristotle solves the problem of producing an infinite number in a finite cosmos: instead of counting actual things, he counts the divisions of the continuum. Here we consider the more difficult question of infinitely increasing lines, on which the very possibility of doing geometry appears to hinge. Among Euclid's postulates, in fact, we find such requests as:

> to produce a finite straight line continuously in a straight line.[35]

Therefore, Greek mathematicians assume the possibility of *e*-increasing, taking the notion of finite straight line as their starting point. This is confirmed by Euclid's definition of parallel lines:

> parallel straight lines are straight lines which, being in the same plane and being produced indefinitely in both directions, do not meet one another in either direction.[36]

Aristotle also refers to parallel lines, and their 'not meeting' is for him a paradigmatic example of an eternal truth.[37] For this reason, we may wonder how Aristotle expected to handle such cases, if lines cannot be infinitely produced.

The answer is predictable: they cannot be produced by a process of *e*-increasing, but they can be produced by a process of *l*-increasing. In Aristotle's words:

> **T7**: This reasoning does not deprive the mathematicians of their study, either, in refuting the existence in actual operation of an untraversable infinite in the direction of increase. Indeed, they do not need the infinite, for they make no use of it[38]; they only need there to be a finite <increase> as great as they want. (*Ph.* III 7, 207b27–31).

The requirement that the increase has to be finite excludes that it may be the (unattainable) result of a process of *e*-increasing, which Aristotle has explicitly rejected. Instead, it will be – and must be regarded as – the (equally unattainable, but finite) result of an *l*-increasing. In an infinite *l*-increasing, in fact, the attribute 'infinite' refers to the procedure, and not the global increase resulting from the procedure.[39]

---

[35] Euclid, *Elements*. Postulate I.2; translations are Heath's, without modifications (Heath 1921).

[36] Euclid, *Elements* Definition I.23.

[37] *APr* II 16, 66a4–9; *APr* II 17, 66a13–15; *APo* I 5, 74a13–16. On the reconstruction of a pre-Euclidean theory of parallel lines see Vitrac (1990, pp. 306–310).

[38] It is worth noting that here Aristotle is not talking about robbing mathematics of the actual infinite per se: it is a plain fact that ancient Greek mathematicians do not use any actually infinite magnitude. The problem is that by removing the actual infinite, on Aristotle's view one also removes the potential infinite which would lead to it, namely, a process of *e*-increasing.

[39] I take the word 'increase' (feminine in Greek) at line 29 as the implied subject of line 31 (for a textual analysis see Ugaglia (2012, p. 84 and 174)). Traditionally, this claim has been referred to the straight line, so that the passage is read as an explicit reference to *El.* VI.10 (Diels 1882, 511.21–24). I prefer to avoid referring to any straight line because given the absence of any mention of lines in the whole preceding argument, this reading seems to me to be a consequence of – rather than evidence for – an interpretation based on *El.* VI.10.

Aristotle's idea is simple: instead of e-increasing lines 'towards the infinite', imagine l-extending them toward the edge of the cosmos, and suppose that what in the first case happened 'at the infinite' is now happening at the edge of the cosmos.

After all, every mathematician knows that the crucial feature of any proof concerning properties of a line that are stated to hold 'at the infinite' is the fact of being framed as a procedure involving, at least in principle, an infinite number of steps: why not suppose that these steps are not related to a process of e-increasing that leads away indefinitely, but to a process of l-increasing, which in infinitely many steps leads to the finite?

In this way, any e-infinite line the mathematician could imagine can be transformed into an l-infinite segment of the size of the cosmos, and this is the core of Aristotle's solution. The subsequent passage contains a minor specification about lines which are e-finite but longer than the size of the cosmos, which is a maximum for existing magnitudes:

> Another magnitude of any size whatever can be divided in the same proportion as the maximal magnitude[40] so that, at least for the purpose of the proof, it will make no difference, and concerning its being it will be among those magnitudes that exist (*Ph*. III 7, 207b31–34).[41]

Although they are finite, such entities cannot be obtained from physical segments simply by leaving change aside, so they are not proper mathematical objects. They can nonetheless be transformed into mathematical objects, without introducing any significant change – no l-increasing is needed, in this case – simply by rescaling them (i.e., reducing them in proportion) to other magnitudes which are less than or equal to the size of the cosmos.

---

[40]It is important to note that this sentence absolutely does not mean that it is sufficient to rescale an infinitely extended line in order to make it finite. While a property that holds at a finite point can be transported to a nearer point by appropriate rescaling – a shrinking of the design, roughly speaking – there is no change of scale that can bring an asymptotic property closer, at a finite point. No matter how small the drawing, the point 'at infinity' will always lie outside, by definition. The possibility of rescaling the e-infinite is maintained in Hussey (1983, p. 95): "Thus, instead of saying that a hyperbola approaches a straight line as it tends to infinity, one may say that, taking sufficiently small scale models of the original hyperbola, we shall find the scale models approaching arbitrarily closely to the corresponding lines". See the first occurrence of the claim in Hintikka (1973, p. 119). Criticisms of Aristotle's system grounded in this kind of 'underestimation' of the mathematical skills of the philosopher are very common: see for example Knorr (1982) and Milhaud (1903). Although endorsing Hintikka-Hussey's interpretation in terms of rescaling, Mendell (2015) offers a more favorable analysis of the passage.

[41]Following the standard interpretation of this passage, which ignores converse increasing, the maximal magnitude is a magnitude as large as the mathematician wants it to be, and every other magnitude can be cut in the same ratio as this maximal magnitude. But a line as large as one wants it to be cannot be a maximum: by definition the maximum element of a set is an element such that any other element of the set is less than or equal to it. If a magnitude can be as large as the mathematician wants, for every given magnitude he can take a larger one, so that no magnitude can be the maximal one. On this point see Ugaglia (2012, p. 174).

14.3.3 Two Caveats

14.3.3.1 We Shouldn't Read Too Much Mathematics into Aristotle

One might be tempted to connect Aristotle's finite cosmos, equipped with the infinite process of converse increasing, with non-Euclidean geometry. No doubt the analogy is suggestive, and I frequently resort to the example of Poincaré's disk, and Escher's strongly evocative pictorial rendering, in order to give an easily graspable idea of what an l-increasing looks like; but one must be very careful, for in no way can the model implied by converse increasing be taken as evidence for Aristotle's envisaging the possibility of an alternative geometry.

There are three sorts of reasons which preclude any possibility of non-Euclidean geometry in Aristotle.

- Mathematical reasons: Aristotle is a Greek philosopher, speaking of Greek mathematics, and not only do the extant works in Greek mathematics actually contain no allusion to non-Euclidean geometry, but they leave no room for its possibility, *pace* Toth.[42]
- Metaphysical reasons: ancient Greek mathematicians do mathematics; they almost never discuss mathematics. This notwithstanding, their *modus operandi*, together with the way in which philosophers like Aristotle speak about it, allows us to make a few rather safe statements. For instance, Greek geometry is concerned with objects, not with space; space itself is a peripheral and unnecessary notion, which only becomes the subject of geometry much later, in the seventeenth century.[43] This means in particular that for Aristotle what we call Euclidean space would be nothing but the 'place' where the properties which will be described in Euclid's *Elements* – which are properties of objects – hold, which is to say the cosmos. Aristotle's assumption that there is only one cosmos,[44] together with his immanentism, implies that only one geometry is possible.
- Textual reasons: it is evident that Aristotle resorts to the procedure of converse increasing for the sole purpose of resolving the impasse he got into in an attempt to reconcile finitism and immanentism with geometric practice. And it is clear – and perfectly understandable – that he is far from perceiving the serious problems which converse increasing would introduce when implemented at a global level.

---

[42]See Toth (1998) and the bibliography quoted there. The reader is referred to Unguru (2013) and Moiraghi (2017), where Toth's statements about the existence of non-Euclidean geometry in ancient Greek mathematics are accurately discussed and definitively dismissed.

[43]The transformations of geometry from a science of objects to a science of space is described in De Risi (2015) (see in particular Mendell 2015 therein, on object, space and spatial relations in ancient Greek geometry), while the meaning of the adjective *Euclidean* as applied to Euclidean period is analyzed in Unguru (2013).

[44]On the uniqueness of the cosmos see in particular *Cael.* I 8–9.

For instance, the statement that the sum of the interior angles of a triangle is equal to two right angles – Aristotle's paradigmatic example of eternal truth – is false on Poincaré's disk.[45]

14.3.3.2   We Shouldn't Read Too Little, Either

On the other hand, reading Aristotle with too little mathematical sophistication can be just as problematic as reading him with too much.

According to the interpretation I have offered in this paper, Aristotle resorts to the procedure of converse increasing in order to provide a viable alternative to the more usual procedures of extending magnitudes, which are incompatible with his finite cosmos.

The existence of a substantial difference between the two kind of procedures – namely the fact that an l-increasing satisfies Aristotle's existence condition **D**, while an e-increasing does not – is a crucial point in favor of my reading.

This difference is easy to express in modern mathematical terms: the existence condition reduces to the requirement that the series of amounts added at every step of the procedure converges. It is immediate that this condition holds only for an l-increasing, where the series tends to the initial segment, while the series connected with an e-increase diverges.

Of course, Aristotle does not put it this way: he explains the difference in natural language (see in particular T2, T5 and T6, in Sect. 14.3.2.1 above), and speaks generically of the conditions of existence **D**, without specifying which requirement fails in the case of an e-increasing. Of course, this is evident from the context, but one must be clear headed about some basic mathematics in order to grasp the point.

I mean in particular the fact that for Aristotle **D1** ("another and another thing is taken, over and over again") is a sufficient condition, while **D2** ("what is taken is finite") is only a necessary one. In particular, both conditions refer to a limit, but while the limit involved in **D1** is the limit of the sum of the series – iff this limit exists, the series converges – the limit involved in **D2** concerns its *partial* sums, i.e., the ones considered at every step of the process. Even if all the partial sums are limited, the series could diverge.

---

[45]In *Ph.* II 9, 200a15–30 Aristotle explains that "because the straight is so and so, it is necessary that the triangle has the sum of the interior angles equal to two right angles, and not the other way round. Still, if the triangle did not have the sum of the interior angles equal to two right angles, the straight is not so and so". Aristotle is here speaking counterfactually, in order to explain what he calls the *necessity ex hypothesis*, namely the fact that, once the first principles are those which are, all the results obtained from them are necessary. But while one can say that the principles are necessary per se, the necessity of the results depends on (the necessity of) the principles. Rather curiously, the passage has been used by Toth in order to prove the exact opposite, namely the possibility of taking a different set of principles.

Of course, it is not necessary to translate Aristotle's statement into mathematical terms, but if one decided to do so, one must be careful: if the translation is done in the wrong way, a completely different conclusion can be obtained – possibly even a contradictory one.

In a recent paper, John Cooper has propounded an alternative interpretation of the procedure of converse increasing, reaching the conclusion that an l-increasing and an e-increasing are ultimately the same – in his terms, that every infinite by addition is based on a division – because the mathematics is 'similar' in both cases.[46]

Now it is not immediately clear what Cooper means when he says that the mathematics of an l-increasing and that of an e-increasing are 'similar', but reading the paper one realizes that similarity comes from the fact that in both cases "there is a limit". The problem is that Cooper does not specify which limit we are talking about, and this is a crucial point. Indeed, he makes a double confusion about limits. On the one hand, he thinks that the existence of the limit mentioned in **D2** is a sufficient condition for the existence of the procedure itself – i.e., that **D2** implies **D1** – while it is at most a necessary one. On the other hand, he confuses the limit of the partial sums of the added amounts – i.e., the limit concerned in **D2** – with the limit of their sequence. Moreover, he thinks that if the sequence of the added amounts is not limited, one can construct another limited sequence – it does not matter how – and this warrants the existence of the procedure.[47]

Cooper may have been misled by the fact that in the case of an l-increasing, it is true *both* that the sequence of amounts added at every step is infinitesimal, *and* that the series converges.[48] Hence, even if he looks at the wrong limit – the limit of the sequence – his conclusion is true: a procedure of l-increasing satisfies **D**, hence it exists.

In the case of an e-increasing, however, the situation is completely different: the sequence of the added amounts is constant, so it does not converge to zero, and the series diverges. Thus, e-increasing should be rejected.[49] But Cooper finds *another* sequence in the same process, an infinitesimal sequence, by looking at the ratios between the amount added at every step and the length reached at the step before. From the fact that this sequence – a sequence with has *nothing to do* with the sum

---

[46]Cooper (2016, p. 172, n. 14).

[47]Cooper speaks generically of the limit of the sequence, but in fact, for the existence of a limit to be a necessary condition, the sequence must be infinitesimal, i.e., this limit must be zero.

[48]In the example chosen by Cooper the sequence is $\frac{1}{3}, \frac{2}{9}, \frac{4}{27}, \frac{8}{81} \ldots$, which tends to zero, and the series is $1+\frac{1}{3}+\frac{2}{9}+\frac{4}{27}+\frac{8}{81} \ldots$, which converges. Indeed, $1+\frac{1}{3}+\frac{2}{9}+\frac{4}{27}+\frac{8}{81} \ldots = 1+\frac{1}{3}\cdot\sum_{n=1}^{\infty}(\frac{2}{3})^n$, and $\sum_{n=1}^{\infty}(\frac{2}{3})^n$ is a geometric series, whose sum is $\frac{1}{1-\frac{2}{3}} = 3$. Then the whole result is $1+\frac{1}{3}\cdot 3 = 2$.

[49]In the case chosen by Cooper, the 'sequence' of the added amounts is $\frac{1}{3}, \frac{1}{3}, \frac{1}{3} \ldots$, and the sum $\frac{1}{3}+\frac{1}{3}+\frac{1}{3} \ldots$.

of the whole added amounts – is limited,[50] Cooper concludes that **D** holds for *e*-increasing too.

In other words, there is a limit, but a limit which does not limit anything; as Cooper himself correctly observes: "though there *is* a limit [...], that does not prevent us from going beyond any given length" (p. 184).[51]

Perhaps because it seems problematic that the limit he finds does not limit anything, Cooper observes that indeed this limit "is not yet *marked off*", and connects the notion of a limit which is not yet marked off with the notion of *indeterminate* finiteness,[52] which he in turn explains in terms of *getting tired*.[53]

Now, if these passages were stated in a clear way, we could immediately reject the argument. But Cooper's pseudo-mathematical language, instead of simplifying the argument, makes it incomprehensible; and that is the only reason his conclusion seems indisputable. What is an indeterminate limit? What is an indeterminate stop? In what respect do they differ? And in what way is a marked-off limit different from one that is not yet marked off?

Although they sound mathematical, none of these terms mean anything in mathematics, including the mathematics of Aristotle's days. Of course, they make it easier to pass from the premises of the argument, which are presented as mathematics, to Cooper's conclusion, which identifies the ultimate reason why the infinite is compatible with a finite cosmos as the fatigue which prevents its actualization.

---

[50]"The amount added remains the same at each step but the proportion (sic) of what was already present immediately before each step continuously diminishes toward a limit *which is never reached or exceeded*." pp. 172-3, n. 14. Italics are mine. Actually, Cooper writes $\frac{4}{3}, \frac{5}{3}, \frac{6}{3} \ldots$, which is far from being infinitesimal. Following the analogous example at pp. 183–184, I assume that it must be corrected to $\frac{1}{3}, \frac{1}{4}, \frac{1}{5} \ldots$. Even then, however, the corresponding series $\frac{1}{3} + \frac{1}{4} + \frac{1}{5} \ldots$, which is to say the harmonic series, diverges.

[51]A further puzzlement is the statement offered in the footnote 16 (pp. 183–184), where Cooper envisages a new kind of *e*-increasing, which in his opinion satisfies **D**. In this case the amounts added *increase* at each step, but the ratio of what was already present immediately before each step remains the same. In this case, Cooper again claims to find "a converging (sic) limit", but for me it has been impossible to reconstruct both where he finds that limit – the sequence of the added amounts diverges and that of the ratios is limited only because it is constant – and what this limit would limit.

[52]"In so far as for Aristotle a spatial magnitude is infinite by addition it is *not* infinite in that it goes on at infinite length from its starting point: instead, it goes on only for some finite extent (no infinite process can be completed), but (and this is his main point) an *indeterminate* one" (pp. 174–5).

[53]"The paths through it that do not come to an end [...] do of course have to come to a stop (anyone who goes any distance along the path will have to stop at some time and place if only *because they get tired* since, again, an infinite process cannot be completed) but where any of them stops is indeterminate" (p. 175, italics are mine). It might be interesting to ask who will get tired in the case of time, which Cooper constructs using the same *e*-increasing, and which is *indeterminately* finite too (p. 186). See also my note 33 above.

References

Cleary, John J. 1995. *Aristotle and mathematics: Aporetic method in cosmology and metaphysics.* Leiden: Brill.

Cooper, John. 2016. Aristotelian infinites. *Oxford Studies in Ancient Philosophy* 51: 161–206.

De Risi, Vincenzo, ed. 2015. *Mathematizing space: The objects of geometry from antiquity to the early modern age.* Basel: Birkhäuser.

Diels, H., ed. 1882. *Simplicii in Aristotelis physicorum libros quattuor priores commentaria.* In *Commentaria in Aristotelem Graeca,* vol. IX. Berlin: G. Reimer.

Heath, Thomas L. 1921. *A history of greek mathematics,* vol. 2. Oxford: Oxford University Press (reprint: New York: Dover 1981).

Hintikka, Jaakko. 1973. *Time and necessity: Studies in Aristotle's theory of modality.* Oxford: Clarendon Press.

Hussey, Edward. 1983. *Aristotle, physics. Books III and IV.* Oxford: Clarendon Press.

Knorr, Wilbour R. 1982. Infinity and continuity: The interaction of mathematics and philosophy in antiquity. In *Infinity and continuity in ancient and medieval thought,* ed. N. Kretzmann, 112–145. London: Cornell University Press.

Lear, Jonathan. 1979. Aristotelian infinity. *Proceedings of the Aristotelian Society* 80: 187–210.

Lear, Jonathan. 1982. Aristotle's philosophy of mathematics. *The Philosophical Review* 91: 161–192.

Mendell, Henry. 2008. Aristotle and mathematics. In *The stanford encyclopedia of philosophy.* Available via DIALOG. http://plato.stanford.edu/archives/win2008/entries/aristotle-mathematics/

Mendell, Henry. 2015. What's location got to do with it? Place, space, and the infinite in classical greek mathematics. In *Mathematizing space: The objects of geometry from antiquity to the early modern age,* ed. V. De Risi, 15–63. Basel: Birkhäuser.

Milhaud, Gaston. 1903. Aristote et les mathématiques. *Archiv für Geschichte der Philosophie* 16: 367–392.

Miller, Fred D. 1982. Aristotle against the atomists. In *Infinity and continuity in ancient and medieval thought,* ed. N. Kretzmann, 112–145. London: Cornell University Press.

Moiraghi, Francesco. 2017. *Geometrie non-euclidee nella matematica greca.* (Dissertation, Università degli studi di Milano a.a. 2016/2017).

Mueller, Ian. 1970. Aristotle on geometrical objects. *Archiv für Geschichte der Philosophie* 52: 156–171.

Mueller, Ian. 1990. Aristotle's doctrine of abstraction in the commentators. In *Aristotle transformed: The ancient commentators and their Influence,* ed. R. Sorabji, 463–479. London: Duckworth.

Rashed, Marwan. 2011. *Alexandre d'Aphrodise, Commentaire perdu la Physique d'Aristote (Livres IV-VIII).* Berlin: De Gruyter.

Toth, Imre. 1998. *Aristotele e i fondamenti assiomatici della geometria. Prolegomeni alla comprensione dei frammenti non-euclidei nel "Corpus Aristotelicum" nel loro contesto matematico e filosofico.* 2nd revised and corrected ed. (Italian) Milano: Vita e Pensiero.

Ugaglia, Monica. 2009. Boundlessness and iteration: Some observation about the meaning of AEI in Aristotle. *Rhizai* 6: 193–213.

Ugaglia, Monica. 2012. *Aristotele, Fisica. Libro III.* Roma: Carocci editore.

Ugaglia, Monica. 2016. Is Aristotle's cosmos hyperbolic? *Educação e Filosofia* 30: 1–21.

Unguru, Shabetai. 2013. Greek geometry and its discontents: The failed search for non-Euclidean geometries in the greek philosophical and mathematical corpus. *International Journal of History and Ethics of Natural Sciences Technology and Medicine* 21: 299–311.

Vitelli, H. 1887. *Ioannis Philoponi in Aristotelis physicorum libros tres priores commentaria* in *Commentaria in Aristotelem Graeca,* vol. XVI. Berlin: G. Reimer.

Vitrac, Bernard. 1990. *Euclide: Les Éléments. Traduction et commentaires par B. Vitrac.* vol. I. Paris: P.U.F.

White, Michael J. 1992. *The continuous and the discrete: Ancient physical theories from a contemporary perspective.* Oxford: Clarendon Press.

Wieland, Wolfgang. 1970. *Die aristotelische Physik.* Göttingen: Vandenhoeck and Ruprecht.

Printed in the United States
By Bookmasters